Host Identity Protocol (HIP)

WILEY SERIES IN COMMUNICATIONS NETWORKING & DISTRIBUTED SYSTEMS

Series Editors: David Hutchison, *Lancaster University, Lancaster, UK*
Serge Fdida, *Université Pierre et Marie Curie, Paris, France*
Joe Sventek, *University of Glasgow, Glasgow, UK*

The 'Wiley Series in Communications Networking & Distributed Systems' is a series of expert-level, technically detailed books covering cutting-edge research, and brand new developments, as well as tutorial-style treatments in networking, middleware and software technologies for communications and distributed systems. The books will provide timely and reliable information about the state-of-the-art to researchers, advanced students and development engineers in the Telecommunications and the Computing sectors.

Other titles in the series:

Wright: *Voice over Packet Networks* 0-471-49516-6 (February 2001)
Jepsen: *Java for Telecommunications* 0-471-49826-2 (July 2001)
Sutton: *Secure Communications* 0-471-49904-8 (December 2001)
Stajano: *Security for Ubiquitous Computing* 0-470-84493-0 (February 2002)
Martin-Flatin: *Web-Based Management of IP Networks and Systems* 0-471-48702-3 (September 2002)
Berman, Fox, Hey: *Grid Computing. Making the Global Infrastructure a Reality* 0-470-85319-0 (March 2003)
Turner, Magill, Marples: *Service Provision. Technologies for Next Generation Communications* 0-470-85066-3 (April 2004)
Welzl: Network Congestion Control: *Managing Internet Traffic* 0-470-02528-X (July 2005)
Raz, Juhola, Serrat-Fernandez, Galis: *Fast and Efficient Context-Aware Services* 0-470-01668-X (April 2006)
Heckmann: *The Competitive Internet Service Provider* 0-470-01293-5 (April 2006)
Dressler: *Self-Organization in Sensor and Actor Networks* 0-470-02820-3 (November 2007)
Berndt: *Towards 4G Technologies: Services with Initiative* 0-470-01031-2 (February 2008)
Jacquenet, Bourdon, Boucadair: *Service Automation and Dynamic Provisioning Techniques in IP/MPLS Environments* 0-470-01829-1 (February 2008)
Minei, Lucek: *MPLS-Enabled Applications: Emerging Developments and New Technologies. Second Edition* 0-470-98644-1 (April 2008)

Host Identity Protocol (HIP):

Towards the Secure Mobile Internet

Andrei Gurtov
Helsinki Institute for Information Technology (HIIT), Finland

A John Wiley & Sons, Ltd, Publication

This edition first published 2008
© 2008 John Wiley & Sons Ltd

Registered office
John Wiley & Sons Ltd, The Atrium, Southern Gate, Chichester, West Sussex, PO19 8SQ,
United Kingdom

For details of our global editorial offices, for customer services and for information about
how to apply for permission to reuse the copyright material in this book please see our
website at www.wiley.com.

The right of the author to be identified as the author of this work has been asserted in
accordance with the Copyright, Designs and Patents Act 1988.

Library of Congress Cataloging-in-Publication Data

Gurtov, Andrei.
  Host Identity Protocol (HIP) : Towards the Secure Mobile Internet / Andrei Gurtov.
    p. cm.
  Includes bibliographical reference and index.
  ISBN 978-0-470-99790-1 (cloth)
1. Wireless Internet–Security measures. 2. Host Identity Protocol (Computer network
protocol) I. Title.
  TK5103.4885.G87 2008
  005.8–dc22

                                                                        2008006092

A catalogue record for this book is available from the British Library.

ISBN 978-0-470-99790-1 (H/B)

Set in 10/12pt Times by Sunrise Setting Ltd, Torquay, UK.
Printed in Great Britain by CPI Antony Rowe, Chippenham, England.

Contents

About the Author

Andrei Gurtov received his M.Sc and Ph.D. degrees in Computer Science from the University of Helsinki, Finland, in 2000 and 2004. At present, he is a Principal Scientist leading the Networking Research group at the Helsinki Institute for Information Technology focusing on the Host Identity Protocol and next generation Internet architecture. He is a co-chair of the IRTF research group on HIP and teaches as an adjunct professor at the Telecommunications and Multimedia Laboratory of the Helsinki University of Technology. Previously, his research focused on the performance of transport protocols in heterogeneous wireless networks. In 2000–2004, he served as a senior researcher at TeliaSonera Finland contributing to performance optimization of GPRS/UMTS networks, intersystem mobility, and IETF standardization. In 2003, he spent six months as a visiting researcher in the International Computer Science Institute at Berkeley working with Dr. Sally Floyd on simulation models of transport protocols in wireless networks. In 2004, he was a consultant at the Ericsson NomadicLab. Dr. Gurtov is a co-author of over 35 publications including research papers, patents, and IETF RFCs. Admitted to the Finnish Union of University Professors at 28, he remains the youngest member at the time of book publication.

Foreword

Jari Arkko

Bob Moskowitz pioneered the idea of the Host Identity Protocol (HIP) in the late 1990s. While the general idea of separating identifiers and locators had been around for a long time, nothing quite as detailed and comprehensive had actually been designed before. The initial designs were little more than sketches but the architecture was appealing. In particular, many people were drawn to the idea of HIP due to the way that security was naturally integrated as a part of the design, or the fact that the same powerful concepts appeared to be capable of solving many different problems, be they about mobility, multihoming, address translation, or stable anchor points.

Despite a lot of interest, HIP was not an overnight success, largely for two reasons. First, in general, deploying any new functionality in the IP layer has proven slow at best. We live in a world with a continuous stream of new Internet innovations, and it may come as a surprise to some that certain parts of the Internet technology are not easy to replace or upgrade. The HIP designers set out to deal with these issues, and today the result is a protocol that does not require upgrades in routers between two communicating hosts, is capable of communicating with legacy hosts, and can support HIP-unaware middleboxes.

Second, the HIP designers took an approach where they wanted to first see their ideas verified and improved in implementations and detailed protocol specifications written, before attempting to acquire any significant deployment. They followed the old principle of the IETF by focusing on running code as opposed to theoretical results or committee agreements. Pekka Nikander, later helped by Thomas Henderson, Andrei Gurtov, a number of implementation teams, and a fair number of other people, started a long process that has now finally been completed.

We are fortunate that Andrei and the rest of the team have decided to write this book. The book describes the HIP architecture and protocols in detail. It also goes beyond the basics, explaining many of the advanced issues and applications, such as middlebox interactions, application programming interfaces, privacy issues, application of distributed hash tables to identifier resolution, and network mobility. Even if you are not deploying or writing code for HIP, the concepts borne out by this protocol design and presented in this book are applicable to future evolution of the Internet architecture in general.

The book comes out at an appropriate time. Official specifications for HIP are coming out of the IETF as RFCs, designs have been verified in many implementations and tests, a number of significant real-world deployments are being discussed, and some are already up and running. A growing body of research on Internet architecture employs ideas from

HIP. Within the set of many identifier–locator separation designs for the Internet, HIP has progressed clearly further than anything else we have so far. It is time to see what HIP can do in larger scale in the real world. In order to make that happen, the world needs a HIP book, and now we have it.

Jari Arkko
Internet Area Director, IETF

Foreword

David Hutchison

Electronic communication, whether using fixed-line or wireless networks, is now a common feature of many activities in everyday life. Enterprises and individuals alike have embraced communication as an indispensable tool. The Internet, the Web and the mobile telephone are three technologies that have made a particular impression. Used in conjunction, these form a powerful combination of technologies that are transforming the ways in which we interact with each other and with information, whether it is for work or leisure. However, these can also be quite literally addictive. There are now reports in the press that many people experience withdrawal symptoms if they are denied access – even temporarily – to their habitual or favourite means of communication. But even more significant is our growing dependency on networks for business use, including for example government, banking, travel, healthcare, monitoring and control: when the network fails – for whatever reason – the consequences may be severe. The working assumption is that the network will act as a utility, i.e. that it will provide an essentially perfect service all of the time. It seems that there is a considerable lack of awareness amongst enterprise owners of the vulnerabilities of networks and a lack of foresight about the impact that failures can have on their operations. Perhaps this is often due to complacency, although ignorance and sometimes unwillingness to invest in appropriate measures cannot be ruled out.

Security and dependability are, then, crucially important aspects of communications and networks that need very much more exposure, as well as considerably more investment by enterprise owners. The Host Identity Protocol (HIP) offers the dual prospects of more secure and more dependable communications through the separation of identity and location, as well as a number of related potential benefits. Although the first HIP draft was submitted in 1999 at the IETF, HIP remained unknown to a wider audience until recently.

This latest – and very welcome – book in the Wiley CNDS Series is written by a co-chair of the HIP Research Group at IRTF. It gives a comprehensive coverage of the subject and should appeal to researchers and practitioners alike in the field of communications and computer networks, not just to those interested in security and mobility.

David Hutchison
Lancaster University

Preface

The main goal of this book is to present a well-structured, readable and compact overview of the Host Identity Protocol with relevant extensions to the Internet architecture and infrastructure. HIP is a new protocol developed by the Internet Engineering Task Force (IETF) to address shortcomings of the current Internet in supporting secure host mobility and multihoming.

As an alternative to Mobile IP, HIP helps to solve security and Denial-of-Service issues that hindered the deployment of IP mobility in an architecturally clean way. In the TCP/IP stack, HIP is positioned between the network (IPv4, IPv6) and transport (TCP, UDP, SCTP, DCCP) layers. The transport protocols use a cryptographic host identity instead of IP addresses.

Since 2000, HIP had been specified in the IETF. Many major telecommunication vendors and operators have started internal activities involving HIP. At the time of writing, an effort is ongoing to create an experimental deployment of HIP in the Internet. HIP is used by several large research projects and often cited in the networking articles. Several public open-source interoperating HIP implementations are available for various platforms, including Linux, BSD, Windows, and Mac OS.

The need for HIP

The Internet has grown tremendously over the past twenty years and become a part of life for millions of people. The basic TCP/IP technology has served us very well. However, important issues such as mobility of Internet hosts over separate IP networks and simultaneous connections to several networks were not a part of the original Internet design. Furthermore, when the Internet grew from a small university network up to a global communication infrastructure, many security issues became apparent. The lack of reliable host authentication has prevented deployment of existing IP mobility extensions. Often, public Internet servers face Denial-of-Service (DoS) attacks that make the service unavailable to other users.

HIP is developed to address these issues in an integrated approach that fits well within the TCP/IP architecture. The original ideas on the separating of host identity and location in the Internet date back to Saltzer in "RFC 1498 On the Naming and Binding of Network Destinations". However, only recent advances in public key cryptography and new requirements of portable terminals have made the actual design and implementation possible.

It is true that HIP is only one of many proposals developed recently in the IETF in the area of security and mobility. Compared with other proposals that often solve only a small part of the problem, HIP integrates host mobility and multihoming in a simple and elegant way.

Independently of which proposal will be deployed in the Internet in the future, we believe that ideas stemming from design and experiments with HIP provide indispensable knowledge for anybody interested in next-generation networking.

Intended audience

We hope the book will be interesting for engineers implementing new Internet protocols, researchers working in the area of next-generation Internet, and students working on Master or Doctoral theses in the area of Internet security, mobility, and multihoming.

For industry engineers, the book includes detailed information on protocol messages and packet headers. Although only the IETF specifications offer complete information necessary for the implementation, the specifications might be difficult to comprehend at first without an overall picture of HIP architecture. The specifications are divided into several RFCs with many cross-references and often repeating parts of the text. Therefore, we hope that the book will be useful to be read before the specifications to obtain a general understanding of the area. In addition to published specifications, the book also includes several experimental or historic proposals from expired Internet drafts that are not easily found otherwise. The ASCII diagrams from the specifications are redrawn to be easily understandable.

For researchers, the book includes an overview of several research articles on HIP infrastructure and design alternatives. The future Internet architecture is a subject of many research projects in the USA such as Future Internet Network Design (FIND) and in the EU (Framework Program 7). The concept of identifier–locator split forms the core of HIP and is often applied in other protocol areas. Several large projects, such as the FP6 Ambient Networks, have adopted HIP as an architectural backbone. Therefore, even the researchers not directly working with HIP may find it useful to understand the concepts and design decisions behind HIP.

University students participating in networking courses and research seminars may find the book useful in covering the Internet security, mobility, and multihoming issues comprehensively. The book can be also used as a text book on a course focusing on networking protocol design, where HIP is taken as a possible example.

The prerequisites for the book include knowledge of TCP/IP networking. Experience with UNIX socket programming is helpful to understand certain parts of the book, such as the socket API interface. Readers would benefit from knowledge of basic cryptography, although an overview of symmetric and asymmetric cryptography, IP security protocols, and DNS is given at the beginning of the book.

Coverage

The book covers all IETF specifications of HIP, several experimental extensions proposed in the IETF, and research results relevant to the development and deployment of HIP. The HIP specifications produced by the HIP Working Group in the IETF include the base specification, IPsec ESP encapsulation, mobility and multihoming extensions, rendezvous server, registration extensions, NAT traversal, and API specification. The experimental proposals from the HIP Research group in IRTF include the DHT interface, opportunistic mode, HIP-aware middleboxes, service discovery and simultaneous multiaccess, and lightweight HIP.

Research results cover the use of overlay networks for routing HIP control packets, HIP privacy extensions, multicast, and integration of HIP with other protocols such as SIP and Mobile IP.

We have considered including a CD-ROM with HIP specifications and implementations together with the book. Instead, for a number of reasons, we decided to create a web page with the book that can be kept up-to-date. First, few HIP specifications are still in the final standardization stages and can change after publication of the book. IETF and IRTF documents are freely available from the Internet. The HIP implementations are constantly improving and supplying a deprecated version on the CD-ROM would only generate obsolete feedback. An overview and setup instructions for HIP on Linux (HIPL) implementation are given in Appendix A towards the end of the book.

The book includes several practical examples of HIP packet exchanges captured with the freeware network analyzer Wireshark. Its latest version, modified to support the HIP packet formats, is available from the book web page (http://www.hipbook.net).

Acknowledgments

This book focuses on the Host Identity Protocol, which has been created by the joint effort of many people. Robert Moskowitz proposed HIP in IETF and served as a coordinator of development efforts in the early days of standardization. Pekka Nikander, Gonzalo Camarillo, and Tom Henderson served as main proponents of HIP architecture and specifications in the following years. Many people have contributed to creating HIP specifications in IETF/IRTF and developing HIP implementations. Many thanks to all of you for doing a great job!

This book would not have been possible without the help of many people. I thank Lars Eggert for helping to draft the book proposal and serving as a co-initiator of the book project. Thanks to anonymous reviewers who have evaluated the proposal and given plenty of useful suggestions for improving it. Tobias Heer contributed two valuable chapters to the book, covering background material and lightweight HIP. Jari Arkko wrote a nice foreword for the book. Samu Varjonen, Abhinav Pathak, Petri Jokela, and René Hummen reviewed various parts of the book. However, any remaining mistakes are those of the author. Assel Mukhametzhanova helped to maintain the project bibliography database. Thanks to Birgit Gruber, Anna Smart, and Sarah Hinton from John Wiley and Sons for help and support in getting the book published.

I am grateful to the Helsinki Institute for Information Technology and Helsinki University of Technology for hosting HIP-related research projects. I thank the Finnish Funding Agency for Technology and Innovation (Tekes), Academy of Finland, and companies that funded research work at HIIT: Elisa, Ericsson, Finnish Defence Forces, Nokia, Secgo, and TeliaSonera.

The book covers a wide range of engineering and research efforts related to HIP. I would especially like to mention a few people whose publications are referenced in the book: Jeff Ahrenholz, Petri Jokela, Andrey Khurri, Miika Komu, Teemu Koponen, Dmitry Korzun, Julien Laganier, Richard Paine, Abhinav Pathak, Laura Takkinen, Hannes Tschofenig, Samu Varjonen, Essi Vehmersalo, Rolland Vida, Ekaterina Vorobyeva, and Jukka Ylitalo. This list is by no means complete and extends to everyone else in the HIP community.

Last but not least, I would like to thank my core and extended family for love and support in getting the book written.

Andrei Gurtov
Helsinki

Abbreviations

ACK	Acknowledgment packet
ACL	Access Control List
ADSL	Asynchronous Digital Subscriber Line
AES	Advanced Encryption Standard
AH	Authentication Header
API	Application Programming Interface
AR	Access Router
ARP	Address Resolution Protocol
ASM	Any Source Multicast
BE	Base Exchange
BEET	Bound End to End Tunnel
BOS	Bootstrap packet
BSD	Berkeley Software Distribution
CA	Certificate Authority
CBA	Credit-Based Authorization
CBC	Cipher Block Chaining
CPU	Central Processing Unit
CRL	Certificate Revocation List
CS	Cryptographic Session
DCCP	Datagram Congestion Control Protocol
DH	Diffie–Hellman
DHCP	Dynamic Host Configuration Protocol
DHT	Distributed Hash Table
DNSSEC	Domain Name System with Security
DoS	Denial-of-Service
DR	Designated Router
DSA	Digital Signature Algorithm
ED	Endpoint Descriptor
ESP	Encapsulated Security Payload
FQDN	Fully Qualified Domain Name
FTP	File Transfer Protocol
GGSN	Gateway GPRS Support Node
GNU	GNU is not UNIX
GPL	General Public License
GPRS	General Packet Radio Service

GRUU	Globally Routable UA URI
GSM	Global System for Mobile communications
GUI	Graphical User Interface
HCVP	Hash Chain Value Parameter
Hi3	Host Identity Indirection Infrastructure
HIP	Host Identity Protocol
HIPD	HIP Daemon
HIPL	HIP for Linux
HISM	Host Identity Specific Multicast
HIT	Host Identity Tag
HMAC	Hash Message Authentication Code
HMIP	Hierarchical Mobile IP
HTTP	Hyper Text Transfer Protocol
i3	Internet Indirection Infrastructure
IANA	Internet Assigned Numbers Authority
ICE	Interactive Connectivity Establishment
ICMP	Internet Control Message Protocol
ID	Identifier
IEEE	Institute of Electrical and Electronics Engineers
IETF	Internet Engineering Task Force
IGMP	Internet Group Management Protocol
IHC	Interactive Hash Chain
IKE	Internet Key Exchange
IMS	IP Multimedia Subsystem
IPv6	Internet Protocol version 6
IRTF	Internet Research Task Force
ISP	Internet Service Provider
L2TP	Layer 2 Tunneling Protocol
LHIP	Lightweight HIP
LRVS	Local Rendezvous Server
LSI	Local Scope Identifier
MAC	Message Authentication Code
MIME	Multipurpose Internet Mail Extensions
MIP	Mobile IP
MITM	Man-In-The-Middle
MKI	Master Key Identifier
MLD	Multicast Listener Discovery
MOBIKE	Mobile Key Exchange
MODP	More Modular Exponential
MR	Mobile Router
MTU	Maximum Transmission Unit
NACK	Negative Acknowledgment
NAT	Network Address Translator
NEMO	Network Mobility
NIC	Network Interface Card
OCALA	Overlay Convergence Architecture for Legacy Applications

ORCHID	Overlay Routable Cryptographic Hash Identifier
OS	Operating System
P2P	Peer-to-Peer
PACK	Pre-Acknowledgment
PDA	Personal Digital Assistant
PGP	Pretty Good Privacy
PIDF	Presence Information Data Format
PISA	P2P Internet Sharing Architecture
POSIX	Portable Operating System Interface
PSIG	Pre-Signature Packet
PSP	Pre-Signature Parameter
RADIUS	Remote Authentication Dial In User Service
RFC	Request for Comments
RFID	Radio-Frequency Identification
PKI	Public Key Infrastructure
ROC	Rollover Counter
RPC	Remote Procedure Call
RR	Resource Record for DNS
RSA	Rivest–Shamir–Adleman algorithm
RTO	Retransmission Timeout
RTP	Real-time Transmission Protocol
RTT	Round-Trip Time
RVA	Rendezvous agent
RVS	Rendezvous server
SA	Security Association
SACK	Selective Acknowledgment
SAD	Security Association Database
SCTP	Stream Control Transmission Protocol
SD	Service Discovery
SDP	Service Discovery Protocol
SHA	Secure Hash Algorithm
SIGMA	Signature and MAC
SIM	Subscriber Identity Module
SIMA	Simultaneous Multi Access
SIP	Session Initiation Protocol
SMTP	Simple Mail Transfer Protocol
SPD	Security Parameter Database
SPI	Security Parameter Index
SRTP	Secure Real-time Transmission Protocol
SSH	Secure Shell
SSL	Secure Sockets Layer
SSM	Source Specific Multicast
SSO	Single Sign-On
SSRC	Synchronization Source
STUN	Simple Traversal of UDP through NATs
SYN	Synchronization packet for TCP

TCP	Transmission Control Protocol
TESLA	Timed Efficient Stream Loss-tolerant Authentication
TLS	Transport Layer Security
TLV	Type-Length-Value
TRIG	Trigger packet
TTL	Time to Live
UA	User Agent
UDP	User Datagram Protocol
UMTS	Universal Mobile Telecommunication System
URI	Universal Resource Identifier
UUID	Universally Unique Identifier
VIGMP	Version-Independent Group Management Protocol
VM	Virtual Machine
VMM	Virtual Machine Monitor
VoWLAN	Voice over WLAN
VPN	Virtual Private Network
Wi-Fi	Wireless Fidelity
WIMP	Weak Identifier Multihoming Protocol
WLAN	Wireless Local Area Network
WPA	Wi-Fi Protected Access
XFRM	Linux IPsec Transforms
XML	Extensible Markup Language

Part I

Introduction

1

Overview

The current Internet architecture, though hugely successful, faces many difficult challenges. The most important ones are the incorporation of mobile and multihomed terminals (hosts) and an overall lack of protection against Denial-of-Service attacks and other lacking security mechanisms. With gradual deployment of IPv6, interoperability between the "old" IPv4 Internet and the "new" IPv6 Internet becomes another challenging problem. Finally, the ongoing convergence on IP-based solutions for telecommunication networks, such as with UMTS/3G and its successors, increases the need to resolve these problems as soon as possible.

Although many of these problems have been widely recognized for some time, a complete and adequate solution is still missing. Most existing approaches are point-solutions that patch support for a subset of the required improvements into the current Internet architecture, but do not cleanly integrate with one another and do not present a stable base for future evolution. As an example, Mobile IP provides some support for host mobility, but still has major security flaws that prevent its widespread deployment.

This book is dedicated to the Host Identity Protocol (HIP), which is a promising new basis for a secure mobile Internet. The cornerstone of HIP is the idea of separating a host's identity from its present topological location in the Internet. This simple idea provides a solid basis for mobility and multihoming features. HIP also includes security as an inherent part of its design, because its "host identities" are cryptographic keys that can be used with many established security algorithms. For example, these cryptographic identities are used to encrypt all data traffic between two HIP hosts by default. Finally, the HIP specifications are patent-free (to our current best knowledge), which is an important criterion for adoption in both commercial and open source Internet platforms.

This chapter starts with Section 1.1 presenting the split in the host identifiers and locations as an important goal for future Internet. The place of HIP in the Internet architecture is discussed in Section 1.2. The origins and history of HIP development in IETF are given in Section 1.3. Section 1.4 walks through the book chapter by chapter to give the reader a general feeling on how the book is structured.

1.1   Identifier–locator split

Routing of IP datagrams in the Internet is based on network prefixes of destination IP addresses. It would be impractical to demand that each Internet router know the next hop for each individual IP address. It would make the routing tables excessively large, increasing the memory costs, and slowing down the lookups. Instead, IP addresses are aggregated and essentially located in a close geographical area. An IP address serves as a *locator* of the host in the Internet topology.

The host *identifier* is something that tells other communicating entities that the host is still the same despite a possible change of its location in the Internet. DNS name is one such widely-deployed form of identifier. In the early days of the Internet, when hosts were stationary, it was possible to use the IP address as a host identifier. In the present Internet architecture, the role of IP addresses as identifier and locators are still mixed, as shown in Figure 1.1, though the old assumptions are not valid anymore. Many hosts are mobile and multihomed. There, each separate service is using own socket as it should, but the endpoint identity is directly attached to the IP address of the interface.

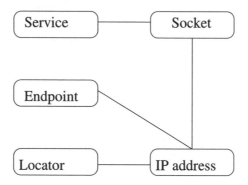

Figure 1.1 Location and identity of hosts are combined in the Internet.

Today many hosts use public IPv4 addresses assigned from a common DHCP pool or ephemeral IPv4 addresses assigned privately behind a NAT. Therefore, an IP address cannot often even be used as a robust identifier. As an example, blacklisting a certain IPv4 address as a source of SPAM messages can in fact disable an innocent host who received the same address later from DHCP or even prevent a large number of hosts using a public IPv4 address of a NAT.

The main architectural theme behind HIP is the split of host identifier and locator (Moskowitz and Nikander 2006). An appropriate security mechanism to enable the host to prove its identity is essential. A long randomly generating string would be sufficient to identify a host in a trusted environment, but not in a public Internet. In HIP architecture, a self-generated public–private key pair becomes the host identity. Figure 1.2 illustrates the positioning of host identity between socket and network interfaces. Now the sockets are bound to the host identity instead of a locator. In fact, there could be several locators associated simultaneously with an identity. The locators can be dynamically added or removed from the active set without any changes on the socket interface.

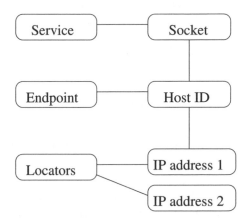

Figure 1.2 Separating location and identity of Internet hosts.

A single host can have multiple identities. Self-asserted unpublished identities are suitable for anonymous communication in a trustworthy way. The peer hosts can reliably verify that the new connection is coming from the same host as before. Host identifiers placed into Secure DNS (DNSSEC), asserted by Public Key Infrastructure (PKI) X.509 certificates, or Pretty Good Privacy (PGP) can be used to authenticate the host.

A single identity can be shared among a group of hosts that possess the same private key. Such hosts can, for example, form a multicast group or perform other distributed operations. Given the challenge of securely distributing private keys over the network, group host identities are still in the research phase and are not supported by the current HIP specifications.

1.2 HIP in the Internet architecture

Since the beginning of the Internet, the IP protocol has been the only routable network-layer protocol in use. Deployment of IPv4 occurred using a "flag day" when the old IP version was not routed anymore. Thanks to its simplicity, the IP protocol is able to run over a wide range of link technologies, including Ethernet (IEEE 802.3), Wireless LAN (IEEE 802.11), Token Ring (IEEE 802.2) and many others. On the other hand, multiple transport protocols can run on top of IP. The TCP and UDP protocols were the only ones widely deployed in practice. The number of applications using the transport protocols is large, with the most important ones being HTTP, SMTP, and FTP. The resulting model of the Internet protocol stack resembles the hourglass as shown in Figure 1.3. The IP is the narrowest part of the stack and sometimes called a waist of the Internet.

A major problem in the original Internet architecture is tight coupling between networking and transport layers. As an example, TCP uses a pseudoheader including IP addresses

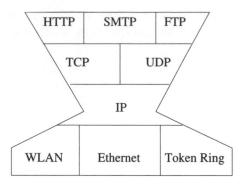

Figure 1.3 IP as a waist of the Internet protocol stack.

in checksum calculations. Therefore, independent evolution of two layers is not possible. Introduction of a new networking or transport protocol requires changes to other layer. In the proposed OSI model of a network stack, different protocol layers are mutually isolated. However, the model has not been realized in the Internet protocol stack.

As many of us have noticed, with passing years it becomes more difficult to control the waist volume. The same happened to the Internet with the introduction of IPv6. The scale of the Internet has grown dramatically and deployment of a new IP version with a flag day is not any more feasible. Now in dual-stack IP networks, both versions of IP protocol need to be simultaneously routed. This creates a challenge for inter-operating the legacy IPv4-only HIP applications with new IPv6 applications. Introduction of the HIP architecture can restore the original Internet hourglass model as shown in Figure 1.4. Then, HIP becomes a new waist of the Internet protocol stack replacing IPv4 in this role. IPv4 and IPv6 run underneath HIP, there as the transport protocols including the new SCTP and DCCP on top of HIP.

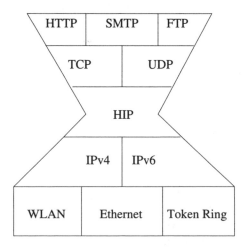

Figure 1.4 HIP as a new waist of the Internet protocol stack.

Another major problem of the present Internet is Denial-of-Service (DoS) attacks. During establishment of a TCP connection, the server creates a significant state by allocating TCP control block structure after replying to a SYN packet with a SYN-ACK packet. At that stage, the server has no assurance even that the SYN has arrived from the same host as stated in the SYN source IP address. Using this exploit, a moderate number of hosts can swamp the server with SYN messages, exhausting the server's memory or other system resources. Therefore, an important requirement of a new interworking protocol is to prevent creating the state at the server before the client is verified to respond from acclaimed IP source address. HIP implements this task and in addition uses cryptographic puzzles to prevent the client generating connection attempts at an overly fast rate. The puzzle is basically asking a client to reverse a hash function of a small length that statistically requires significant computational resources. On the other hand, verifying the puzzle at the server is a short operation.

Figure 1.5 shows a well-known structure of the present IP stack. The application creates a Berkeley socket using the IP address and transport protocol family of the peer. The state created at a transport layer is also using the IP address in addition to transport protocol-specific port numbers to deliver data segments to a correct application. The network layer uses the destination IP address to determine a right transmission link for a packet. The Network Interface Card (NIC) address is added on the link for transmission.

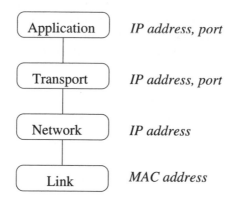

Figure 1.5 The IP protocol stack.

HIP takes place as a shim sub-layer between the network and transport layers, as shown in Figure 1.6. Now the applications and transport protocols use the host identity tag in their messages. The HIP sub-layer maps HITs to the best IP address of the destination internally before passing a packet to the networking layer. Transmission of the packet afterward follows the same pattern as in a regular IP stack.

1.3 Brief history of HIP

The problem of naming of hosts and data in the Internet has been discussed for a long time in the Internet Engineering Task Force (IETF). As an example, RFC 1498 from 1993 reprints the paper on naming and binding of network destinations originally from 1982 (Saltzer 1993).

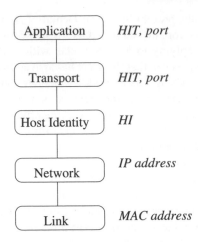

Figure 1.6 The protocol stack of HIP.

The paper starts from considering a resource name, address and route as the basic definitions used by John Shoch. The definitions are extended to cover services and users, network nodes, network attachment points, and paths. The paper identified three successive and changeable bindings of a service to a node, a node to an attachment point, and an attachment point to a route.

The Name Space Research Group (NSRG) has been active in IRTF from 1999 to 2003. It produced a report on the use of other namespaces than the 32-bit IPv4 addresses.

The original inventor of HIP was Robert Moskowitz from ICSA, Inc. The draft-moskowitz-hip-00 was published as an individual submission in the IETF in May 1999. Quoting its acknowledgment section: "The drive to create HIP came to being after attending the MALLOC meeting at IETF 43. It is distilled from many conversations from the IPsec mailing list and the IPsec workshops. Particularly Rodney Thayer should be mentioned for giving this protocol its initial push. Steve Bellovin assisted on some of the public key and replay concerns. Baiju Patel and Hilarie Orman gave extensive comments on the initial format, resulting in the present document. Hugh Daniels and IPsec implementers have kept after me to see that HIP moved beyond concept to spec".

From 1999 to 2002, Robert Moskowitz has held several informal meetings during the IETFs. Several revisions of the HIP architecture and protocol specifications were published as individual submissions.

In 2002, Pekka Nikander became interested in HIP and took over the leading of the standardization effort from Robert Moskowitz, who had other commitments. Together with a number of people at Ericsson NomadicLab, Boeing and HIIT he developed new packet structure, the state machine and the protocol details. The specifications were published as individual submissions until 2004.

After a period of background development, an IETF working group on HIP was created and draft-ietf-hip-base-00 was published in June 2004. The HIP WG is chaired by David Ward (Cisco) and Gonzalo Camarillo (Ericsson). The purpose of the WG was "to define the minimal elements that are needed for HIP experimentation on a wide scale".

The first outcome of the group was the overview of HIP architecture. Originally the group focused on creating the HIP base exchange and ESP encapsulation specifications, mobility and multihoming extensions, DNS and rendezvous, and registration extensions.

In late 2006, the WG was re-charted to include legacy NAT traversal, the application support and native API as WG items. At the time of writing (early 2007) the HIP specifications are under review process in IESG and their publication is expected shortly. The document quality meets the requirements for standard-track RFCs, but their status is "experimental" due to unknown implications on HIP deployment to the Internet. It is expected that after more experience with the use of HIP in real networks is obtained, the specifications could be elevated to the standard-track status. The document specifying an IPv6 prefix for HIP is published as an independent submission.

A HIP Research Group was chartered at the Internet Research Task Force (IRTF) in 2004. From the establishment Pekka Nikander (Ericsson) and Tom Henderson (Boeing) had served as the chairs of HIP RG. In 2005, Andrei Gurtov (HIIT) replaced Pekka Nikander after his election to the Internet Architecture Board (IAB). The task of HIP RG is to evaluate the impact of wider HIP deployment on the Internet and develop experimental protocol extensions that are not yet ready for standardization in the IETF.

1.4   Organization of the book

The book is organized into four parts: introduction, protocol specifications, infrastructure support, and applications. The first part includes a general HIP overview and background on cryptography and Internet protocols. The second part covers the base protocol, the core extensions for mobility and multihoming, rendezvous and DNS, as well as advanced extensions such as the HIP opportunistic mode. The third part describes changes to the Internet infrastructure useful for wider HIP deployment, such as middlebox traversal and name resolution. The last part covers practical use cases for HIP and its possible use with other protocols, such as SIP and Mobile IP.

In the overview (Chapter 1), we outlined the problems that the current Internet faces and provided the motivation for HIP, as well as the range of existing solutions and their shortcomings. Chapter 2 presents the necessary background on public–private key and symmetric cryptography and Internet protocols (IPsec, Internet Key Exchange). Readers that are familiar with all or some of this background information can skip these sections.

Part II starts with Chapter 3 on the place of HIP in the Internet architecture. It presents the concept of Internet namespaces, position of HIP within the TCP/IP stack, and construction of host identifiers. Chapter 4 describes the core parts of the HIP specifications that establish a secure connection between two hosts in a way that is DoS-resistant and makes spoofing difficult. The HIP base exchange packets I1, R1, I2 and R2 are explained in detail, as well as NOTIFY, UPDATE, CLOSE, and CLOSE_ACK HIP control packets. The IPsec encapsulation of HIP data packets within a Bound End-to-End Tunnel (BEET) is presented.

Chapter 5 then proceeds to describe extensions to HIP that enable host mobility and multihoming (the UPDATE packet and the LOCATOR parameter), the role of a rendezvous server and registering with it, and DNS extensions. Chapter 6 covers advanced extensions. The HIP opportunistic mode enables a HIP host to establish a HIP association to another host without prior knowledge of its identity. Piggybacking of data to HIP control packets can reduce the HIP association establishment time. HIP service discovery extensions enable a HIP host to locate available service, such as a rendezvous server, in the local network. Simultaneous multiaccess enables efficient utilization of multiple network paths between the HIP host. The SIMPLE presence protocol can be used to exchange host identities for HIP. Chapter 7 presents implementation and performance results of using HIP on lightweight hardware, a Linux PDA. Chapter 8 concludes the second part with a description of lightweight HIP that suits well for Internet hosts with modest hardware resources.

Part III describes extensions to the Internet infrastructure that enable wide-scale HIP deployment. The part starts with Chapter 9, which focuses on the interactions between the end-to-end HIP and middleboxes located on the path between two HIP hosts. Because the base HIP has difficulties in passing through NATs and firewalls in the Internet, the NAT traversal extensions for HIP form an important part of this chapter. The chapter also considers HIP-aware middleboxes that are designed to explicitly support HIP.

Chapter 10 covers the process of mapping of host name to its current locator, also known as the name resolution. The chapter starts with describing the requirements for the name resolution service and an overview of Distributed Hash Tables (DHTs). The interface between OpenDHT and the HIP host is described to insert and lookup HIT-IP mappings. Next, the use of overlay networks for routing of HIP control packets is introduced. The Host Internet Indirection Infrastructure is described in detail. In the initial HIP deployment phase, many of the infrastructure services are running on PlanetLab, a distributed research testbed of roughly 700 servers worldwide. Chapter 11 introduces extensions to HIP architecture that enable host micromobility. The third part is concluded with Chapter 12 outlining protocol and infrastructure extensions to support location privacy of HIP hosts.

Part IV starts with Chapter 13 listing several practical applications that benefit from the use of HIP protocol. As a real-world case study, we describe the HIP deployment in a wireless testbed on a Boeing airplane factory. Chapter 14 presents an interface between HIP and applications. First, compatibility mechanisms that let existing applications benefit from communicating over HIP without the need for application modifications are described. The Native API enabling the use of wider functionality for new applications is presented next.

Chapter 15 discusses the integration of HIP with other protocols, such as SIP and Mobile IP. Experimental proposals for replacing the HIP base exchange or IPsec encapsulation with IKEv2 or SRTP are presented. The chapter concludes with description of a HIP proxy that enables the legacy hosts to benefit from HIP.

Appendix A describes how HIP can be gradually deployed on the readers' own computers and what the benefits are. In particular, this chapter includes practical advice and tutorial-style sections on downloading and installing an existing HIP implementation, utilizing PlanetLab's HIP services, and simple "first steps" experiments with HIP using the Wireshark network analyzer.

2

Introduction to network security

Tobias Heer[1]

The Host Identity Protocol uses a wide range of cryptographic mechanisms to secure the Host Identity (HI) namespace, to securely establish a protected channel, to defend against Denial-of-Service attacks, and to protect the mechanisms that, among other features, enable mobility and multihoming. This chapter introduces some basics that are necessary to understand the design and rationale of the security mechanisms employed by HIP. After stating some goals for secure communication protocol, we will discuss various attacks that aim at undermining these goals. In the following, we introduce some basic cryptographic techniques that serve as building blocks for security and key-exchange protocols. We conclude the chapter with an introduction to some security protocols that are closely related to HIP and its design.

2.1 Goals of cryptographic protocols

The most obvious reason to use cryptography in communication is to protect data from being read by unauthorized persons. Besides providing confidentiality, several other properties are desirable for secure communication protocols. Depending on the application scenario, some of these goals may appear more or less important. However, when designing security protocols it is important to consider all of these goals and to carefully judge before giving up one in favor of another. Essential or desirable properties of secure communication protocols are:

Authentication: A receiver of a message should be able to determine the origin of the message. This implies that no attacker should be able to send a message with forged source information.

Authorization: Only authorized network entities or users should have access to restricted resources, services, and data.

---

[1]This chapter, with the exception of Section 2.7, is contributed by Tobias Heer (RWTH Aachen University, Distributed Systems Group).

Host Identity Protocol (HIP): Towards the Secure Mobile Internet   Andrei Gurtov
© 2008 John Wiley & Sons, Ltd

Accountability: It should be possible to identify the user of a service unambiguously in order to account for that service.

Data integrity: A receiver of a message should be able to verify that the contents of the message have not been altered on the communication path.

Confidentiality: Data should be protected from unauthorized access. This also applies to data in transit that must be encrypted in some way to achieve confidentiality. Confidentiality is not necessarily restricted to the protection of data but may also comprise meta-information about the communicating entities.

Reliability: A host that provides services should not be vulnerable to attacks that affect the availability or the quality of the services in a negative way.

Non-repudiation: A network entity or user should not be able to falsely deny its participation in a communication.

Privacy: The identity of a user or a network entity should not be revealed to unauthorized parties. This requirement is also called identity protection.

Consistency: If two honest hosts establish a communication context, both should have a consistent view of the parties involved in the communication process.

Some of these security goals build on each other. For example, achieving accountability or authentication may be difficult without the requirement that both hosts need a consistent view of the parties involved in the communication. However, these dependencies may differ from one use-case to another. For some cases, like when identities are transmitted over insecure channels, privacy requires confidentiality while for other uses it may not be required.

2.2 Basics and terminology

In a communication process, two or more hosts are exchanging data. These hosts are called *peers*. In order to distinguish certain peers we use a popular naming scheme throughout this chapter. The first host that acts is denoted as *A* or *Alice* while the reacting host is denoted as *B* or *Bob*. Depending on the role of a peer in the communication process, peers are also denoted as *Initiator* and *Responder*. The Initiator initiates a process (e.g., a key exchange) and the Responder responds to the Initiator's request. It is difficult to discuss security protocols without considering attackers. Following the same naming convention as for Alice and Bob, we denote an attacking party *Mallory*.

When applying cryptography to communication channels, all peers involved must share the same *security context*. A security context is the set of information that is required to apply security measures. For instance, two hosts that use a shared secret to encrypt and decrypt data must be in possession of the same secret. Thus, the secret is part of their shared security context.

Algorithms that encrypt and decrypt data are called ciphers. The unprotected text that serves as input for the encryption function is denoted as *plaintext* and the encrypted text that was transformed by the cipher is denoted as *ciphertext*.

2.3 Attack types

To understand the rationale of security mechanisms in communication protocols, it is necessary to understand possible attack techniques. This section gives a brief overview of the most relevant attacks and countermeasures for these attacks. This overview is not comprehensive but it comprises the most important attacks. These attacks aim at undermining one or several of the security goals stated above. The discussed attacks can be categorized as attacks that target communication protocols, hosts, and ciphers.

2.3.1 Eavesdropping

Eavesdropping denotes the process of overhearing a private communication. An eavesdropper can read all or some messages that a group of communicating peers exchange. Depending on the network topology and the employed transmission technology, eavesdropping can be either simple or difficult. Eavesdropping on unprotected wireless communication channels is simple while eavesdropping on wired communication is more difficult as it requires physical access to the medium. Without adequate cryptographic protection, eavesdropping can compromise the confidentiality of data in transit. Data encryption is the most common way to deal with eavesdroppers.

2.3.2 Impersonation

Impersonation attacks take place whenever an attacker aims at maliciously pretending to represent another host or user. If, for example, Mallory succeeds to pretend to be Bob while communicating with Alice she impersonates Bob. Therefore, besides consistency, impersonation undermines a range of security goals such as authentication, authorization, non-repudiation, accountability, and possibly data integrity and confidentiality. Impersonation often requires to forge authentication data or to send messages with forged source addresses. A special case of an impersonation attack is the Man-In-The-Middle attack presented below.

2.3.3 Man-In-The-Middle attacks

A *Man-In-The-Middle* (MITM) attack is a special form of impersonation attack. It can take place whenever an attacker is situated on the network path between two hosts and can delay, modify, or drop packets. Assume Alice initializes a communication process with Bob and that Mallory performs a MITM attack. When Alice tries to contact Bob, Mallory intercepts the messages between Alice to Bob and modifies the messages to make Alice and Bob believe that they are directly communicating with each other while they are communicating with Mallory.

MITM attacks are one way to compromise public key cryptography (cf. Section 2.4.2) if an attacker can modify the exchange of cryptographic identities unnoticeably. Instead of attempting to break the public keys of Alice or Bob, Mallory replaces their public keys in the messages with her own public keys. If this manipulation is not recognized, both peers establish a security context with Mallory. Figure 2.1 illustrates the situation. The attacker forwards all messages between Alice and Bob. This enables her to act exactly like Alice and Bob would. As Alice and Bob have established a security context with Mallory, she can decrypt, read, and possibly modify all messages exchanged between them.

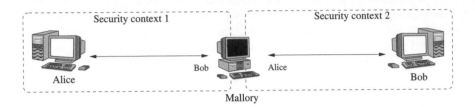

Figure 2.1   Mallory uses a Man-In-The-Middle attack on Alice and Bob.

2.3.4   Delay and replay attacks

The *replay attack* is a form of attack in which the attacker uses validly encrypted or integrity protected data in a fraudulent way to subvert a communication protocol. In order to use a replay attack, the attacker must be able to eavesdrop on the communication channel and to send forged messages. Such an attacker can retransmit previously sent messages. An attacker who is situated on the communication path between the victims can, in addition, delay, reorder, or drop packets. The attack can take place immediately after a message has been recorded or at any later point in time. The goal of a replay attack is to trick the receiver of the replays into triggering duplicate transactions. Recorded messages can also be used to subvert communication processes other than the one from which the messages were recorded. An attacker could, for example, try to use a recorded protocol handshake to impersonate one of the parties that have been involved in the original communication process.

2.3.5   Denial-of-Service attacks

Another attack is the *Denial-of-Service* (DoS) attack, which typically targets hosts that provide some sort of service (servers). The goal of a DoS attack is to consume the resources of a victim to an extent that it is not able to provide adequate or any service to legitimate hosts. Thus, this attack aims at the security goal of reliability.

To maximize the impact of the attack, attackers try to invoke processes on the server that consume much CPU time, memory, bandwidth, or other scarce resources. Processes that are exposed to the network and do not require authentication are especially prone to DoS attacks. Moreover, the authentication process itself can be vulnerable. To amplify the impact of the attack, the attacker often mounts parallel attacks, causing one, several, or hundreds of hosts to simultaneously initiate as many processes as possible. An attacker can, for example, waste resources of the victim by connecting numerous times until the resources that are used for the connection establishment are depleted. In this situation, neither the attacker nor legitimate clients can establish new connections with the server, which, therefore, cannot provide its service any more.

2.3.6 Exhaustive key space search

Besides attacks that try to exploit weaknesses in communication protocols, there are attacks that specifically try to break the ciphers employed by the protocol. As most cryptographic algorithms require the possession of a key, finding the secret key is the natural way of breaking the applied protection.

Encryption schemes typically use keys for encrypting and decrypting data. One of the least sophisticated ways to break such a key-based cryptographic algorithm is to try different keys until one of the keys is usable for decrypting encrypted data. This kind of attack is possible for all encryption schemes in that only one sensible plaintext is possible for every ciphertext otherwise an attacker cannot decide whether or not the decryption was successful. However, most cryptographic schemes minimize the effectiveness of the exhaustive key space search by using large key-spaces, leading to an enormous computational cost for an exhaustive key search. Due to the simplicity and the high cost, the exhaustive key space attack is often called *brute force* attack.

2.3.7 Cryptoanalysis

Cryptoanalysis is the means of deciphering a message without knowledge of the encryption key. Different ciphers and cryptographic mechanisms are vulnerable to different cryptoanalytical attacks that exploit mathematical findings and shortcuts that break or decrease the security of a cipher. Cryptoanalysis can either support brute-force attacks by reducing the size of the probable key-space or make key-search unnecessary by providing alternative ways of deciphering the ciphertext.

A special kind of cryptoanalytical attack is the side channel attack, exploiting weaknesses in the physical implementation of the cipher. Such attacks observe timing, energy consumption, or memory usage of the cipher and deduct information about the key or the state of encryption or decryption processes from these.

2.4 Defense mechanisms

Secure network protocols are built from algorithms and techniques that can be seen as complementary building blocks. Each measure or technique provides protection against one or several threats discussed above ensuring one or several of the security goals stated in Section 2.1. However, for many purposes it is necessary to combine several mechanisms to reach a security goal.

2.4.1 Symmetric cryptography

Algorithms that use the same key for encryption and decryption are called symmetric ciphers. If the keys of symmetric ciphers are not the same, one key can be generated from the other. This property requires the communicating entities to keep the key secret in order to prevent attackers from decrypting encrypted data. All communication peers must be in possession of the same secret key in order to use symmetric key cryptography to secure communication between them. Symmetric cryptography is widely used because it is less CPU-demanding than public-key cryptography.

There are two classes of symmetric ciphers: *block ciphers* and *stream ciphers*. Block ciphers encrypt plaintext blocks of a fixed length l into identical sized blocks of ciphertext while stream ciphers take streams of plaintext and encrypt the stream symbol by symbol. The following sections give a brief overview of block and stream ciphers. As block ciphers are most relevant for HIP, stream ciphers are only discussed briefly.

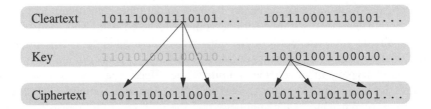

Figure 2.2 The principles of confusion and diffusion.

Block ciphers

Block ciphers operate on fixed-size blocks of input data. Typical block lengths are 64 bits, 128 bits, and more. The plaintext is segmented into blocks. As the length of the plaintext is often not a multiple of the block size, the last block is filled with additional symbols, so-called padding data.

Secure ciphers follow the concepts of *confusion* and *diffusion* as proposed by Shannon (Shannon 1949) as depicted in Figure 2.2. The goal of these concepts is to make the relationship between plaintext, ciphertext, and key as complex as possible to prevent attacks on the cipher. In practice, confusion means that changing one bit of the plaintext changes multiple bits in the ciphertext whereas diffusion means that the ciphertext is not related to the key in a trivial way. Thus, changing one bit of the key changes several bits of the ciphertext.

Feistel ciphers   Feistel ciphers are block ciphers that follow an architectural concept first introduced by Horst Feistel. They are the basis for many popular block ciphers, such as the *Digital Encryption Standard* (DES) (Schneier 1996), triple DES (3DES), and *Blowfish* (Schneier 1996). Feistel networks follow the principles of confusion and diffusion in order to transform plaintext into ciphertext. A Feistel network consists of several encryption rounds in which a number of operations are applied to the plaintext or the intermediate results from the previous round. Typically each round consists of a substitution and a transposition step. Substitution means that symbols of the input data are replaced with different symbols while transposition means that the order of the symbols is changed.

Figure 2.3 depicts a Feistel network. To encrypt data, n *round keys* (K_n) are generated from a shared key. The input block of size l is split into two blocks of $l/2$. The right-hand sub-block R_0 is combined with the first round key K_0 in a substitution function F. The result is combined with the left-hand sub-block L_0 by using the *xor* operation. A transposition step that flips the order of the intermediate results R_1 and L_1 follows. The same combination of substitution and transposition steps is repeated n times with the corresponding round keys.

The result of the last encryption round is the ciphertext. The decryption process is identical to the encryption process with the difference that the round keys are applied in reversed order beginning with R_n.

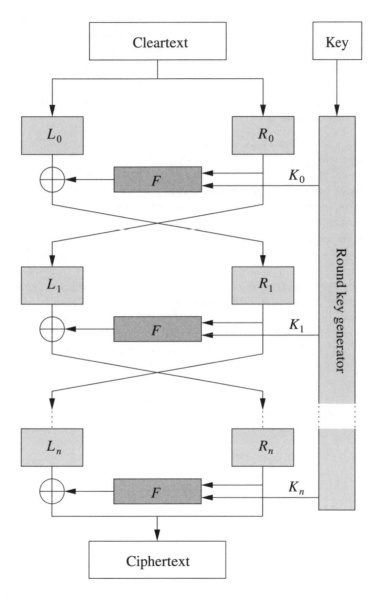

Figure 2.3 The structure of a Feistel network.

The operations that are performed during each round have a low computational complexity, resulting in efficient encryption and decryption. Moreover, the low complexity allows hardware implementations,which further reduces the computation time on specialized computer systems. The fact that the operations for the encryption and decryption are identical also supports hardware implementations as the same routines can be used for both.

The Advanced Encryption Standard: AES   The official successor of DES is the *Advanced Encryption Standard* (AES). It is finalized as *Federal Information Processing Standard* approved encryption algorithm (Standard 2001). Like for Feistel ciphers the operations of the AES algorithm are applied in several rounds. AES operates on a fixed block length but allows 128-, 192-, and 256-bit keys to be used. The number of rounds (10, 12, or 14) is determined by the length of the key. AES is considered to be secure and has undergone thorough cryptographic analysis.

Stream ciphers

In contrast to block ciphers, stream ciphers encrypt a stream of plaintext symbol by symbol. The most common method of stream-cipher encryption is to generate a key stream from a shared key. The data stream is added to the key stream symbol by symbol. The cipher stream resembles a stream of random numbers and, thus, is hard to remove from the ciphertext. The way of operation of stream ciphers approximates the use of a *one-time pad*. The one-time pad encryption uses a random bit stream to encrypt and decrypt data. The one-time pad is an unbreakable cipher, provided the source of the key stream is truly random. The difference to the one time pad is that stream-ciphers use a pseudo-random key stream.

A popular stream cipher is RC4 (Schneier 1996), being used in protocols such as IEEE 802.11 *Wired Equivalent Privacy* (WEP) and *Wi-Fi Protected Access* (WPA) or the *Secure Sockets Layer* (SSL) (Frier *et al.* 1996). It generates a pseudo-random key stream that is added to the data stream with an *xor* operation. The cipherstream can be decrypted by reversing this operation by using *xor* and the key stream. Therefore, the sender and the receiver must compute the same key stream. The key-stream generation is based on an initial key that sets the initial internal state for the RC4 algorithm. Figure 2.4 depicts how a stream is encrypted and decrypted with a stream cipher.

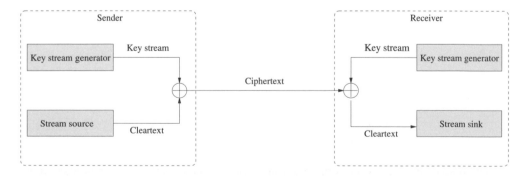

Figure 2.4 Stream ciphers use an identical key stream to encrypt and decrypt a data stream.

Key exchange for symmetric cryptography

The question of how to exchange the symmetric keys arises whenever two hosts employ symmetric cryptography. Many security protocols utilize public-key cryptography to securely agree on shared secrets. The Diffie–Hellman key exchange can be used to generate the necessary symmetric keys during the base exchange. The following sections give a brief introduction to public-key cryptography and related key exchange mechanisms.

2.4.2 Public-key cryptography

Unlike symmetric key cryptography, public-key cryptography uses encryption and decryption keys that are related in a non-trivial way. This allows to make the encryption key publicly accessible as a *public key* while the decryption key is kept secret as a *private key*. This enables hosts to send encrypted data to any host without exchanging shared secrets first, provided the sender knows the public key of the receiver.

Easy, hard, feasible and infeasible Public-key encryption algorithms are based on so-called *trapdoor functions*. These trapdoor functions can be calculated easily but are hard to reverse without additional information in the form of a private key. In this context, the terms hard and easy, as well as feasible and infeasible, relate to the complexity of the calculation. Computations that take up to millions of years are denoted hard or infeasible. On the contrary, computations that can be calculated in a matter of milliseconds, seconds, or hours, depending on the computational power that is used, are denoted easy or feasible.

Properties of trapdoor functions Functions that are suited for public key cryptography must provide three properties:

- The function must be easy to compute and therefore the encryption is computationally feasible.

- It must be hard to reverse the function without the possession of additional information. This provides that the encrypted data cannot be decrypted without the private key (the additional information).

- It must be computationally hard for an attacker to calculate the secret if it is in possession of the cryptographic algorithm, the public key, and an arbitrary amount of encrypted data, and corresponding plain text.

Public-key cryptography is based on mathematical problems that are hard to solve. Prime factorization and the discrete logarithm are such problems. HIP uses PK cryptography to authenticate hosts. The Host Identities are essentially public keys. Furthermore, HIP generates the keys for the symmetric encryption by utilizing PK cryptography in the Diffie–Hellman key exchange.

Diffie–Hellman key exchange

The first PK cryptosystem was published by Whitfield Diffie and Martin Hellman (Diffie and Hellman 1976). It provides a way to securely exchange symmetric keys over insecure channels. It is based on the discrete logarithm problem. The original version of the

Diffie–Hellman (DH) key exchange uses the multiplicative group of integers modulo p. Given the primitive g modulo p, with p as a large prime number and g as generator of the modular group, computing x from g^x is considered infeasible.

The key exchange The discrete logarithm problem can be used as the basis for a key exchange protocol. In order to exchange a secret shared key, two parties agree on p and g. Both numbers may be exchanged in public. In practice, only few combinations of g and p are in use. These *Diffie–Hellman groups* do not need to be kept secret and are often publicly defined in protocol specifications as for the *Internet Key Exchange protocol* (IKE) (cf. Section 2.5.1).

The DH key exchange consists of four steps. A secret key is publicly negotiated during these steps. However, an attacker can neither read nor create the shared secret due to the computational complexity of the discrete logarithm problem. We give a short description of the four steps. Further explanations and analysis are provided in Schneier (1996).

1. The Initiator (I) of the key exchange selects a random number $x \in \{1, \ldots, p-2\}$ and calculates $g^x \bmod p$. It sends g^x to the Responder R and keeps x secret. It is computationally infeasible to calculate x from g^x due to the complexity of the discrete logarithm problem. The value g^x is used as the Initiator's public key whereas x is its private key.

2. The Responder (R) randomly selects a secret number $y \in \{1, \ldots, p-2\}$. It calculates $g^y \bmod p$ and $k = g^{x^y} \bmod p = g^{xy} \bmod p$. The result k is the shared secret that can be used as a symmetric key. The Responder sends back g^y as its public key. The Responder may destroy its private key y afterwards.

3. The Initiator calculates the shared secret $k = g^{y^x} \bmod p = g^{y^x} \bmod p = g^{x^y} \bmod p$. It may discard x now.

Both hosts are in possession of the secret key k after the key exchange. They generate symmetric keys of the desired key length from this secret by selecting some bits or combinations of bits from this key (cf. Section 2.5.2). Typical key lengths l that are assumed to be safe for longer terms are 1024 bits or more. The values p, g, x, y, and k are all large numbers in the order of magnitude of l. Arithmetic computations on large numbers are typically not supported in hardware. Especially the exponentiation of large numbers is costly in terms of CPU cycles.

Using the Diffie–Hellman Key exchange allows for *Perfect Forward Secrecy* (PFS) because it is typically not used to authenticate hosts and, thus, the secret key can be deleted after the communication ended. To achieve PFS, both communicating peers delete their secret values x and y and the generated session key $g^{xy} = k$. After deletion of these values, there is no way to reproduce the shared secret and, thus, it is impossible to easily decrypt encrypted data without breaking the Diffie–Hellman key exchange or the symmetric cipher. This protects the encrypted data even in cases in which an attacker gains full control over one of the communicating peers after the communication process has ended.

The values $g^x \bmod p$ and $g^x \bmod p$ can be pre-calculated before the key exchange. This reduces the operations that need to be computed online to one exponentiation per host, which reduces the time for the exchange.

The RSA algorithm

The RSA algorithm is another PK encryption algorithm, published by Rivest, Shamir, and Adelman (Rivest *et al.* 1978). The RSA algorithm is named after their initials. It is the first PK algorithm that was suitable for encryption as well as for message authentication.

The RSA algorithm builds on the fact that though multiplying two large prime numbers is easy, factorizing their product into the two original prime factors is computationally hard. The prime factorization problem is used as a trapdoor function, which enables encryption and decryption with a public-key pair.

Assume that Alice wants to send an encrypted message m to Bob over an insecure channel. Bob first generates a public-key pair k_{pub} and k_{priv} and publishes its public key k_{pub}. Alice uses Bob's public key to encrypt the message with the RSA encryption function E and sends the encrypted message $E_{k_{pub}}(m)$ to Bob. Bob can use its private key and the decryption function D to decrypt the message $D_{k_{priv}}(E_{k_{pub}}(m)) = m$.

RSA key generation The RSA key generation process finds a public key and a private key that can be used for encryption and decryption, respectively. The key generation process consists of four steps. Mathematic preliminaries and a detailed description are provided by Schneier (1996).

1. Find two large prime numbers p and q of similar size. This is typically done with probabilistic primality tests like the Miller–Rabin test (Miller 1975; Rabin 1980).

2. Calculate $n = pq$ and $\varphi(n) = (p - 1)(q - 1)$. φ is Euler's φ function, which gives the amount of numbers x with $x < n$ and $\gcd(x, n) = 1$. $\gcd(x, y)$ denotes the *Greatest Common Divisor* of x and y.

3. Choose a random number e, $1 < e < \varphi(n)$ with $\gcd(e, \varphi(n)) = 1$.

4. Calculate $d = e^{-1} \bmod \varphi(n)$ by using the *extended Euclidean algorithm.*

The pair (n, e) is used as the public key k_{pub} and d is used as the private key k_{priv}.

RSA encryption and decryption Alice can use the public key (n, e) of Bob to send an encrypted message. The encryption process can be summarized in three steps:

1. Alice looks up the public key of Bob $k_{pub} = (n, e)$ by using some public-key infrastructure or by asking Bob for its public key. Note that the second variant in that Bob gives Alice its public key can be subject to a MITM attack if an attacker is situated on the communication path between Alice and Bob.

2. Alice divides the message into pieces and transforms each into a number $m \in \mathbb{Z}_n$.

3. Alice calculates the cipher text $c = m^e \bmod n$.

Alice can send the cipher text c to Bob over an insecure channel as only Bob is able to decrypt c by using its private key k_{pub}. The decryption process repeats the encryption process with d used instead of e:

$$c^d \bmod n = (m^e \bmod n)^d \bmod n = m \qquad (2.1)$$

RSA signatures are considered secure provided that suitable numbers for $n = pq$ are used[2] and that advances in cryptography do not yield algorithms that are able to efficiently calculate prime factors or to solve other equivalent problems. Typical sizes for $n = pq$, d, and e are 1024 bits and more. Note that short RSA keys are considered unsafe since a 576-bit RSA key has been broken (decomposed into its prime factors) (Cavallar *et al.* 2000).

Typically, small numbers are chosen for e, making encryption more efficient than decryption because d depends on the choice of e and d cannot be chosen freely.

RSA signatures   The RSA encryption and decryption process can be used to digitally sign messages. A valid signature ensures that the content of a message has not been modified or forged by third parties that are not in possession of the private key components. The signature process works similarly to the encryption process. The only difference is that the encryption is performed with the private key of the sender and the decryption with its public key. The encryption process with the private key is called signature process while the decryption is the signature verification.

Like the DH key exchange, RSA encryption and decryption require to compute CPU-intensive exponentiation of large numbers, slowing down the encryption and decryption process. Therefore, RSA is mainly used to transmit short encrypted messages like symmetric keys or signatures of short message digests (cf. Section 2.4.3).

The public key of a host or user can be seen as its identity because only the host in possession of the private key can use it to sign and decrypt data. *Public key infrastructure* (PKI) is a system that allows to map the public key of a host or a user to its real world identity in a secure way, thus allowing to verify the binding between the user or host and its key. Using PKI solves the problem of MITM attacks because a mis-binding between an attacker's public key and a victim's identity can be recognized by consulting the PKI. To constitute the correctness of a binding, a *Certificate Authority* (CA), which is part of the PKI, issues a certificate that holds the public key and the identity information of a host or user. The CA signs this certificate to protect it from manipulations. This allows every host that is in possession of the CA's public key to verify certificates and, therefore, the binding between a public key and an identity.

The certificates can be used whenever a host needs to prove its real world identity by means of public key cryptography. The public keys of trustworthy CAs are often already included in the operating system or other software like web browsers. This allows to verify certificates without the problem of getting the CA's public key first. HIP uses public keys as identifiers for hosts, allowing it to operate without CA's because the mapping between the identity and the public key is trivial and cannot be forged.

DSA signatures   The *Digital Signature Algorithm* (DSA) has been specified to be used with the *Digital Signature Standard* (DSS). It is a variety of the *El Gamal* signature scheme and requires the use of a cryptographic hash function. Like the Diffie–Hellman key exchange, DSA relies on the complexity of the discrete logarithm problem. The properties, application, and performance of DSA are similar to RSA. Therefore, we do not provide a detailed description here. However, interested readers can find a detailed description of the DSA standard in Schneier (1996).

---

[2]The selection and the size of the numbers matter. For details see Schneier (1996).

2.4.3   One-way cryptographic hash functions

Besides symmetric and asymmetric cryptography, *one-way cryptographic hash functions* form a third class of cryptographic mechanisms. These functions transform a given value of variable length into a value of length n in a way that does not allow to compute the original value from the result. Thus, the computation of $H(x)$ should be feasible, while reversing the function should be infeasible. Hash functions inherit their name from their way of operation because the characteristics of the hashed result resembles a random number and cannot be correlated to the input any more. Cryptographic hash functions should be:

Second preimage resistant A hash function is second preimage resistant if for a given x it is computationally hard to find $x' \neq x$ with $H(x') = H(x)$.

Collision resistant A hash function is collision resistant if it is computationally hard to find any two values x and x' with $x \neq x'$ and $H(x) = H(x')$.

Preimage resistant A hash function is called preimage resistant if for a given $z = H(x)$ it is computationally hard to find a preimage x' with $H(x') = z$.

There are two ways of designing cryptographic hash functions. For one, their design can be based on the same algebraic problems as public-key cryptosystems. These hash functions have the same computational complexity but also the same performance issues. Alternatively, block-cipher-based hash functions, following the Merkle–Damgård construction (Damgård 1990; Merkle 1989), can be used as building blocks for hash functions. The Merkle–Damgård construction is the most common design for cryptographic hash functions. It constructs hash functions with arbitrary input length by cascading a compression function. Since the computational complexity of these hash functions is very low compared with the complexity of public-key operations, they are widely used for many purposes. Examples of popular cryptographic hash functions are the SHA hash function family (National Institute of Standards and Technology 1995) and MD5 (Rivest 1992).

Message digests

In practice, cryptographic hash functions are often used for message authentication. Signing long messages with RSA or DSA is a CPU intensive operation and produces signatures of the size of the messages, thus doubling the required space for storage and the bandwidth for transmission. Cryptographic hash functions can solve this problem because they can produce a short *message digest* of fixed length. The second preimage- and collision resistance of the hash function ensures that signing the short message digest is sufficient to securely sign the message. Although signing the fixed length digest signs an infinite number of messages that map to the same hashed output, finding one of these alternative messages is considered infeasible.

Hashed message authentication codes: HMACs

Cryptographic hash functions provide a possible method to verify the authenticity and the origin of a message without using public-key signatures. In order to sign messages with hash functions, all communicating peers must be in possession of a shared secret. This shared secret can be exchanged manually, by using the Diffie–Hellman key exchange, or in any other secure way. Given the shared secret s, a message m, and a cryptographic hash function $H(x)$, the peers compute a *Message Authentication Code* (MAC) $MAC(x)$ to sign a message. The symbol $\|$ denotes concatenation in this context.

$$MAC(m) = H(k \parallel m)$$

This simple scheme allows hosts that know the secret key to authenticate the message m by recalculating the MAC. Other hosts are not able to forge this signature as they are not in possession of the secret key. The *preimage resistance* of the cryptographic hash function prevents the secret k from being calculated from $MAC(m)$ and the property of *collision resistance* prevents an m' with $h(m') = h(k \parallel m)$ from being found.

However, this is only true when an ideal hash function is used. Using practical hash functions leads to security problems. Iterative hash functions that are designed according to the Merkle–Damgård construction principles are susceptible to length extension attacks, which cause serious problems with this naive MAC. An attacker can forge the MAC for a suitable m. Mironov describes this attack and further security issues for practical hash functions (Mironov 2005). Descriptions and practical implications of further attacks on hash functions were provided by Wang and Yu (2005), Joux (2004) and Kaminsky (2004), and others.

The HMAC message authentication mechanism is designed to provide a safe way to authenticate messages that is not susceptible to length extension attacks. It was standardized by the IETF in RFC 2104 (Krawczyk *et al.* 1997). The HMAC is calculated with the following formula:

$$HMAC_k(m) = H((k \oplus opad) \parallel H((k \oplus ipad) \parallel m)).$$

The constants $opad = 0x5c5c5c..5c$ and $ipad = 0x3c3c3c..3c$ are of the same size as the input block length of the hash function (512 bits for SHA-1 and MD5). *Ipad* and *opad* are mnemonics for inner and outer padding. The symbol $\oplus$ denotes *exclusive or*.

The HMAC computation performs three steps. It concatenates a secret key with the message. It adds noise to the key to complicate attacks on the mechanism and uses two hash operations to generate the *keyed-hash message authentication code*. Signature verification is analogous to the signature generation process. A receiver computes the HMAC from a given signed message and the shared key and compares the result to the HMAC attached to the message. If both match, both HMACs have been created for the same message with the same shared key. This allows to verify the origin of the message, i.e. the holder of the shared key, and the integrity of the message.

A disadvantage of HMAC signatures is that only messages exchanged between hosts that share a secret key can be authenticated. Therefore, other measures are needed to identify a host and to verify messages from a host for which no security context in form of a shared key exists. Typically the number of hosts that share the secret is kept very small. Therefore, middleboxes that inspect and relay packets usually have no knowledge of the shared keys that were used for the HMAC authentication of packets they forward. Thus, they cannot verify the HMAC signatures.

The HMAC generation and verification is very cheap in terms of CPU cycles. It only requires two hash computations and, therefore, outperforms signature schemes that are based on public-key algorithms by magnitudes.

Hash chains

In 1981, Leslie Lamport proposed to use chains of hashes to protect remote password authentication over insecure channels from replay and eavesdropping (Lamport 1981). Since then, hash chains have been used in several other fields of application like electronic cash (Sander and Ta-Shma 1998) and micropayments (Rivest and Shamir 1996).

The basic idea behind hash chains is the iterated application of a cryptographic hash function on some random or pseudo-random seed value r, which we call *seed* of the hash chain. The hash function H is applied to this initial value. The result of the hash function is used as input for the next round until the hash chain has reached the desired length. The intermediate results form a chain of hashes, which is called the *hash chain*.

$$h_0 = r \qquad (2.2)$$

$$h_n = H(h_{n-1}) \qquad (2.3)$$

$$= H(H(H(\cdots H(r)\cdots))) \qquad n \text{ times} \qquad (2.4)$$

The sequence $(h_i, h_{i-1}, h_{i-2}, \ldots, h_1, r)$ is called the hash chain. Its first element h_i is called the *hash chain anchor*. In order to use a hash chain, the anchor is published while the rest of the chain is kept secret. The elements of the hash chain are disclosed in reverse order of their generation beginning with the anchor value. Thus, the first element that is revealed is h_i, the second one is h_{i-1}, and so on. Used in this way, hash chains provide a number of appealing properties, enabling their use for authentication in several ways.

- Given h_{i-1} and h_i, it is easy to verify that h_i belongs to the same hash chain as h_{i-1}. This property results from the equation 2.4 and the basic requirement for hash functions, that it should be feasible to calculate a hash function.

- It is computationally hard to find h_{i-1} if only h_i is given. This property is provided by the preimage resistance of the hash function. It ensures that someone who is not in possession of the hash chain cannot calculate unrevealed elements from already disclosed ones.

- Given h_{i+1}, it is hard to find an h' with $H(h') = H(h_i) = h_{i+1}$. This property also arises from the preimage resistance of the hash function. It ensures that no-one can forge a single unrevealed element of the hash chain.

- Given h_i, it is possible to verify that h_{i-n} is part of the same hash chain if $0 < n \le i$. This property comes from the iterated construction of the hash function. The check requires iterated application of the hash function to h_i. It ensures that hash chain values can be verified even when some values get lost due to transmission errors or attacks.

The owner of a hash chain can exchange an anchor with its peer. It can disclose the next undisclosed hash chain element later. The peer can verify that the owner of the hash chain has disclosed this element.

2.4.4 One-time signatures

Lamport proposed one-time signatures based on hash functions. A host uses two random values r and r' and applies the hash function H to both. The results $H(r)$ and $H(r')$ are published as a public key. A one-bit message can be signed by publishing r as signature s if the bit is 1 and r' otherwise.

Comparing the hash of the signature $H(s)$ with the public key values $H(r)$ and $H(r')$ verifies or falsifies the signature of the message.

This signature can be used to sign l-*bit* messages by using $2l$ hashes. The signature allows instant verification, provided the public key is known. However, $2l$ hashes must be calculated per signature.

An improvement of this scheme (Merkle 1988) reduces the number of hashes to $l + \lfloor \log_2(l) \rfloor + 1$ and the average number of hashes per signature to $(l + \lfloor \log_2(l) \rfloor + 1)/2$. This is achieved by only signing the bits with value 1. The number of bits with the value 1 is appended to the message and signed as well to prevent signature forgery. This improves the efficiency almost by a factor of two.

Winternitz proposed a scheme that reduces the signature size even further but his approach requires more computation. The scheme uses two hash chains of length l to sign a $\log_2(l)$ bit long message. Only two hashes need to be transmitted to sign a message but a multitude of hashes must be computed in order to generate the public keys.

Zhang (1998) proposed two protocols for signing routing messages. These protocols use hash chains to provide the hashes for the signature. This enables the use of one public key consisting of a set of hash chain *anchors* to sign several messages. The signature size for these protocols is about 2 to 4 KB per signature and has the same computational overhead as Merkle's scheme.

2.4.5   Sequence numbers

As discussed in Section 2.3.4, attackers can try to manipulate network protocols by replaying, reordering, or dropping packets. Sequence numbers are a simple measure to avoid such attacks. A monotonically increasing number is assigned to every message. This allows the receiver of a message to determine if a message has already been received and processed. Messages without the next expected sequence number can be identified as duplicate messages or as part of a replay attack. The decision whether to consider a message as part of an attack can be based on a strict or relaxed matching of sequence numbers. Strict matching means that all preceding packets must have been received before a packet with a higher sequence number is accepted or processed. Alternatively, hosts can use a window of legitimate sequence numbers, enabling less restrictive ways of matching. A range of sequence numbers is considered valid and any of these can be accepted next allowing to use sequence numbers in cases where sequential delivery of packets is not guaranteed.

The use of sequence numbers alone does not provide a sufficient defense against attackers that can modify the contents of a packet. Therefore, the sequence numbers must be integrity protected in some way. This can be done by using keyed message authentication codes (cf. Section 2.4.3), digital signatures (cf. Section 2.4.2), or data encryption (cf. Section 2.4.1).

2.4.6   Cryptographic nonces

Cryptographic nonces are a way to verify the reachability of a host and to defend against replay attacks. A nonce is a number or a string that is used only once. Nonces are often used in request–response mechanisms, e.g. when Alice sends a nonce to Bob who is expected to send the nonce back, possibly in an encrypted or otherwise integrity protected way. This mechanism allows Bob to verify that Alice can respond to the nonce, proving that she is able to respond to packets addressed to a certain network address. Moreover, nonces can be used to prove the possession of a secret. Alice can send a nonce to Bob with the request to apply a cryptographic technique like encryption or signatures to the nonce. Bob can only solve this task if he is in possession of the secret. The fact that the nonce is only used once forestalls replay attacks as it cannot be reused by an attacker.

Nonces are often derived from random or pseudo-random numbers but they may also include information that the host wants to embed in encrypted or unencrypted form. The sender of a nonce can use such embedded information to identify the nonce in the response of its peer as originated by itself. Furthermore, the embedded information can be used to ease the verification of the uniqueness of the nonce.

2.4.7 Client puzzles

Attackers that perform DoS attacks often target mechanisms that do not require many resources from the attacker to invoke them but require much effort or especially scarce resources from the victim to perform them. The rate at which an attacker can invoke vulnerable processes determines how fast the resources of the victim are depleted and, therefore, how effective the attack can be. Client puzzles are a way to level this imbalance by allowing a host to artificially generate CPU load on its peer, and thus, the potential attacker. Requiring a correct solution of the puzzle as a precondition before allocating scarce resources helps to make such vulnerable mechanisms less attractive to attackers and forces attackers to reduce their attack rate.

To not increase the load on the victim, the puzzle solution must be easy to verify while difficult to solve. Many puzzles are based on cryptographic hash functions (cf. Section 2.4.3). The peer or attacker must repeatedly apply a cryptographic hash function to the concatenation of a server-generated nonce and a varying client-chosen value until the result of the hash function exhibits a certain property.

Assume Bob is a potential victim who wants to defend against a DoS attack. If Bob uses the puzzle mechanism, he sends a random or pseudo-random value i to Alice. Alice is required to find a number j for which the binary representation of the result of $H(i|j)$, has k bits with the value 0 as lowest-order bits. H is a cryptographic hash function and | denotes concatenation. Alice must vary j and apply the hash function to the concatenation with the new j until a suitable solution for the puzzle is found. This kind of puzzle requires Alice to try 2^{K-1} values for j on average while Bob can verify the validity of a solution within constant time.

As increased puzzle difficulties force an attacker to spend more CPU time on solving the puzzle, simultaneous attacks by a single attacker are slowed down to a degree that the server can handle. However, legitimate clients with limited CPU resources, such as mobile devices, suffer from difficult puzzles as well. These clients may not be able to solve the puzzle within an acceptable time.

2.5 Security protocols

Security protocols employ selected cryptographic algorithms to achieve some or all of the security goals mentioned above. These protocols specify how two or more communicating parties must act in order to reach the goals. The security protocols presented in this section either establish a security context or use an already established security context to secure communication. A security context consists of parameters for the employed security mechanisms such as the choice of security algorithms, keys, and the state of the communicating peers.

The life cycle of a security context or a security association typically consists of an establishment phase during which all communicating peers agree on a set of algorithms and the required keys. After the establishment phase, the security context is used to secure the communication between the involved hosts. Eventually the security context is modified and other algorithms or keys are used. The life cycle ends with the closing of the security association and the deletion of the security context. There are secure communication protocols that manage all or certain parts of this life cycle. The SIGMA protocol family and with it the Internet Key Exchange (IKE) are examples for protocols that establish a security context and authenticate the peers involved in the communication process. The IPsec security architecture secures communication channels and, therefore, is concerned with the use of security contexts after their creation. HIP manages all of the mentioned phases including the establishment, modification, and tear-down of a security context and secure communication channel. The following section will give a brief introduction to SIGMA, IKE, and IPsec because HIP is closely related to these protocols or uses them.

2.5.1 Modular exponential Diffie–Hellman groups

Many security protocols use the Diffie–Hellman key exchange to generate a shared secret. In order to successfully generate the secret, both hosts must use the same modular exponential groups. To simplify the key exchange, RFC2412 (Orman 1998) and RFC3526 (Kivinen and Kojo 2003) specify a number of well known Diffie–Hellman modular exponential (MODP) groups. These groups have been published for the IKE but other security protocols use these groups as well. The prime numbers p of the different groups vary from 768 bits to 8192 bits. In order to allow devices with few CPU resources to use HIP, a 384-bit group is defined by Moskowitz *et al.* (2008). However, due to the short key length, this group is insecure and should only be used if a host cannot use DH groups with longer key sizes or if the security requirements are low. Each group is identified by a group identifier.

As described in RFC2412, the groups have been selected based on a certain pattern. The high order and low order 64 bits are forced to 1 to simplify the computation of different remainder algorithms. The middle section of each group is taken from the binary expansion of the number Pi (π). This means that the bits are selected from a portion of pi that with the 64 1-bits attached at the beginning and at the end is a prime number. Pi is suited as a basis for the modular groups because it inherits a sufficient amount of randomness. Moreover, using pi as the source eliminates the possibility that a modular group could be chosen in a way that intentionally would lead to weak primes and, therefore, to insecure keys. The 768-bit *well-known modular group 1* as defined in Orman (1998) and Moskowitz *et al.* (2008) is depicted in Figure 2.5 as an example. The generator g for this group is 22.

2.5.2 Keying material

The Diffie–Hellman key exchange does not generate symmetric keys directly but generates keying material of the length of the Diffie–Hellman group that was used during the exchange.

```
FFFFFFFF  FFFFFFFF  C90FDAA2  2168C234  C4C6628B  80DC1CD1
29024E08  8A67CC74  020BBEA6  3B139B22  514A0879  8E3404DD
EF9519B3  CD3A431B  302B0A6D  F25F1437  4FE1356D  6D51C245
E485B576  625E7EC6  F44C42E9  A63A3620  FFFFFFFF  FFFFFFFF
```

Figure 2.5 768-bit well-known modular group for HIP.

The length of the keying material is typically greater than the length of the symmetric keys that are required by the security protocol. Therefore, single bits or combinations of bits must be selected from the keying material as a symmetric key. IKE and HIP use a cryptographic hash function for this purpose. The hash function reduces the length of the keying material and also allows to derive several keys by adding extra information to the input string. These derived keys have the advantage that they cannot be related to each other without knowledge of the keying material. This ensures security for the remaining keys even if one of the keys is compromised.

To create a binding between the session keys and certain additional parameters of the key exchange, these parameters can be included into the derivation of the symmetric key. This is done by concatenating the Diffie–Hellman shared key with the additional parameters before deriving the symmetric keys. HIP, for example, bases the generation of its session key on the Diffie–Hellman shared key, the HITs of the Initiator and the Responder, the puzzle and its solution, and a two-byte counter. Using not only the Diffie–Hellman-generated keying material but also other information as the basis for the shared key allows both hosts to verify that this additional information is identical for both. Any malicious modification of one of the relevant parameters would lead to mismatching symmetric keys and would be recognized by all involved peers.

2.5.3   Transforms

There is a large variety of cryptographic algorithms and mechanisms that can be employed by a security protocol. In many cases, the algorithms are interchangeable. Examples for this are AES, DES, 3DES, and Blowfish or RSA and DSA signatures. Moreover, different key-lengths can be used for symmetric and asymmetric ciphers. In order to keep a security protocol suitable for many application scenarios, the choice of algorithms and key-sizes should not be fixed but negotiated from case to case. These negotiations typically take place during the handshake. A host offers a set of applicable algorithms in a so-called *transform* parameter. This parameter typically contains a number of *transform suites*. Each transform suite describes the combination of several cryptographic algorithms and key-lengths. Suites are identified by an index that allows a host to simply select a set of algorithms and key-lengths by referring to the corresponding index.

The transform parameter contains a set of indices to applicable transform suites. The order of the transform suites in the transform parameter expresses the preferences of a host in decreasing order. Accordingly, the most preferred transform suite is listed first. A host receiving a transform suite parameter can choose any of the contained transform suites according to its own preferences. After selecting an appropriate suite it sends back its choice to its peer to inform it about the transform they have agreed on. Both hosts use the algorithms and key-lengths, indicated by the chosen transform suite, throughout the communication process.

2.5.4 IP security architecture: IPsec

The Internet Protocol (IP) itself does not offer sufficient security for many application scenarios. It provides no defenses against eavesdroppers or attackers that modify or forge IP packets. In 1998, the IPsec security architecture was published as RFC2401 (Kent and Atkinson 1998b) and later replaced by RFC4301 (Kent and Seo 2005). IPsec allows for authenticating the source and the integrity of IP packets and can ensure confidentiality for IP payload. According to the layering principle, it adds a new header, distinguishing unprotected IP traffic from IPsec traffic. Support for IPsec is mandatory for standard compliant IPv6 implementations.

Security Associations To use IPsec, both communication endpoints must agree on a set of algorithms and keys. These negotiations can either be pre-configured or can take place during a separate handshake. The choice of how to negotiate the keys and algorithms is not specified by IPsec. Therefore, hosts can employ a range of protocols (e.g. IKE (cf. Section 2.5.8) or HIP) for this purpose. IPsec assumes that a shared symmetric key is present at both IPsec endpoints. These keys are part of the security context two hosts must share to communicate securely over IPsec. The shared security context is also called a *Security Association* (SA). According to RFC 4301 (Kent and Seo 2005), IPsec security associations are simplex connections that afford security services to the traffic carried by them. This means that two SAs – an incoming and an outgoing SA – are necessary to protect a duplex channel.

Security Association and Security Policy databases Hosts typically manage security associations in two data structures: the *Security Policy Database* (SPD) and the *Security Association Database* (SAD). The SAD holds information about established security associations. It contains two security associations for every secured duplex communication to a distant host – one association for incoming and one for outgoing traffic. Among other information, the SAD contains information about the IP address of the distant host, the IPsec protocol that is used to transmit the payload, the encryption and authentication algorithms, and the keys that are used. The SAD contains the contexts, needed to process incoming and outgoing IP packets.

The *Security Policy Database* (SPD) contains information about the services that are offered to IP datagrams and about the fashion in which they are offered. It also defines the kind of security mechanisms, i.e. algorithms and protocols, that must be used. Furthermore, it defines traffic that stays unprotected and traffic that may not leave the host at all. In the case that an SPD entry defines that IPsec processing is to be applied to a class of IP packets, it must contain the security association specification that is applicable, too.

It is possible that several IPsec security associations exist between a pair of hosts. In order to process the incoming packets with the right algorithms and keys, some demultiplexing information is needed. Port numbers as used by the transport layers are not suited for this purpose because the transport layer headers may be encrypted and therefore not be accessible before decryption. IPsec uses the *Security Parameter Index* (SPI) as demultiplexing information in the IPsec header. Each IPsec packet contains an SPI, identifying the right security context on the destination host. Every security association uses a different SPI. Therefore, both hosts must agree on two SPIs to send data back and forth. For outgoing traffic, security associations are referenced by the matching SAD entries. Incoming traffic is assigned to a security association by the SPI contained in the packet.

2.5.5   IPsec modes

In its basic form, IPsec offers two modes of operation. Firstly, IPsec can protect traffic between hosts in an end-to-end fashion. IPsec uses the *transport mode* for this purpose. Secondly, IPsec allows to securely connect several networks over a potentially insecure communication link to form a *virtual private network*. IPsec uses the *tunnel mode* for this purpose. In this mode, complete IP packets are tunneled from one network gateway to a gateway of a different network. A third mode is the *Bound End-to-End Tunnel* (BEET) mode, currently being standardized by the IETF. It combines aspects of the tunnel and transport mode.

Transport mode   The *transport mode* is used when traffic between two end-hosts has to be protected. To achieve this, an additional protocol header structure is added between the IP header and the transport layer header. This header structure contains all information required to verify or decrypt the packet at the legitimate receiver. The structure of the IP packet after adding the *IPsec header* is depicted in Figure 2.6. The transport mode cannot encrypt the IP header because it is used for routing the packet between the end-hosts. However, it can integrity protect parts of the IP header to detect unauthorized modifications. Figure 2.6 depicts how the IPsec header is inserted into an IP packet when the transport mode is used.

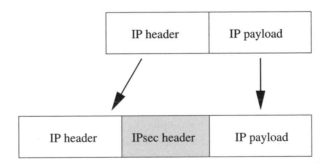

Figure 2.6 The transformations in IPsec transport mode.

Tunnel mode   The *tunnel mode* is mainly used to securely connect networks. When using tunnel mode, not the communication endpoints but two intermediate gateways, connecting two networks, establish an IPsec association and forward protected packets from one network to the other. Thus, the traffic is securely tunneled between these gateways.

As depicted in Figure 2.7, the tunnel mode adds an additional IP header before the IPsec header. The additional header is marked in grey. The original IP header stays the same when the packet is tunneled because the addresses of the gateways are provided in the additional IP header. The receiving gateway, i.e., the endpoint of the tunnel, removes the outer IP and IPsec header and forwards the unencrypted packet to its actual destination. Each IPsec tunnel mode packet consists of the outer IP header, the IPsec header, the inner IP header, and the payload. The inner IP header can be encrypted as it is not needed for routing the packet between the gateways. Therefore, the identity of the sender and the receiver can be protected from eavesdroppers as only the IP addresses of the gateways are present in the outer IP header. Parts of the outer IP header can be integrity protected to detect fraudulent modifications.

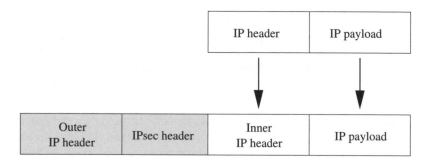

Figure 2.7 IPsec tunnel mode encapsulates the original IP packet into a new IP packet.

Bound End-to-End Tunnel mode   The *BEET* mode offers semantics similar to the *tunnel mode* with a *transport mode* like packet format. The way the BEET mode operates resembles a *Network Address Translator* (NAT). Like in tunnel mode, IP packets are tunneled from one gateway to another. However, the outgoing gateway does not attach a new header to the IP packet but modifies the existing IP header. Therefore, it effectively performs a network address translation step. The address translation is reversed by the incoming gateway of the destination network. This replacement procedure requires both tunnel endpoints to maintain a mapping between the SPI in the IPsec headers and the destination IP address. The address translation rules are bound to the SAs. Consequently, the incoming gateway can use the SPI in the IPsec header to identify the right translation rules. The payload is protected by IPsec while it is in transit between the gateways. Figure 2.8 depicts how the BEET mode protects IP traffic. The modified IP header is marked in light grey.

The BEET mode is less flexible than the tunnel mode because only one SPI to IP mapping can be used whereas the tunnel mode allows to send packets to different hosts in the destination network. The communicating peers are, therefore, fixed for the BEET mode. On the contrary, the tunnel mode can deliver packets from and to a range of source and destination IP addresses.

The BEET mode can be used to implement HIP payload encryption in a very efficient way if the HITs are used as source and destination IP addresses while the actual IP addresses of the hosts are the tunnel endpoints. This means the hosts and the gateways are the same network entities. The tunnel is used for encrypted IP packet delivery while the translation maps the IP addresses to the corresponding host identifiers.

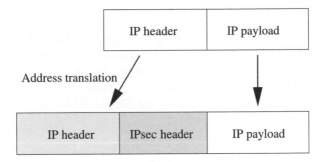

Figure 2.8 The IPsec BEET mode transforms the addresses in the IP header.

2.5.6   IPsec security protocols

IPsec offers two security protocols for protecting IP payload. The *Authentication Header* (AH) protocol (Kent 2005a) ensures the authenticity of an IP packet and its source whereas the *Encapsulating Security Payload* (ESP) protocol (Kent 2005b) additionally applies encryption.

AH is applicable whenever IP packets must be integrity protected but not necessarily encrypted. In order to protect an IP packet, AH employs checksums, based on cryptographic hashes as described in Section 2.4.3. Standard compliant IPsec implementations must support HMAC with SHA-1 (Eastlake 3rd 2005). AH protects the IP packet payload as well as certain fields of the IP header that may not be modified by routers and middleboxes. Examples for unprotected fields are the *Time to Live* (TTL) and the *Header Checksum* fields.

ESP protects IP payload against eavesdroppers by employing symmetric cryptography. It also allows the receiver of an ESP protected packet to authenticate the sender. ESP neither encrypts nor integrity protects the IP header. Instead of integrity protecting the source address, the source authentication is performed indirectly by using the key that is defined in the SA and, therefore, bound to the IPsec peer.

A sequence number field in the IPsec header protects both protocols from replay attacks. Instead of expecting the packets to arrive in strict sequential order, IPsec defines a range, a so-called window of sequence numbers that are considered valid. This window is called a *replay protection window*. This loose matching of sequence numbers is necessary because the IP protocol does not necessarily deliver IP packets in the right order. The range of valid sequence numbers (the position of the window) is determined by the size of the window and the highest valid sequence number received by a host. This highest sequence number sets the "right" edge of the window. All packets with lower sequence numbers are dropped if their sequence number is lower than the "left" edge of the window. For packets with sequence numbers within the range of the window, a host must verify that no packet with the same sequence number was received and processed before. Thus, the replay protection windows allow out of order delivery of IP packets in certain bounds. The minimum window size that must be supported is 32. A typical window size is 64 while for some rare cases, requiring reordering of packets or certain packet scheduling, sizes up to 1024 are used in practice.

2.5.7   SIGMA

The *SIGn and MAc* (Krawczyk 2003) family of key exchange protocols describes a general pattern on which many of today's key exchange protocols are based. The thorough crypto-graphic analysis of this pattern and well-known security properties of the SIGMA protocol family allow protocol developers to derive new key exchange protocols with the same core security properties from it. A very popular example for a SIGMA based protocol is IKE. The basic design elements of SIGMA protocols are:

- The use of the Diffie–Hellman key exchange for creating a session secret. This allows perfect forward secrecy (cf. Section 2.4.2).

- The use of public key authentication based on digital signatures to allow mutual authentication.

- The use of MACs (cf. Section 2.4.3) to bind the session keys to the identities of the communicating peers.

- The use of encryption to protect the identities of the protocol peers from being learned by attackers.

- The ability to delay the authentication of one peer to protect its identity from active attackers.

The SIGMA protocol does not specify a concrete communication protocol with all details like the choice of cryptographic algorithms. It abstracts from these details by assuming that its building blocks represent classes of interchangeable algorithms. SIGMA uses public-key cryptography for signing and verifying digital signatures, MACs, and symmetric encryption. During the SIGMA key exchange, both hosts mutually authenticate their communication peers and generate a shared secret that can be used to establish a secure channel.

SIGMA uses asymmetric cryptography to represent the identities of peers. The public keys are used as identity information and the private keys are used to prove the ownership of an identity. This corresponds to the use of public-key cryptography in PKI and the HI namespace.

Basic SIGMA protocol   The basic SIGMA protocol as illustrated in Figure 2.9 consists of a three-way handshake. The first message from Alice to Bob contains her Diffie–Hellman public key (g^x). The Diffie–Hellman public keys serve two purposes in the basic SIGMA design. For one, they are the basis for the Diffie–Hellam key exchange. For another, they serve as nonces that guarantee the freshness of the key exchange. However, using the public keys as nonces requires that Alice and Bob use different Diffie–Hellman exponents for every key exchange. A concrete instantiation of SIGMA may also use additional nonces to explicitly verify the freshness of a key exchange.

Bob generates his own Diffie–Hellman public key g^y and uses Alice's Diffie–Hellman public key to create a shared secret K. It sends back a message containing his identity B, g^y, a digital signature of g^x and g^y, and a MAC of his identity ($MAC_K(B)$). With the signature, Bob proves that he is in possession of the private key belonging to B. The inclusion of g^x under the signature allows Alice to verify that Bob's reply is fresh and not a replay of a previous key exchange. The inclusion of g^y ensures that g^y is Bob's Diffie–Hellman key and that it was not modified or exchanged during the transmission. The MAC over B binds the shared key to the identity of Bob.

Initiator   Responder

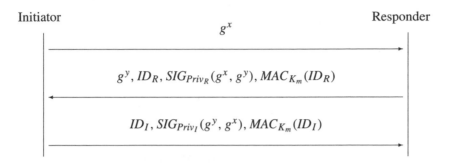

Figure 2.9 The basic SIGMA key exchange without identity protection.

On receipt of Bob's message, Alice can generate the shared secret g^{xy} from which she derives the session key K. Alice replies with her own identity A, a digital signature of g^x and g^y, and a MAC of her identity ($MAC_K(B)$). Alice, therefore, proves her identity by using her private key. Moreover she proves that she has computed the shared secret by appending the MAC, binding her identity to the shared session key.

The exact use of the SIGMA protocol may vary from application to application. In particular, SIGMA compliant key exchange protocols may need to add further auxiliary information (like sets of supported cryptographic algorithms for later use) to the signed or MAC protected parts of the message. Even variants in which the MAC is also covered by the signature are possible. This does not affect the security of the SIGMA protocol as long as the identity of the hosts is covered by the MAC and the freshness of the signature is proven by a unique nonce or Diffie–Hellman key.

The basic SIGMA key exchange protects neither the identity of Alice nor the identity of Bob. A passive eavesdropper can learn both identities by only observing the communication channel.

Integrity protection The requirements for SIGMA compliant protocols contain two seemingly contradicting goals: privacy and authentication. On one hand, to communicate securely, hosts should be able to verify the identity of their peers, but on the other hand, the identity of a host should not be revealed to unauthorized parties. There are two types of attackers from which the identities must be concealed: active and passive ones. Active attackers interfere with the key exchange protocol and execute or modify some or all steps of the protocol whereas passive attackers can only observe the key exchange. Encrypting the identities in the key exchange with the symmetric keys as shown in Figure 2.10 protects against passive attackers. However, to allow secure communication a host must reveal its identity to its peer at some point in the protocol. Therefore, active attackers can execute the protocol until they learn the identity of the victim. The SIGMA protocol allows to delay the authentication of the Initiator or the Responder, depending on which of the two entities requires identity protection. This allows a peer to verify the identity of its peer before it reveals its identity. These two variants of SIGMA are called *SIGMA-I* for Initiator protected SIGMA and *SIGMA-R* for Responder protected SIGMA. We will discus two examples in which either Initiator or Responder protection is important.

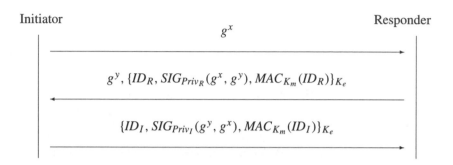

Figure 2.10 The SIGMA-I key exchange protects the identity of the Initiator.

If, for example, a client connects to a server, it is often not necessary to protect the identity of the server as it is already known before the connection. However, it may be necessary to delay the authentication of the client until the server has authenticated to allow the client to verify that it communicates with the right host. SIGMA-I as depicted in Figure 2.10 is applicable in such a case. The Responder authenticates before the Initiator reveals its identity. Note that all identities are encrypted, which protects both identities from passive attackers.

We consider a mobile device as a usecase for SIGMA-R. Knowledge about the point of network attachment of a mobile client typically allows to roughly estimate the location of the mobile device and its owner. It is clear that no unauthorized parties should be able to learn about the identity of a mobile host to avoid that anyone with such a mobile device can be traced. If the mobile host would authenticate first during the key exchange, an attacker could easily sniff the identity of a host at a certain address by simply initiating a SIGMA-I key exchange. The attacker would not have to authenticate itself if it canceled the key exchange after the Responder's reply. The SIGMA protocol defines the SIGMA-R variant to allow for identity protection for the Responder. The SIGMA-R key exchange is depicted in Figure 2.11.

SIGMA-R reverses the authentication process in the sense that the Initiator must authenticate first. The Responder can decide whether it is willing to authenticate to the Initiator or if it prefers to cancel the key exchange. Like SIGMA-I, SIGMA-R encrypts all packet contents that contain identities. In contrast to SIGMA-I, the Responder does not send its identity in the first message but it only sends its Diffie–Hellman public key. The third and fourth message of the SIGMA-R key exchange resemble the second and third message in SIGMA-I, with the difference that the direction (Responder to Initiator) is reversed and that no Diffie–Hellman public keys are contained because SIGMA-R exchanges these keys within the first and second message.

Initiator   Responder

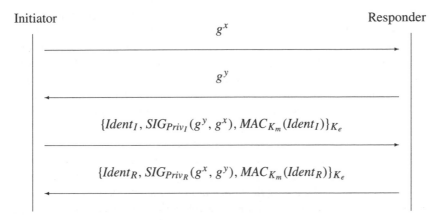

g^x

g^y

$\{Ident_I, SIG_{Priv_I}(g^y, g^x), MAC_{K_m}(Ident_I)\}_{K_e}$

$\{Ident_R, SIG_{Priv_R}(g^x, g^y), MAC_{K_m}(Ident_R)\}_{K_e}$

Figure 2.11 The SIGMA-R key exchange protects the identity of the Responder.

2.5.8   Internet Key Exchange: IKE

IPsec defines certain security mechanisms that protect communication on the network layer. However, the authentication, integrity protection, and encryption features of IPsec rely on a shared state, present at both peers. IPsec does not specify how this state – including the symmetric keys for integrity protection and encryption – is established. The state can either be configured manually or it can be delegated to specialized key exchange protocols that allow mutual authentication, negotiating security algorithms and keys for these algorithms, and managing and maintaining IPsec SAs.

The Internet Key Exchange (IKE) protocol as specified in RFC2407, RFC2408, and RFC2409 (Harkins and Carrel 1998) and its successor IKEv2 (Kaufman 2005) perform these tasks. IKEv2 simplifies the security mechanisms in IKE and summarizes the three standards in one RFC document. Moreover, it closes some security loopholes and weaknesses of IKE. Therefore we only discuss IKEv2. The key negotiations and the management of IPsec security associations as done by IKEv2 take place in two phases. During the first phase, a host verifies the identity of its peer, generates the keying material for the symmetric keys, and establishes a secured channel: the IKE_SA. During the second phase, new SAs – so called child SAs – are negotiated over this channel.

Authentication and key negotiation

The IKE negotiations start with the establishment of the IKE_SA and the mutual authentication of the peers. This phase requires a four-way handshake, as depicted in Figure 2.12. A host takes the role of the Initiator by initiating the key exchange by sending a packet, containing the IKE header, a list of supported algorithms (cf. transform suites), a nonce, and its Diffie–Hellman public key to the *Responder*. The list of supported algorithms allows the Initiator to propose several Diffie–Hellman groups. Therefore, the first message may contain several Diffie–Hellman public keys. The Responder selects one of the offered transform suites and completes the Diffie–Hellman key exchange with its private Diffie–Hellman key. At this point, the Responder has created the keying material that is used for creating the symmetric key for encrypting and integrity protecting the secure channel.

The Responder sends back a message containing the index of the selected transform suite, a random nonce, and its public Diffie–Hellman key. It may also attach a list of trust anchors – a list of certificate authorities – that it accepts as trustworthy. On receiving this second message, the Initiator can compute the keying material for the shared secret as well. As both peers are in possession of the symmetric keys, the following message exchanges are sent encrypted.

The next two messages are used to mutually verify the identities of the hosts and to setup the IKE_SA. The Initiator proves its identity by using its private key to sign the Responder's nonce, parts of the message exchange, and a special key derived from the keying material and the identity of the Initiator. Including the nonce in the signature ensures that the signature is fresh and not a replay of an old signature. It attaches the signed data to a message to the Responder. It may also attach a list of accepted trust anchors. Additionally, it attaches other parameters that are used to negotiate the child SAs. The child SA negotiation is described in the next paragraph. The Responder can verify the identity of the Initiator by verifying the signature. It sends back a signed message that similarly to the Initiator's signature contains parts of the message exchange, the Initiator's nonce, and a key derived from the keying material and the identity of the Responder.

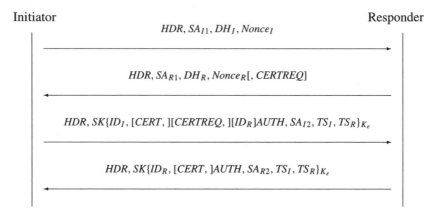

Figure 2.12 The IKE handshake.

Setup of new child SAs

Setting up a new child SA, which is used by IPsec to transmit integrity protected or encrypted IP payload, requires a two-way message exchange. As an optimization, the negotiations for child SAs already take place during the four-way IKE handshake. The necessary information is piggybacked on the third and fourth handshake message. The whole process of creating a new child SA is protected by the algorithms and keys being used for the IKE_SA.

Either of the endpoints may initialize the creation of a new SA. Therefore, the host requesting the new SA is called *Initiator* and the host responding to the request is called

Responder regardless of which roles the hosts inherited in the handshake. In order to create a new SA, both hosts must agree on the symmetric keys and the algorithms that should be used to protect the traffic belonging to the SA. First, the Initiator sends a packet that contains a proposal for suitable algorithms and a nonce. Moreover, the message can contain a Diffie–Hellman public key, which can be used to achieve *perfect forward secrecy* for every child SA if the Diffie–Hellman keys are deleted after the key-exchange. The *create child* request can be used to rekey an existing SA. In this case, the Initiator must attach a notification of which SA should be rekeyed. Rekeying is used to replace an existing SA with a new one to change the secret symmetric keys.

The Responder selects one of the proposed algorithms and generates the Diffie–Hellman shared secret if a Diffie–Hellman public key is present. The symmetric keys for the SA are derived from the keying material and the nonces contained in the messages. The Responder sends back its selection of algorithms and a nonce. The Initiator can now compute the symmetric keys from the nonces and the keying material as well, to complete the creation of the child SA.

IKEv2 uses *traffic selectors* to communicate rules of what classes of traffic should use the newly created SAs. Thus, the traffic selectors are used to synchronize the SPDs of both hosts. The traffic selectors set the rules for source or destination IP address ranges, port numbers, and transport layer protocol numbers. The Initiator of a child SA setup process adds traffic selectors for the incoming and outgoing traffic of the SA. The Responder can narrow this selection or signal its compliance with the proposed set by replying the same selection.

2.6 Weak authentication techniques

Arkko and Nikander (2002) compare different weak authentication techniques. Weak authentication relates to peers that use strong authentication techniques without trusted third parties. This authentication is called *weak* because a small possibility of an attack remains.

Weak authentication is used when the costs for using strong authentication with third parties like PKIs surpass the benefits, for instance, if cheap lightweight devices become big and expensive to support strong authentication.

There are three classes of weak authentication techniques:

Temporal separation describes the ability of hosts to relate communication at a certain time t_1 to an earlier communication at time t_0. This means that a host can verify that the host it is talking to at t_1 is the same host it has been talking to at t_0. Consequently, no-one is able to impersonate the peer for a certain time span after t_0. Temporal separation can typically only be guaranteed under certain circumstances. For instance, the first contact at t_0 must be authentic and there must not be an active attacker on the communication path during this contact.

Spatial separation refers to the ability of a host to ensure that its peer is located on a certain communication path. In practice, this often means that the host is located at a certain network address. This property is often verified with challenge response mechanisms. For instance, the HIP *echo request–echo response* test ensures that a host is located at a certain IP address. However, hosts in a similar location might be able to pass the test if they are able to eavesdrop on traffic to a certain IP address and if they are able to spoof packets from this IP address.

Asymmetric costs refer to the case in which the costs of an attack surpass the benefits for the attacker. This can be achieved by making it harder for the attacker to find the right target. This forces the attacker to perform multiple attacks, increasing the costs of an attack.

Arkko and Nikander argue that sometimes incomplete security is enough for the task at hand. They state that weak authentication is not suitable for applications that require real-world identities. Multihoming, routing, and mobility are given as typical examples for such applications. An ephemeral identity is sufficient for these applications. Real-world identities that are ensured by a third party are, in most cases, too much overhead for these applications.

2.7   Secure DNS

The Internet critically depends on the Domain Name System (DNS) for resolving human-friendly host names to IP addresses. Yet, the present DNS is insecure, susceptible to attacks forging DNS replies for a host name or even entire subdomains. A most common attack on DNS is cache poisoning, when a DNS server is maliciously supplied with incorrect information on the IP addresses of a DNS subdomain. An honest DNS server then unwillingly propagates bogus data further to its child subdomains.

A common way to implement cache poisoning is to make an honest DNS server query the attacker's DNS server for some data. The attacker's DNS server, after a correct response, can include "additional" DNS fields claiming it is a responsible server for a domain the attacker chose to subvert. The honest server could believe that and redirect all future queries for the subverted domain to the attacker's DNS server. The attacker's server can return IP addresses of hosts containing advertisements or viruses instead of real IP addresses of a queried host name. RFC3833 contains a comprehensive threat analysis of the DNS system.

While HIP can be used in many useful scenarios without any infrastructure, to exploit its security properties fully, a host needs to store its public key to the DNS server. In this case, secure DNS extensions (DNSSEC) ensure that the public key indeed comes from a trusted DNS server and has not been modified in transit. As the DNSSEC is important for HIP deployment, this section describes its current design in detail.

Current DNSSEC specifications (also known as DNSSEC-bis) are contained in three RFCs. RFC4033 gives an overview and requirements of DNSSEC architecture. New Resource Records (RR) for DNSSEC are described in RFC4034 and the DNS protocol extensions in RFC4035. In addition, RFC3833 contains a comprehensive threat analysis of the DNS system that has influenced the current DNSSEC architecture. These RFCs obsolete the previous design of DNSSEC that was found unsuitable for Internet-wide deployment due to the scalability issues. In particular, in the old architecture, a change of a domain key would require updating all DNS servers in its subdomain. In practice, for domains such as ".com" it would require tens of millions of update messages. In the present architecture, this problem is removed at the expense of extra work the DNS clients have to do to verify legitimacy of a DNS reply.

The DNSSEC architecture is based on storing a DNS server's public key in a DNSKEY Resource Record. For clients that indicate the support of DNSSEC in a query, the DNS server signs its replies with its private key placing the signature into RRSIG RR. The client also obtains the Delegation Signer (DS) parameter from the parent DNS server. The DS parameter contains the digest of the child DNS server public key. By verifying the digest, the DNS client can establish a trust link from the parent DNS server to the child DNS server that replied to the client. To establish authenticity of the reply, the client has to construct a chain of trust links up to the DNS server for which the client has a trusted public key stored. Typically it would be the root server of the DNS system.

After the authenticity of the reply is established, the client knows that the DNS information for a queried host indeed exists and has not been tampered with. However, DNSSEC does not provide confidentiality of data (DNS messages are not encrypted) nor does it provide protection against DoS attacks. In fact, because cryptographic operations are computationally intensive, there are new opportunities for DoS attacks against DNSSEC servers.

DNSSEC supports authentic denial of existence for a host name. In other words, if a DNS reply tells that the given host name does not exist, it is true and is not an attempt to block an existing host. The Next Secure (NSEC) resource record implements denial of existence. All host names in a DNSSEC server are ordered according to a canonical representation. The NSEC RR contains host names previous to and following the queried host name.

NSEC introduces an issue of zone enumeration when a client querying non-existing names can obtain a full list of DNS names registered at a DNSSEC server. While some people argue that DNS information should be public, a malicious host can exploit a full list of host names, e.g. to probe for security vulnerabilities in all hosts. Additionally, revealing DNS information can be in conflict with privacy protection laws, for example in Germany. To resolve the controversy, a new resource record NSEC3 is introduced that contains hashes of host names instead of full names.

DNSSEC deployment is slowly progressing. In addition to policy issues, such as zone enumeration, deployment difficulties include updating the DNS servers and clients, and initial distribution of DNS public keys. As of 2006, DNSSEC support in root DNS servers is still missing, though there are promises to enable it soon. From country-level domains, Sweden has deployed DNSSEC.

```
[gurtov@hippy ~]$ dig +dnssec www.se any

; <<>> DiG 9.3.2 <<>> +dnssec www.se any
;; global options:  printcmd
;; Got answer:
;; ->>HEADER<<- opcode: QUERY, status: NOERROR, id: 8237
;; flags: qr rd ra; QUERY: 1, ANSWER: 6, AUTHORITY: 3,
;; ADDITIONAL: 2

;; OPT PSEUDOSECTION:
; EDNS: version: 0, flags: do; udp: 4096
;; QUESTION SECTION:
;www.se.                         IN      ANY

;; ANSWER SECTION:
www.se.           3260   IN      A       212.247.204.135
www.se.           6860   IN      RRSIG   NSEC 5 2 7200
20060821182406 20060816070558 54245 se.
```

```
Oex8Tj20u9Q9f7l5Idepb53acoFLbNvwX6xcNBpUezEnxKAze+0vrQvM
7i0X5QYq/EQ8EtPEhLH/SXjJS4+mmYXV3mNVdT96rYSeKX7oh+N82+VY
vCaKPxZLZy22JYhEPxs5CZxiSGEgPq3ncYw6+mhzKjxm3AA5XtpUhZ3B tW4=
www.se.              6860    IN      NSEC    www-3.se. NS RRSIG NSEC
www.se.              3260    IN      NS      ns.nic.se.
www.se.              3260    IN      NS      ns2.nic.se.
www.se.              3260    IN      NS      ns3.nic.se.

;; AUTHORITY SECTION:
www.se.              3260    IN      NS      ns.nic.se.
www.se.              3260    IN      NS      ns2.nic.se.
www.se.              3260    IN      NS      ns3.nic.se.

;; ADDITIONAL SECTION:
ns3.nic.se.          17256   IN      A       212.247.3.80

;; Query time: 406 msec
;; SERVER: 193.210.18.18#53(193.210.18.18)
;; WHEN: Thu Aug 17 12:41:52 2006
;; MSG SIZE  rcvd: 358

[gurtov@hippy ~]$
```

Part II

The Host Identity Protocol

3

Architectural overview

In this chapter, we describe the concept of a namespace in the Internet (Section 3.1), the generation of Host Identity Tag (Section 3.2), and the overlay-routable IPv6 prefix for HITs (Section 3.3), and outline the role of IPsec (Section 3.4). Section 3.5 concludes the chapter with an overview of IETF activities related to HIP.

3.1 Internet namespaces

Namespace in the Internet allows to uniquely identify an entity such as a host or a service. At the moment two namespaces for hosts are globally deployed in the Internet: IP addresses and DNS names. The IP addresses also serve as host locators in the Internet as we will describe in the following sections. DNS names provide hierarchical human-friendly host names. DNS names can be location independent (such as from .net domain) or limited to a certain geographical area (such as .fi for Finland). The DNS namespace has several limitations. Updating the current IP address in DNS can be too slow to support mobility (Walfish *et al*. 2004). Furthermore, most hosts do not even have modification access to the DNS servers they are using.

The basic DNS service is not secure and can be easily spoofed. DNSSEC offers an improvement in security, but is still not universally used. Many DNS names are bound to a specific organization or country. For example, if a user changes employer or school, the host DNS name suffix will almost certainly change to reflect the new administrative location. Such updates can last for many hours due to caching of DNS information. To make things worse, some applications such as Internet Explorer ignore DNS Time-to-Live information and can cache DNS entries longer than their lifetime.

Current namespaces have three shortcomings. First, changing the host address is not directly possible without breaking transport layer connections. Second, authentication of the host is not supported; spoofing of a source IP address is a common problem in the Internet. Third, privacy-preserving communication is not provided.

Host Identity Protocol (HIP): Towards the Secure Mobile Internet   Andrei Gurtov
© 2008 John Wiley & Sons, Ltd

3.2   Methods of identifying a host

In HIP, a pair of self-generated public and private keys provides the Host Identity. The length of the public key can be 512, 1024 or 2048 bytes and is generated with the RSA algorithm by default. Most HIP implementations also support the DSA algorithm as it was the default before. Generation of new keys is a relatively time-consuming operation and occurs only infrequently, e.g. when the old keys have been compromised. Current Unix HIP implementation store public and private keys in the file system at /etc/hip directory. In this chapter, we assume that a single identity per host is sufficient. In reality, often several identities are needed to protect the privacy of the user; we return to this issue in Chapter 12.

Using the public key as a host identifier in packets and the application interface is inconvenient due to large and variable size. A typical Maximum Transmission Unit of 534 bytes would not even fit the shortest public key. For this reason, and to maintain compatibility with existing applications using the Berkeley socket interface, two additional forms of host identity are introduced as shown in Figure 3.1.

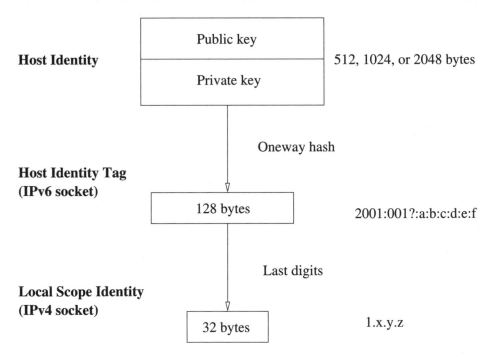

Figure 3.1 Methods of identifying a host.

The Host Identity Tag (HIT) is a 128-byte hash of the public key. HIT has deliberately the same length as an IPv6 address and can be used instead of it by applications. The hash is one-way, it is not possible to restore the original public key from it. HITs are statistically unique given their sufficient length. The probability of a collision when two different public keys map to the same HIT is negligible. HITs have a prefix 2001:0010::/28 that enables to distinguish them from currently allocated IPv6 addresses. Having fixed-length identifiers gives an additional benefit of protocol independence from the cryptographic algorithm used to generate public–private keys.

The Local-Scope Identifier (LSI) is a 32-bit identifier that can be constructed taking the last bytes of the HIT. LSIs have shorter lengths than HITs and the probability of their collisions is significant. Therefore, LSIs have only local meaning and cannot be assumed to be globally unique. The LSI have the same length as IPv4 addresses and can be used by the legacy IPv4-only applications in the socket interface. LSIs have a prefix 1. to distinguish them from publicly allocated IPv4 addresses.

Each HIP implementation is required to support multiple HIs. One HI should be reserved for anonymous communication. Initiators are expected to utilize anonymous communication more often than Responders. HIP privacy extensions are described in detail in Chapter 12.

3.3 Overlay Routable Cryptographic Hash Identifiers

IPv6 Prefix for Overlay Routable Cryptographic Hash Identifiers (ORCHIDs) (Nikander *et al.* 2007b) reserves a part of the IPv6 address space to serve as identifiers in the socket APIs. Internet Assigned Numbers Authority (IANA) allocated a prefix for ORCHIDs. ORCHIDs appear as IPv6 addresses but are not routable at the IP layer, although are expected to be routable at the overlay layer on top of IP. Applications can transparently use ORCHIDs in place of regular IPv6 addresses. ORCHIDs can contain, for example, HIP HITs or Temporary Mobile Identifiers for the Mobile IP Privacy Extensions.

3.3.1 The purpose of an IPv6 prefix

The main goal of introducing a special format for ORCHIDs is to prevent confusion with regular IPv6 addresses. Naturally, an application can use a subset of IPv6 addresses in place of identifiers. That, however, can cause leaking of non-routable addresses to unaware applications for example through referrals. In addition, different applications can select different prefixes for identifiers that can potentially prevent interoperability in the Internet.

ORCHIDs are meant to be used as identifiers in the legacy application APIs. Newly developed applications are expected to use "native" API utilizing identifiers such as a public key in the interface instead of 128-bit ORCHIDs. However, in the near future it is unreasonable to expect that all applications and host OS are updated to support the new model. Instead, ORCHIDs offer a possibility to experiment with new network architectures in a reasonable way.

ORCHIDs have the following properties. They are generated using a hash function that provides secure binding to the input parameters and statistical uniqueness. ORCHIDs are compatible with an IPv6 global unicast address format.

3.3.2 Generating and routing an ORCHID

ORCHIDs are produced by taking an SHA1 hash over a 128-bit context ID concatenated with an input bitstring. The bitstring must be statistically unique with the given context, and can often be a public key. It is proposed that a context ID is a randomly generated value that defines the type of ORCHID. The final ORCHID is formed by concatenating an IANA-allocated 28-bit prefix with a 100-bit bitstring extracted from the middle of the hash output.

While it is desirable to create as long a hash as possible to prevent collisions, another goal for ORCHIDs was to support several different protocols (HIP, MIP) with the same allocated IPv6 address prefix. It was decided that the context ID is not explicitly present in ORCHID and all remaining space in the 128-bit IPv6 address is used as a hash output. It is possible to verify that a given ORCHID and a bitstring belong to a given context by applying the hash function.

By design, ORCHIDs are not routable at the IP layer and present a location-independent end-point identifier. Therefore, routers may be configured to drop an IP packet containing ORCHID as a source or destination IPv6 address. Furthermore, if a source address is a regular unicast IPv6 while the destination is an ORCHID, the router can reply with an ICMP "Destination Unreachable" packet. However, routers should not include any hard-wired handling of ORCHIDs and special handling, if any, should be configurable in software.

ORCHIDs can be routed on an overlay layer, using a lookup service, such as a DHT (Section 10.2), or forwarding service, such as Hi3 (Section 10.5). Overlay routing is more flexible in providing advanced services such as anycast or multicast. It is also more expensive in terms of the latency and bandwidth, as overlay hops typically stretch over multiple IP hops. To enable overlay routing, a host should be able to map the ORCHID identifier to the IP address of the next overlay hop. Such mapping is performed by a shim protocol, such as HIP.

3.3.3   ORCHID properties

ORCHIDs are statistically unique meaning that collisions are theoretically possible although very improbable. Two types of collision are possible. First, two hosts generated within the same context but with different bitstrings can collide. In this case, the shim protocol (HIP) would detect an attempt to use duplicate ORCHIDs and prevent an attempt to establish an association with one of duplicates. Second, two ORCHIDs from different contexts can collide. To resolve this case, the specification recommends for hosts supporting multiple contexts to support a database containing ORCHIDs in use and their context. Then, if a context generates an ORCHID that matches any existing ORCHID in another context, the generation attempt would fail.

The collision resolution measures remain largely theoretical. Even if two hosts do have same ORCHIDs, a conflict would be created only if some host attempts to communicate with both colliding hosts. Otherwise, the collision would not be noticeable. To ensure a negligible collision probability, a host must include to an input bitstring a randomly-generated public key or an incremental counter to ensure statistically unique input to the hash function.

For security reasons, it is required that all contexts use the same hash function to generate an ORCHID from the input bitstring. Allowing different hash functions without their explicit encoding in ORCHIDs would allow *bidding down* attacks. In particular, if one hash function gets compromised, attacks even against ORCHIDs generated with other secure hash functions would be possible. Consequently, the use of a different hash function implies allocation of a new prefix.

The ORCHID prefix (2001:0010::/28) is allocated from the IANA Special Purpose Address Block (2001:0000::/23) meant for experimental purposes.

3.4   The role of IPsec

By default, HIP data packets are carried using IPsec utilizing the Encapsulated Security Payload (ESP) transport mode. The HIP control messages essentially provide a session key exchange between two hosts. While it would have been possible also to use other key exchange protocols, such as IKE or IKEv2, the new HIP architecture was created to be friendly with middleboxes. Middleboxes, such as firewalls, are able to examine the process of HIP key negotiation.

The ESP transport mode spans from an end host to another end host encrypting IP packet payload. Packets are multiplexed using the Secure Parameter Index (SPI) value to identify a *Security Association* (SA) between two hosts. An SA is established between HITs of two end points; only a single SA between a pair of HITs is supported. As the SA is bound to HITs, not IP addresses, the active set of IP addresses at the SA end point can change dynamically. The SA is identified using the SPI in ESP packets that are mapped to HITs at the end point. This property is sometimes called the HIT compression. There is no need to transmit HITs in data packets, which reduces the packet size. In summary, there is no HIP-specific data packet format defined, but the standard IPsec ESP mode is used.

While the use of IPsec is preferable in all cases to enhance the security level, it is not always feasible e.g. for busy web servers or lightweight devices. Furthermore, encryption could be also implemented on the upper layer such as SSH or TLS. Therefore, a different transport mode than IPsec might be needed in HIP in some cases. While some activities on developing *lightweight HIP* are ongoing, there is not yet a standardized solution.

3.5   Related IETF activities

Figure 3.2 shows the position of HIP with regard to other relevant proposals in the IETF. The figure shows relevant groups in the areas of Internet architecture, security, mobility, and multihoming. Most proposed protocols are placed in one or two areas and thus require combination with other protocols to achieve all desired properties. Complexity of such combinations can result in poor performance and implementation errors. HIP, on the other hand, provides secure mobility and multihoming with a simple and architecturally sound approach. In this section, we present a brief overview of IETF Working Groups relevant to HIP.

Mobility for IPv4 (mip4) and *Mobility for IPv6 (mip6)* Working Groups continue to develop the Mobile IP protocol based on the use of Home Agent for providing a stable IP address to a mobile host. The WGs document the existing deployment experience and examine interoperability issues between the implementations. The current goals of WGs include adoption of IKEv2 for establishing IPsec Security Associations, dual-stack support and reducing the configuration burden per mobile node. Switching of the Home Agent and bootstrapping a mobile node after powering on, as well as firewall traversal and location privacy, are also within the scope.

The *IKEv2 Mobility and Multihoming Protocol (MOBIKE)* WG concluded in 2006 after publishing protocol specifications in RFC 4555. The MOBIKE protocol enables the IP addresses of IPsec tunnel mode Security Associations to change and can be used for mobile VPN or site multihoming. The MOBIKE protocol removes the need to create new Security

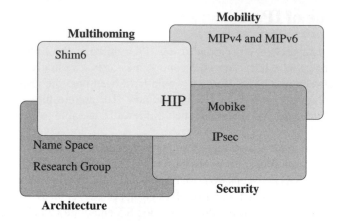

Figure 3.2 HIP relation to other IETF activities.

Associations, which reduces the computation overhead and can save the user from entering codes from a token card.

The *Site Multihoming in IPv6 (multi6)* WG documented the ways that multihoming is currently implemented in IPv4 networks and evaluated several approaches for advanced multihoming. The security threats and impact on transport protocols were covered during the evaluation. The work continued in another WG Site Multihoming by IPv6 Intermediation (shim6) focusing on specifications of one selected approach. This WG uses the approach of inserting a shim layer between the IP and the transport layers that hides effects of changes in the set of available addresses. The applications are using one active address that enables referrals. Shim6 relies on cryptographically generated IPv6 addresses to solve the address ownership problem. The shim6 host can benefit from multihoming properties even when its peer host does not support shim6 extensions.

The current IPsec and IKE protocols provide strong security guarantees but require the use of pre-existing credentials that can be validated. *Better-Than-Nothing Security (btns)* WG specifies extensions to the IPsec architecture and IKE to support unauthenticated Security Associations. The use of self-signed certificates and self-generated public keys with BTNS would enable simpler and faster deployment of IPsec. The WG also studies how to use IPsec to secure upper-layer protocols. As an example, if a user accesses a sensitive web site over IPsec, there must be some indicator in the browser confirming that the traffic is protected by IPsec.

The WG *Mobility for IP: Performance, Signaling and Handoff Optimization (mipshop)* develops Hierarchical Mobile IPv6 (HMIPv6, RFC 4140) and Fast Handovers for Mobile IPv6 (FMIPv6, RFC 4068) to reduce the delay and packet loss for a mobile host. The goal of the group is to advance the specifications from an experimental to the standard-track status. The group also develops on optimization of return ratability test in terms of security and performance by using Cryptographically Generated Addresses and Credit-based Authorization.

Other WGs possibly related to HIP include Mobile Nodes and Multiple Interfaces in IPv6 (Monami6), Network Mobility (nemo), and Network-based Localized Mobility Management (netlmm).

4

Base protocol

This chapter describes HIP control packets I1, I2, R1, R2 (Section 4.1), UPDATE, NOTIFY, CLOSE, CLOSE_ACK (Section 4.2) and the protocol for establishing HIP associations. Section 4.3 describes the IPsec encapsulation of HIP data packets.

4.1 Base exchange

The HIP base exchange establishes a security association between two hosts (Moskowitz *et al.* 2008). The base exchange consists of four messages in two round-trip times, as shown in Figure 4.1. The host that starts the base exchange is called the Initiator and the other host is a Responder. Resilience to DoS and replay attacks has been taken in all aspects of the protocol design, most importantly with a cryptographic puzzle that the Initiator must solve before the Responder creates any state for the HIP connection. The base exchange distributes Diffie–Hellman keys and authenticates the hosts.

4.1.1 I1 packet

The base exchange is initiated when the application passes a datagram for transmission or tries to open a TCP connection. The HIP implementation should buffer such a first packet, complete the base exchange and then transmit the packet over the established HIP association. Some implementations, however, discard the packet that triggers the base exchange and rely on the transport protocol or the application to retransmit the data packet. In the case of TCP, the initial timeout value for the SYN packet is three seconds.

There can be only one HIP association between a pair of HITs. Therefore, the only way to support multiple associations between two hosts is to have several HITs per host. Naturally, a host can have multiple associations to separate hosts simultaneously.

Initiator transmits the I1 packet to the Responder's IP address. Responder's address and HITs can be obtained from the DNS server or using another distributed repository service. We return to this issue in Chapter 10. The I1 packet contains the Initiator and Responder HITs. The Responder HIT can be also NULL, if it is unknown to the Initiator. In that case the base exchange operates in the *opportunistic mode* described in detail in Section 6.1.

Host Identity Protocol (HIP): Towards the Secure Mobile Internet   Andrei Gurtov
© 2008 John Wiley & Sons, Ltd

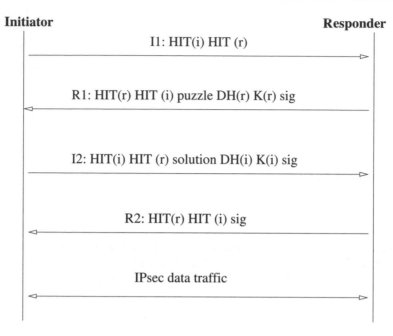

Figure 4.1 The HIP base exchange protocol.

The opportunistic mode is susceptible to the Man-In-The-Middle attacks and should be used only in a trusted environment.

When an I1 packet is sent, a retransmission timer is started. After a timeout, the host retransmits the I1 packet and restarts the timer. The same I1 packet can be sent in parallel to up to three different IP addresses of a peer host to reduce the latency of association establishment.

All HIP control packets are transmitted after a basic IPv4 or v6 header. The protocol number assigned by IANA for HIP is 139; early HIP implementations were using the number 253, which is reserved for testing purposes. The basic HIP header common to all HIP messages is shown in Figure 4.2. The I1 message contains only the basic header and no parameters. Its size, therefore, is fixed to 20 bytes plus the size of the IP header. The HIP header follows the same structure as an IPv6 extension header. However, processing of any subsequent headers is currently unsupported. Figure 4.3 shows an example of an I1 packet captured from a network. The protocol analyzer Wireshark was enhanced by the OpenHIP project to understand HIP control packets.

Currently the only defined HIP control flag is the lowermost bit for Anonymous identifiers. Such identifiers are not publicly distributed in any repository and should not be stored by the Responder. The rest of the control bits are set to zero, as well as three reserved bits (RES) after the HIP version. The two other bits are set as shown for compatibility with SHIM6 IPv6 multihoming protocol.

The HIP checksum is calculated over a pseudoheader including source and destination IP addresses, HIP packet length, and protocol number. Since the checksum includes the IP addresses, a middlebox re-writing the addresses must also recalculate the HIP checksum.

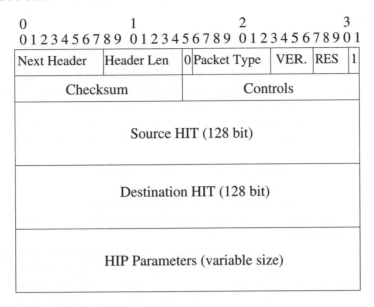

Figure 4.2 A general packet format of HIP messages. The I1 message has no parameters attached, and the destination HIT can be NULL.

```
Internet Protocol, Version: 4
    Header length: 20 bytes
    Differentiated Services Field: 0x00 (DSCP 0x00: Default; ECN: 0x00)
    Total Length: 60
    Identification: 0x0000 (0)
    Flags: 0x04 (Don't Fragment)
    Fragment offset: 0
    Time to live: 63
    Protocol: Unknown (0xfd)
    Header checksum: 0x4087 [correct]
    Source: 193.167.187.106 (193.167.187.106)
    Destination: 193.167.187.132 (193.167.187.132)
Host Identity Protocol
    Payload Protocol: 59
    Header Length: 4
    Packet Type: 1
    Version: 1, Reserved: 1
    HIP Controls: 0x0000
        .... .... .... ..0. = Certificate (One or more CER packets follows)
        .... .... .... ...0 = Anonymous (Sender's HI is anonymous)
    Checksum: 0x6b06 (correct)
    Sender's HIT: 200100150A097C449CA1257176DD0872
    Receiver's HIT: 20010014766EFBEEF74DEC73D6C528C0
```

Figure 4.3 HIP I1 packet captured with Wireshark.

4.1.2   R1 packet

When Responder receives the I1 packet, it does not store any information (called state) yet about the Initiator. Responder replies with an R1 packet to the Initiator using the IP source address of the I1. The R1 packet contains the Diffie–Hellman key, a cryptographic puzzle, and the public key of Responder each as a separate HIP parameter. The packet is signed by the Responder using its public key. The obligatory fields of R1 packet are shown in Figure 4.4. Parameters are encoded in the Type-Length-Value (TLV) format. Figure 4.5 shows an example of an R1 packet captured from a network.

All parameters have the length multiple of 8 bytes for proper alignment. Padding data is added if a parameter ends earlier. The order of parameters in HIP messages is fixed as shown in Figure 4.4. Each parameter has a critical bit to indicate its importance. If a host does not recognize a critical parameter, the packet is not processed further, but a NOTIFY or ICMP message is sent to the peer host.

The cryptographic puzzle exists to protect the Responder from Denial-of-Service attacks. Before creating state, the Responder verifies a puzzle solution transmitted by Initiator in I2 packet. The puzzle is a task to find a value that produces zeros when passed through an SHA-1 hash function. The number of zeros determines the puzzle difficulty; it takes longer for Initiator to find a right value when K increases. Puzzle verification at Responder is a single fast hash computation. Responder can dynamically adjust the puzzle difficulty for Initiators based on its current load. It is expected that typically Responders ask for a short puzzle and only increase its difficulty when under a DoS attack.

The field K in the puzzle parameter in R1 determines the puzzle difficulty as the number of bits to be verified. The puzzle lifetime field specifies the maximum number of seconds the Initiator may use for finding a solution. The opaque field contains data that is copied back by the Initiator in a Puzzle Solution parameter. The random number field is used as an input to the RHASH function concatenated with HITs of Initiator and Responder, and a variable J. The task of the Responder is to find such J that K lowest-order bits of the hash are zeros. The K value should be less than 20.

The Diffie–Hellman parameter in R1 contains the public key from a certain group. Mandatory for all HIP implementations is 384-bit and 1536-bit More Modular Exponential (MODP) groups. The parameter can include a maximum of two groups. The idea is that the Initiator that is capable of supporting a stronger group selects the one with a longer bit key. However, a lightweight Initiator, such as a mobile phone, can select a weaker key.

The HIP_TRANSFORM parameter in R1 contains a list of cryptographic algorithms (transforms) supported by the Responder. The list can contain up to six transform identifiers (Suite-IDs). All HIP implementations support at least AES-CBC with HMAC-SHA1 and NULL-ENCRYPTION with HMAC-SHA1. Recently discovered vulnerabilities of the AES-CBC encryption may result in the change of the default algorithm for HIP.

The HOST_ID parameter includes a host public key of the length given in the HI Length field. Additionally, the parameter includes a domain identifier, a Fully Qualified Domain Name (FQDN) or a Network Access Identifier. FQDN is a host DNS name up to the top domain terminated by a period. It is sent in a binary representation.

The R1 packet is terminated by a HIP_SIGNATURE parameter that includes a signature. Either RSA or DSA is used to generate the signature over an entire HIP packet up to the signature parameter itself. The HIP checksum field is set to zero during the computation.

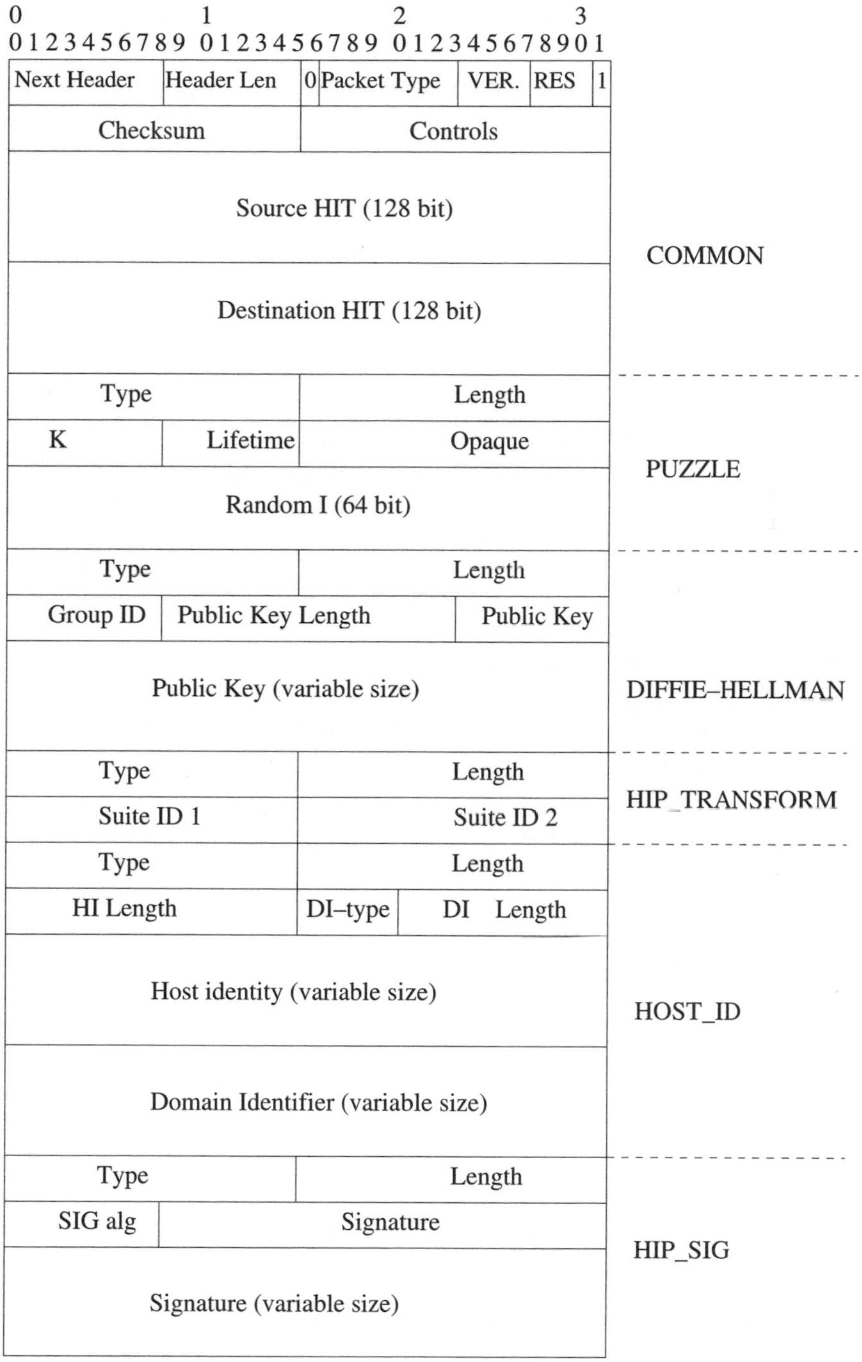

```
0                   1                   2                   3
0 1 2 3 4 5 6 7 8 9 0 1 2 3 4 5 6 7 8 9 0 1 2 3 4 5 6 7 8 9 0 1
```

| Next Header | Header Len | 0 | Packet Type | VER. | RES | 1 |

| Checksum | Controls |

Source HIT (128 bit)

— **COMMON**

Destination HIT (128 bit)

| Type | Length |
| K | Lifetime | Opaque |

Random I (64 bit)

— **PUZZLE**

| Type | Length |
| Group ID | Public Key Length | Public Key |

Public Key (variable size)

— **DIFFIE–HELLMAN**

| Type | Length |
| Suite ID 1 | Suite ID 2 |

— **HIP_TRANSFORM**

| Type | Length |
| HI Length | DI–type | DI   Length |

Host identity (variable size)

Domain Identifier (variable size)

— **HOST_ID**

| Type | Length |
| SIG alg | Signature |

Signature (variable size)

— **HIP_SIG**

Figure 4.4 Obligatory fields of HIP R1 packet.

```
[Time delta from previous packet: 0.001762000 seconds] Internet
Protocol, Version: 4
    Header length: 20 bytes
    Differentiated Services Field: 0x00 (DSCP 0x00: Default; ECN: 0x00)
    Total Length: 668
    Identification: 0x0000 (0)
    Flags: 0x04 (Don't Fragment)
    Fragment offset: 0
    Time to live: 64
    Protocol: Unknown (0xfd)
    Header checksum: 0x3d27 [correct]
    Source: 193.167.187.132 (193.167.187.132)
    Destination: 193.167.187.106 (193.167.187.106)
Host Identity Protocol
    Payload Protocol: 59
    Header Length: 80
    Packet Type: 2
    Version: 1, Reserved: 1
    HIP Controls: 0x0000
            .... .... .... ..0. = Certificate (One or more CER packets follows)
            .... .... .... ...0 = Anonymous (Sender's HI is anonymous)
    Checksum: 0x2ce7 (correct)
    Sender's HIT: 20010014766EFBEEF74DEC73D6C528C0
    Receiver's HIT: 200100150A097C449CA1257176DD0872
    HIP Parameters
        PUZZLE (type=257, length=12)
            Puzzle Difficulty K: 10
            Puzzle Lifetime: 42
            Opaque Data: 0x4849
            Puzzle Random I: A70EB9051DB3BD8A
        DIFFIE_HELLMAN (type=513, length=246)
            3 (1536-bit MODP group)
            Public Value: 00C0EC125E884D4F813834F038ACF09AF7C0EB344386715D...
        HIP_TRANSFORM (type=577, length=6)
            1 (AES-CBC with HMAC-SHA1)
            2 (3DES-CBC with HMAC-SHA1)
            5 (NULL with HMAC-SHA1)
        ESP_TRANSFORM (type=4095, length=8)
            Reserved: 0x0000
            1 (AES-CBC with HMAC-SHA1)
            2 (3DES-CBC with HMAC-SHA1)
            5 (NULL with HMAC-SHA1)
        HOST_ID (type=705, length=163)
            Host Identity Length: 136
            Domain Identifier Type: 1
            Domain Identifier Length: 23
            Host Identity flags: 0x0202ff05
                0000 0010 0000 0010 .... .... .... .... = Flags: key is associated with
                                                          non-zone entity (0x00000202)
                .... .... .... .... 1111 1111 .... .... = Protocol: key is valid for
                                                          any protocol (0x000000ff)
                .... .... .... .... .... .... 0000 0101 = Algorithm: RSA (0x00000005)
            RSA Host Identity e_len (exponent length): 3
            RSA Host Identity e (exponent): 010001
            RSA Host Identity n (public modulus):
                B35BC70CF07E82649FD954A7E7837B0FDD279E266D1EAD56...
            FQDN: hipserver.infrahip.net
        HIP_SIGNATURE_2 (type=61633, length=129)
            5 (RSA)
            Signature: 1BBA36EF3CE5F20837156C39CE0564AA8E70F3AD8645ADA9...
```

Figure 4.5 HIP R1 packet captured with Wireshark.

Furthermore, the random number I is not included to the signature and is replaced with zeros. This operation allows the Responder to use pre-generated R1 packets.

The R1 packet can optionally include the ECHO_REQUEST parameter, which the Initiator should return unmodified in I2. This parameter can either be located before the signature and covered by it or placed after the signature.

If an optional R1_COUNTER parameter is present in R1 packet, it is placed before any other parameter. The counter gives the currently valid generation of puzzles in an 8-byte number. It is incremented periodically to protect against an attack with a large number of pre-computed puzzles.

4.1.3   I2 packet

The format of I2 packet is shown in Figure 4.6. I2 has similar obligatory fields as R1, except the puzzle parameter is replaced by the solution parameter and an HMAC parameter is added before the signature. The solution parameter includes the random number and a J value that should hash into K zero bits as described for the puzzle parameter in R1. The hash message authentication code (HMAC) is calculated over the entire HIP packet excluding SIGNATURE and ECHO parameters. HMAC is faster to verify than a HIP signature, and is checked before the signature as an additional protection against replay attacks.

The host identity can be sent in plain or encrypted. A plain version facilitates traversal of some middleboxes that may want to verify host identities before permitting further communication. I2 may also include unmodified R1_COUNTER and ECHO_REPLY parameters copied from R1 packet. Figure 4.7 shows an example of I2 packet captured from a network.

4.1.4   R2 packet

Figure 4.8 shows the content of R2 packet. Its only fields are HMAC and HIP_SIGNATURE parameters that must be verified by Initiator. Figure 4.9 shows an example of R2 packet captured from a network.

The data can flow on a HIP association after the fourth packet in the base exchange. The transport protocol, such as TCP, further requires exchange of own control packets, which additionally delays the point when an application can actually transmit user data. In the TCP case, the three-way handshake takes an additional round-trip time after two round-trips of the base exchange. To reduce the delay observed by applications, it may be possible to transmit upper-layer data already in I2 and R2 HIP packets. Earlier transmission is not possible due to security concerns. Data transmission in HIP control packet is not a part of the base specification, but is described in a separate document (Lindqvist 2006b).

Figure 4.10 illustrates an ICMP echo request and echo reply packets transported in an ESP bound end-to-end tunnel. The packets were captured with Wireshark immediately after a HIP base exchange. Both packets are encrypted in ESP so that only SPI value and ESP sequence numbers are visible to a third party on the wire.

Figure 4.11 shows a simplified state machine for HIP association establishment. Transitions between states are marked with a condition for transition at the top and the resulting action at the bottom. It is assumed that packets causing transitions have correct checksums and MAC; otherwise they are discarded without a change in the state.

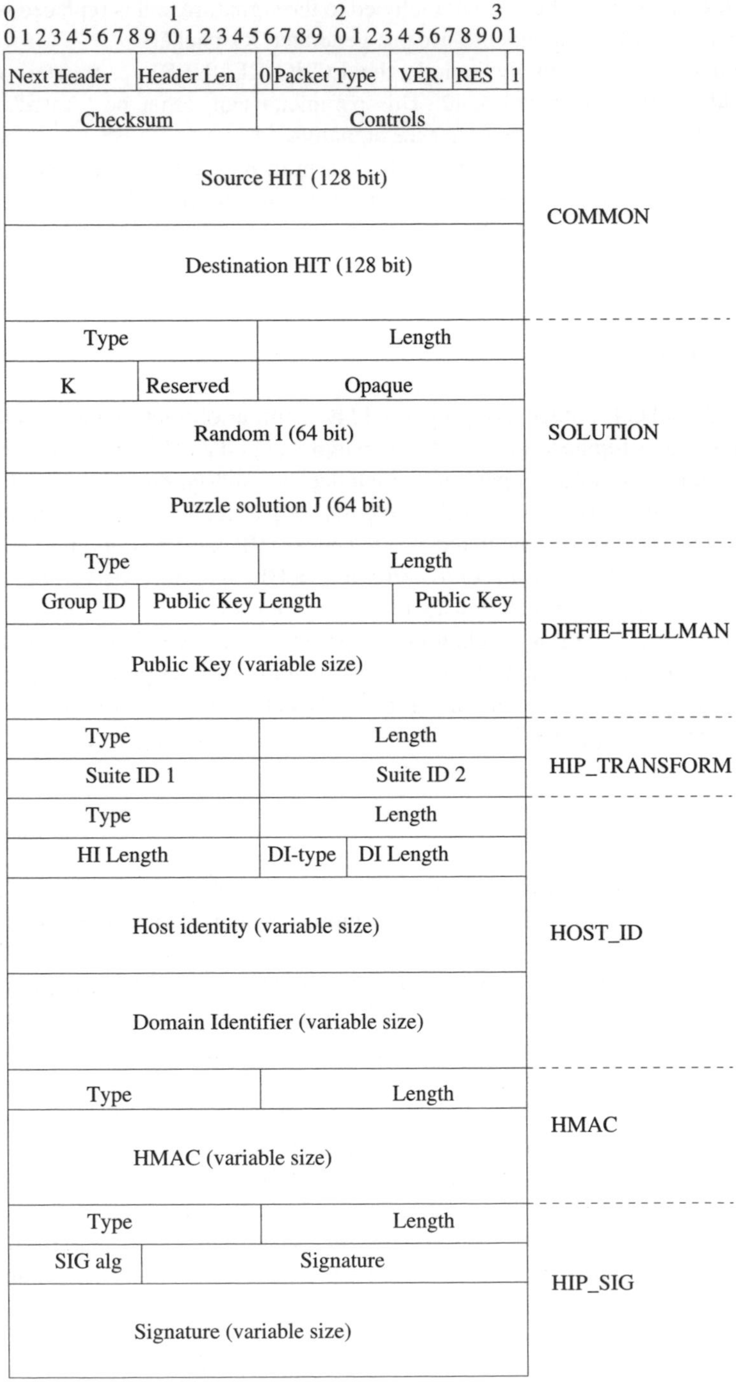

```
 0                   1                   2                   3
 0 1 2 3 4 5 6 7 8 9 0 1 2 3 4 5 6 7 8 9 0 1 2 3 4 5 6 7 8 9 0 1
┌─────────────┬─────────────┬─┬───────────┬────────┬─────┬─┐
│ Next Header │ Header Len  │0│Packet Type│  VER.  │ RES │1│
├─────────────┴─────────────┼─┴───────────┴────────┴─────┴─┤
│        Checksum           │            Controls          │
├───────────────────────────┴──────────────────────────────┤
│                                                           │
│             Source HIT (128 bit)                          │
│                                                           │              COMMON
│                                                           │
│          Destination HIT (128 bit)                        │
│                                                           │
├───────────────────────────┬──────────────────────────────┤  - - - - - - - - - -
│          Type             │            Length            │
├─────────┬─────────────────┼──────────────────────────────┤
│    K    │    Reserved     │            Opaque            │
├─────────┴─────────────────┴──────────────────────────────┤
│             Random I (64 bit)                             │              SOLUTION
├───────────────────────────────────────────────────────────┤
│             Puzzle solution J (64 bit)                    │
├───────────────────────────┬──────────────────────────────┤  - - - - - - - - - -
│          Type             │            Length            │
├─────────┬─────────────────┼──────────────┬───────────────┤
│ Group ID│ Public Key Length              │   Public Key  │
├─────────┴────────────────────────────────┴───────────────┤             DIFFIE–HELLMAN
│          Public Key (variable size)                       │
├───────────────────────────┬──────────────────────────────┤  - - - - - - - - - -
│          Type             │            Length            │
├───────────────────────────┼──────────────────────────────┤
│        Suite ID 1         │         Suite ID 2           │              HIP_TRANSFORM
├───────────────────────────┼──────────────────────────────┤  - - - - - - - - - -
│          Type             │            Length            │
├───────────────────┬───────┼──────────────────────────────┤
│     HI Length     │DI-type│         DI Length            │
├───────────────────┴───────┴──────────────────────────────┤
│          Host identity (variable size)                    │              HOST_ID
├───────────────────────────────────────────────────────────┤
│          Domain Identifier (variable size)                │
├───────────────────────────┬──────────────────────────────┤  - - - - - - - - - -
│          Type             │            Length            │
├───────────────────────────┴──────────────────────────────┤             HMAC
│          HMAC (variable size)                             │
├───────────────────────────┬──────────────────────────────┤  - - - - - - - - - -
│          Type             │            Length            │
├───────────────┬───────────┴──────────────────────────────┤
│    SIG alg    │              Signature                   │
├───────────────┴──────────────────────────────────────────┤             HIP_SIG
│          Signature (variable size)                        │
└───────────────────────────────────────────────────────────┘
```

Figure 4.6 Obligatory fields of HIP I2 packet.

```
[Time delta from previous packet: 0.088695000 seconds] Internet
Protocol, Version: 4
    Header length: 20 bytes
    Differentiated Services Field: 0x00 (DSCP 0x00: Default; ECN: 0x00)
    Total Length: 844
    Identification: 0x0000 (0)
    Flags: 0x04 (Don't Fragment)
    Fragment offset: 0
    Time to live: 63
    Protocol: Unknown (0xfd)
    Header checksum: 0x3d77 [correct]
    Source: 193.167.187.106 (193.167.187.106)
    Destination: 193.167.187.132 (193.167.187.132)
Host Identity Protocol
    Payload Protocol: 59
    Header Length: 102
    Packet Type: 3
    Version: 1, Reserved: 1
    HIP Controls: 0x0000
        .... .... .... ..0. = Certificate (One or more CER packets follows)
        .... .... .... ...0 = Anonymous (Sender's HI is anonymous)
    Checksum: 0x3b64 (correct)
    Sender's HIT: 200100150A097C449CA1257176DD0872
    Receiver's HIT: 20010014766EFBEEF74DEC73D6C528C0
    HIP Parameters
        ESP INFO (type=65, length=12)
            Reserved: 0x0000
            Keymat Index: 0x0048
            Old SPI: 0x00000000
            New SPI: 0x545434d4
        SOLUTION (type=321, length=20)
            Puzzle Difficulty K: 10
            Puzzle Lifetime: 0
            Opaque Data: 0x4849
            Puzzle Random I: A70EB9051DB3BD8A
            Puzzle Solution J: F7037C04772CB48C
        DIFFIE_HELLMAN (type=513, length=195)
            3 (1536-bit MODP group)
            Public Value: 00C0C6C51DD129B6170DE89E3063F07610F34C24B7036014...
        HIP_TRANSFORM (type=577, length=2)
            1 (AES-CBC with HMAC-SHA1)
        ESP_TRANSFORM (type=4095, length=4)
            Reserved: 0x0000
            1 (AES-CBC with HMAC-SHA1)
        ENCRYPTED (type=641, length=452)
            Reserved: 0x00000000
            IV: C0AA92024E015B75
            Encrypted Data (440 bytes)
        HMAC (type=61505, length=20)
            HMAC: 26700FB555915065FC1C40587D071778C6827C19
        HIP_SIGNATURE (type=61697, length=42)
            3 (DSA)
            Signature: 08082426AD2D594A669C8AFA5BC3FA0827660A7D883909B6...
```

Figure 4.7 HIP I2 packet captured with Wireshark.

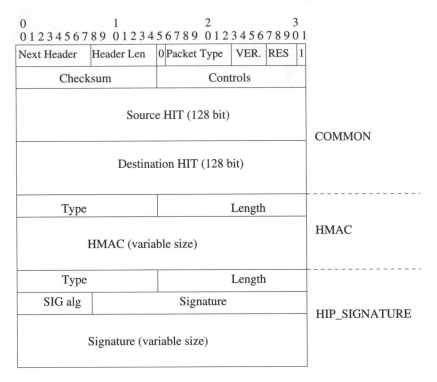

Figure 4.8 Fields of HIP R2 packet.

4.2   Other HIP control packets

This section describes the remaining HIP packets meant to control the HIP association: UPDATE, NOTIFY, and CLOSE. Finally, HIP-specific ICMP messages are discussed.

The UPDATE packet has a similar format to R2. Before the HMAC parameter, the UPDATE packet can include also SEQ and ACK parameters. If the UPDATE packet includes the SEQ parameter, it is acknowledged by the peer host by sending an UPDATE packet with the ACK parameter. The SEQ number is incremented by one starting from zero in each new UPDATE packet. The number has scope within a single HIP association. The ACK parameter can include several sequence numbers. Unlike TCP, HIP acknowledgments are not cumulative, but each individual SEQ is acknowledged by a separate ACK parameter. A single UPDATE packet can include both SEQ and ACK parameters.

The optional NOTIFY packet may be used by the HIP host to inform its peer on protocol errors. Due to security concerns, the HIP host should not make any state changes based purely on NOTIFY packets. The NOTIFY packet includes one or more NOTIFY parameters, one or more HIs, and HIP_SIGNATURE. The following errors types can be used in the NOTIFY packet: UNSUPPORTED_CRITICAL_PARAMETER_TYPE, INVALID_SYNTAX, NO_ DH_PROPOSAL_CHOSEN, INVALID_DH_CHOSEN, NO_HIP_PROPOSAL_CHOSEN, INVALID_HIP_TRANSFORM_CHOSEN, AUTHENTICATION_FAILED, CHECKSUM_

```
[Time delta from previous packet: 0.129714000 seconds]
Internet Protocol, Version: 4
    Header length: 20 bytes
    Differentiated Services Field: 0x00 (DSCP 0x00: Default; ECN: 0x00)
    Total Length: 236
    Identification: 0x0000 (0)
    Flags: 0x04 (Don't Fragment)
    Fragment offset: 0
    Time to live: 64
    Protocol: Unknown (0xfd)
    Header checksum: 0x3ed7 [correct]
    Source: 193.167.187.132 (193.167.187.132)
    Destination: 193.167.187.106 (193.167.187.106)
Host Identity Protocol
    Payload Protocol: 59
    Header Length: 26
    Packet Type: 4
    Version: 1, Reserved: 1
    HIP Controls: 0x0000
        .... .... .... ..0. = Certificate (One or more CER packets follows)
        .... .... .... ...0 = Anonymous (Sender's HI is anonymous)
    Checksum: 0xb358 (correct)
    Sender's HIT: 20010014766EFBEEF74DEC73D6C528C0
    Receiver's HIT: 200100150A097C449CA1257176DD0872
    HIP Parameters
        ESP INFO (type=65, length=12)
            Reserved: 0x0000
            Keymat Index: 0x0048
            Old SPI: 0x00000000
            New SPI: 0x940e5f06
        HMAC_2 (type=61569, length=20)
            HMAC: 4A30965252D4A96A626587817DDED0243B1C37DD
        HIP_SIGNATURE (type=61697, length=129)
            5 (RSA)
            Signature: 2D18C7CCEF2591258F5579B80A68EA9DC0BE7DB4A3E51E1F...
```

Figure 4.9 HIP R2 packet captured with Wireshark.

FAILED, HMAC_FAILED, ENCRYPTION_FAILED, INVALID_HIT, BLOCKED_BY_POLICY, SERVER_BUSY_PLEASE_RETRY, I2_ACKNOWLEDGEMENT.

HIP associations are terminated by a CLOSE packet, typically after a period of inactivity of 30 minutes. The CLOSE packet is acknowledged by a CLOSE_ACK packet. The CLOSE packet includes ECHO_REQUEST, HMAC, and HIP_SIGNATURE parameters. A peer host sends CLOSE_ACK containing ECHO_REPLY with data sent in ECHO_REQUEST. Hosts must verify both HMAC and HIP_SIGNATURE in CLOSE and CLOSE_ACK packets.

An ICMP message may be used by a HIP host to indicate unsupported protocol version, invalid puzzle solution, non-existing HIP association, or a malformed parameter. Due to security risks, ICMP messages must be transmitted at a limited rate. Either ICMPv4 or v6 can be used with a message type "Parameter Problem" with a pointer to the packet field that caused the problem. However, if the checksum fails, the HIP packet is silently dropped without sending ICMP messages. ICMP messages can be misused by an attacker to trick the host that its peer does not support HIP or simulate a protocol error. Therefore, a host must be

```
[Time delta from previous packet: 0.003767000 seconds]
Internet Protocol, Version: 4
    Header length: 20 bytes
    Differentiated Services Field: 0xa7 (DSCP 0x29: Unknown DSCP; ECN: 0x03)
    Total Length: 136
    Identification: 0x0000 (0)
    Flags: 0x04 (Don't Fragment)
        0... = Reserved bit: Not set
        .1.. = Don't fragment: Set
        ..0. = More fragments: Not set
    Fragment offset: 0
    Time to live: 63
    Protocol: ESP (0x32)
    Header checksum: 0x405f [correct]
    Source: 193.167.187.106 (193.167.187.106)
    Destination: 193.167.187.132 (193.167.187.132)
Encapsulating Security Payload
    SPI: 0x940e5f06
    Sequence: 1

[Time delta from previous packet: 0.000186000 seconds]
Internet Protocol, Version: 4
    Header length: 20 bytes
    Differentiated Services Field: 0x00 (DSCP 0x00: Default; ECN: 0x00)
    Total Length: 136
    Identification: 0x0000 (0)
    Flags: 0x04 (Don't Fragment)
        0... = Reserved bit: Not set
        .1.. = Don't fragment: Set
        ..0. = More fragments: Not set
    Fragment offset: 0
    Time to live: 64
    Protocol: ESP (0x32)
    Header checksum: 0x4006 [correct]
    Source: 193.167.187.132 (193.167.187.132)
    Destination: 193.167.187.106 (193.167.187.106)
Encapsulating Security Payload
    SPI: 0x545434d4
    Sequence: 1
```

Figure 4.10 Captured headers of ICMP ping echo request and reply packets inside ESP.

cautious about incoming ICMP messages. As an example, after transmitting an I1 message, the host must wait for a full timeout value for an R1 packet even if an ICMP message arrives.

4.3   IPsec encapsulation

The HIP base specifications do not give details of any encapsulation protocol to be used to carry HIP data packets, but state that the ESP support is mandatory. The ESP encapsulation is described in a separate document (Jokela *et al.* 2008). The ESP part of IPsec is used to set up a pair of Security Associations (SAs), one in each direction, to encapsulate HIP data packets.

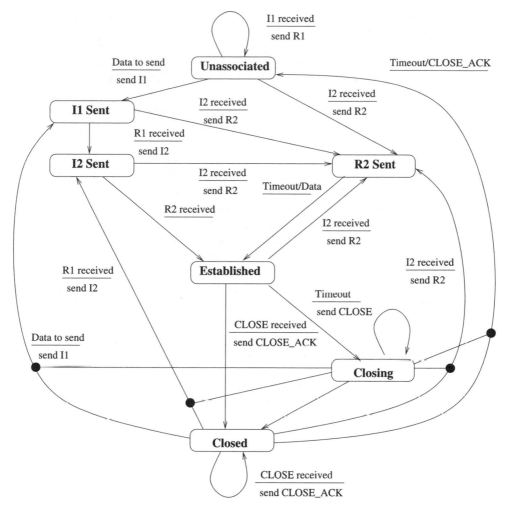

Figure 4.11 A simplified state machine of a HIP association.

The data packets are authenticated and encrypted by ESP using the symmetric keys agreed during the HIP base exchange.

4.3.1   ESP transforms

On-the-wire HIP ESP packets look exactly the same as IPsec ESP packets (Kent and Atkinson 1998a). Host identifiers are not transported explicitly in the HIP ESP packets, instead SPIs at both ends are used to locate a correct HIP association. To support the ESP mode for data packets, the HIP base exchange messages include ESP_TRANSFORM and ESP_INFO parameters. The ESP_TRANSFORM parameter included into I2 by the Responder suggests a set of supported IPsec encryption and hashing algorithms (transforms). ESP_INFO is sent in I2 and R2 and includes the SPI value to be used in an SA. AES and

HMAC-SHA-1-96 are transforms supported by all HIP implementations. In addition, ESP NULL encryption with SHA1 or MD5 authentication is supported.

There are two ways to implement ESP support: a standard IPsec ESP transport mode and a new Bound End-to-End Tunnel (BEET) mode (Nikander and Melen 2006). The standard transport mode uses IP addresses in setting Security Parameter Database (SPD) and Security Association Databases (SAD). The upper protocol layers and transmitted packets also use the IP addresses. However, the HIP layer calculates and verifies upper-layer checksums using HITs in place of IP addresses. The BEET mode provides "native" IPsec support for HIP and is described in more detail in Section 4.3.2.

The Initiator determines the SPI value to be used by the Responder in its outgoing Security Associations, while the Responder gives the SPI for Initiator. The SPI value should be assigned randomly and changed for each new SA between two hosts to prevent replay attacks. One possible method of generating an SPI is to take a hash of HIT concatenates with an incremental counter and take 32 high-order bits.

If a host cannot successfully process a request to establish an SA, HIP NOTIFY parameter is used to inform the peer host of a problem. NO_ESP_PROPOSAL_CHOSEN or INVALID_ESP_TRANSFORM_CHOSEN error types in the NOTIFY message correspond to the cases when there is no acceptable cryptographic algorithm to choose and when the chosen algorithm is not the one offered by the Responder. When receiving an ESP message with an SPI that does not correspond to any established SA at the host, it may send ICMP an error message "parameter problem" with a pointer to the SPI field in the packet header.

The process of creating a new pair of SAs and removing the old SAs is called *rekeying*. Rekeying may be necessary, for example, when wrapping up the 64-bit sequence number field in the ESP header. Either the Initiator or Responder can initiate this procedure. The SPI value for a new SA is sent in the ESP_INFO parameter using HIP UPDATE message. Once a host receives data on the new SA, it can safely remove the old SA. The new SPI value in the ESP_INFO parameter can be used by middleboxes to update their state. For this reason, ESP_INFO is signed but not encrypted. The HIP UPDATE message is transmitted reliably using the underlying retransmission mechanism.

Idle SA time out after an inactivity timer expires in the host. The timeout value is determined locally; the default recommended value is 15 minutes. This means that if no data is sent over a HIP connection, it is terminated and a new base exchange is necessary before new data can be transmitted.

4.3.2   ESP Bound End-to-End Tunnel

The previous section has described supporting ESP encapsulation for HIP using packet header rewriting. This section concentrates on the new BEET IPsec mode, which is a recommended way of implementing HIP. Using BEET, all IPsec state information is kept in one place, there as using the transport IPsec mode, synchronization between SADB and a separate database is necessary.

The BEET mode is a combination of transport and tunnel IPsec modes. As the name suggests, BEET creates a tunnel between two end hosts. However, it is more efficient than the standard tunnel mode as it eliminates the need to transmit a pair of IP addresses in packet headers. We distinguish two separate pairs of IP addresses: *outer* addresses that are used on the wire and *inner* addresses that the applications see. For the BEET mode, inner addresses are HITs that are fixed for the lifetime of the SA and thus can be left out of the transmitted packet headers. The outer addresses can change dynamically during the SA lifetime.

The BEET mode suits ideally for using with HIP. Other possible applications of BEET can be Mobile IP v4 and v6, and multihoming protocols. The main advantage of BEET over the transport IPsec mode is the ability to traverse middleboxes. As the addresses seen by applications may differ from addresses on the wire, NATs can safely rewrite packet headers without breaking an SA. If an IPsec transport mode were used, changes in IP addresses would invalidate any transport-layer checksums. This fact has hindered the use of transport IPsec mode given that even UDP encapsulation is not a sufficient solution. The tunnel mode is free from the above-mentioned problem at the expense of larger header overhead.

In the BEET mode, inner and outer addresses can be from different address families. As an example, the inner addresses can be IPv6 while outer addresses IPv4. Recall that HITs have the same format as IPv6 addresses. That allows using IPv6 applications connecting to HITs over the IPv4 network.

Figure 4.12 illustrates the case of inner IPv4 and outer IPv6 addresses. The outer IPv6 address is the one transmitted on the wire followed by IPv6 extension headers. The body of the packet includes TCP (or other transport protocol) segment encapsulated into ESP. Normally, no inner header is present in the transmitted packet. However, in the illustrated case the inner IPv4 header included some IP options. To support them, a pseudoheader is added after the ESP header. The pseudoheader includes the header type and length fields, as well as any of original IP options. Fortunately, IP options are infrequently used and this case is shown only for completeness.

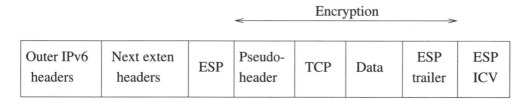

Figure 4.12 BEET headers with inner IPv4 addresses with IP options and outer IPv6 address.

The parameter values from the inner IP header, such as the Time-To-Live counter and the Don't-fragment bit are copied to the outer header upon transmission and back to the inner header upon reception of the packet. That required extensions to the PF_KEYv2 interface are SADB_X_IDENTTYPE_ADDR and SADB_IDENTTYPE_MAX parameters.

In principle, it would be possible to emulate the BEET mode using the transport IPsec combined with two NAT address mappings in hosts. However, this is a more complex and error-prone solution that also requires correct host firewall configuration to prevent leaking of inner addresses. Therefore, the BEET mode is the recommended way of HIP implementation. The BEET mode is currently implemented to the BSD KAME stack as well as to Linux 2.6 XFRM architecture. The BSD implementation is a little over 500 lines of code and took six person-days to realize. The Linux BEET code proved to be more complex, but it is included in the official kernel release starting from version 2.6.18.

5

Main extensions

In this chapter, we present HIP mobility and multihoming extensions (Section 5.1), the rendezvous server (Section 5.2), the DNS extensions (Section 5.3), and the registration protocol (Section 5.4).

5.1 Mobility and multihoming

The base protocol specification for HIP describes setting up an association between two hosts, each having a single IP address. Mobility and multihoming support for HIP are given in a separate IETF document (Henderson 2007). These extensions cover the basic scenarios of mobility and multihoming, while advanced cases are left for future study. In particular, the document does not consider localized mobility management (similar to hierarchical MIP) but focuses on end-to-end mobility. Furthermore, scenarios when two hosts move simultaneously (double jump), mobility behind a NAT or location privacy are outside the scope of this section. The specification assumes that HIP uses the ESP transport format.

5.1.1 Mobility and multihoming architecture

The *locator* is a more general concept than a network address. A HIP host may need additional information to demultiplex a HIP data packet to a correct security association. One currently defined locator format combines a prefix of SPI with an IP address. Other format may include IPv6 Flow Labels or transport protocol numbers.

The mobility and multihoming extensions described in this section modify the set of locators at a peer using an UPDATE message. Therefore, they require that the peer is still reachable after a mobility event. This is the reason why double jump and other scenarios where two hosts lose direct end-to-end connectivity are not directly supported.

For the reasons of observing the ESP anti-replay window explained below, it may be necessary to establish a SA between each pair of IP addresses of two hosts. The job of the HIP layer is to map arriving ESP packets to a HIT using the SPI value in the packet and select the source address and interface according to the SPI value set by ESP.

Mobility with a single SA pair without rekeying is the simplest scenario. Two hosts have a single SPI pair between them. When one of the hosts moves to another subnet, renews its DHCP lease or receives a new IPv6 router advertisement message changing the local network prefix, it needs to change its IP address. After obtaining the new IP address, the host sends an UPDATE message to the peer containing the new address in the LOCATOR parameter. The message also includes ESP_INFO parameter with the Old SPI and New SPI values both set to the current outgoing SPI and a sequence number.

Upon receiving the UPDATE message, the peer constructs its own UPDATE to acknowledge its reception and verify host reachability at the claimed new address. The second UPDATE message includes ESP_INFO with Old and New SPIs set to the current outgoing SPI. The message includes an acknowledgment for the first UPDATE, a sequence number, and an ECHO_REQUEST to verify the new peer address.

When a host that initiated a mobility update receives the UPDATE message, it replies with the third UPDATE message containing ECHO_RESPONSE to prove that it indeed receives data on the new address. It also acknowledges the second UPDATE message. The third UPDATE is not acknowledged. Instead, the peer host retransmits the second UPDATE if the third UPDATE was lost.

A mobility scenario with a single SA pair with rekeying additionally includes a change of SAs as illustrated in Figure 5.1. A mobile host initiates the UPDATE including a New SPI field set to a value to be used for incoming SPI and an index in the key material to generate a new ESP session key. The peer host replies with an UPDATE containing its own new SPI and,

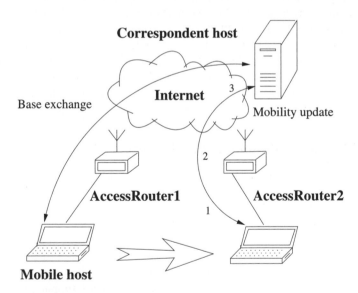

1)UPDATE(ESP_INFO, LOCATOR, SEQ)

2)UPDATE(ESP_INFO, SEQ, ACK, ECHO_REQUEST)

3)UPDATE(ACK, ECHO_RESPONSE)

Figure 5.1 Mobility update signaling.

finally, the mobile host acknowledges it as described for update without rekeying. Optionally, instead of retrieving the keys from the existing keying material, the mobile host may force a new keying material to be generated. To accomplish the change, the first and second UPDATE messages include DIFFIE_HELLMAN parameters for calculating a new shared secret to be used to generate the new keying material.

As a practical example of HIP mobility signaling, we present a handover of HIP on Linux (HIPL) implementation captured with Wireshark protocol analyzer. During the handover that followed the base exchange presented in Figures 4.3–4.10, the IPv4 address of the Initiator is changed from 193.167.187.106 to 193.167.187.107. After receiving a notification from the Linux kernel via the NETLINK interface, the Initiator starts an update without rekeying.

Figure 5.2 illustrates the first UPDATE packet. The LOCATOR parameter contains the new IPv4 address and the available IPv6 address of the Initiator. The Responder replies with the second UPDATE packet as shown in Figure 5.3. The packet contains opaque data that is echoed by the Initiator in the third UPDATE packet in Figure 5.4. This check ensures that the Initiator indeed is available at the claimed address and prevents possible replay attacks. After the handover, data continues to flow inside of the ESP tunnel with same SPI values but a different IP address of the Initiator. Figure 5.5 shows headers of ICMP echo request and reply packets captured after the handover.

5.1.2   Multihoming as extension of mobility

The basic multihoming scenario with adding a new IP address for a host corresponds to mobility without rekeying where the old address is not deprecated but can be used afterwards. Multihoming with a single pair of SAs is feasible, for example, when IP addresses belong to a single subnet. For a general case, when addresses correspond to different topological places in the Internet, a possibility of packets violating the ESP anti-replay window requires setting separate SAs for a new address. To inform the peer host of a new SA, the mobile host sets the old SPI value in ESP_INFO parameter to zero and the new SPI value to the desired incoming SPI value. The peer host uses the destination address of the UPDATE message to create a new SA. Otherwise, adding a new address to the host corresponds to mobility with rekeying. The peer host must verify reachability of a newly added address before disabling Credit-Based Authentication for it. See Section 5.1.6.

An advanced scenario may combine mobility and multihoming. A mobile host having several network interfaces can enable and disable them based on its current location. The host is reachable by any address in its current active locator set. A practical example includes a mobile phone having UMTS, GPRS, and WLAN interfaces. While GPRS has almost universal coverage in Europe, UMTS is often available in urban areas. WLAN has limited coverage in public hotspots. The phone could keep the GPRS data connection active at all times, and add UMTS and WLAN IP addresses if their wireless network coverage is available at a spot. Although the current LOCATOR in UPDATE mechanism should be flexible to accommodate such scenarios, they have not been sufficiently experimented with.

In addition to sending the LOCATOR parameter in UPDATE messages, hosts can send LOCATOR in R1 and I2 during the base exchange. Such practice has the effect of notifying the peer host of other locally available IP addresses. However, hosts must verify address reachability before sending a significant data amount to it. The source address used in R1 and I2 packets remains the preferred locator after the base exchange.

```
Internet Protocol, Version: 4
    Header length: 20 bytes
    Differentiated Services Field: 0x00 (DSCP 0x00: Default; ECN: 0x00)
    Total Length: 220
    Identification: 0x0000 (0)
    Flags: 0x04 (Don't Fragment)
    Fragment offset: 0
    Time to live: 63
    Protocol: Unknown (0xfd)
    Header checksum: 0x3fe6 [correct]
    Source: 193.167.187.107 (193.167.187.107)
    Destination: 193.167.187.132 (193.167.187.132)
Host Identity Protocol
    Payload Protocol: 59
    Header Length: 24
    Packet Type: 16
    Version: 1, Reserved: 1
    HIP Controls: 0x0000
        .... .... .... ..0. = Certificate (One or more CER packets follows)
        .... .... .... ...0 = Anonymous (Sender's HI is anonymous)
    Checksum: 0x97a2 (correct)
    Sender's HIT: 200100150A097C449CA1257176DD0872
    Receiver's HIT: 20010014766EFBEEF74DEC73D6C528C0
    HIP Parameters
        ESP INFO (type=65, length=12)
            Reserved: 0x0000
            Keymat Index: 0x0090
            Old SPI: 0x545434d4
            New SPI: 0x545434d4
        LOCATOR (type=193, length=56)
            Traffic Type: 0
            Locator Type: 0
            Locator Length: 4
            Reserved: 0xb7
            Locator Lifetime: 0x00000000
            Address: 2001:708:140:220:216:41ff:fee5:232a
            Traffic Type: 0
            Locator Type: 0
            Locator Length: 0
            Reserved: 0x0
            Locator Lifetime: 0x00000400
            Address: ::ffff:c1a7:bb6b (193.167.187.107)
        SEQ (type=385, length=4)
            Update ID: 0x00000002
        HMAC (type=61505, length=20)
            HMAC: 15A73BCFE17D7584F314A0C37DC86B5D4B6B867B
        HIP_SIGNATURE (type=61697, length=42)
            3 (DSA)
            Signature: 082FBD0B96E39898AD03D4C2801C617D3359F6092C3D67A1...
```

Figure 5.2 First UPDATE packet.

```
[Time delta from previous packet: 0.305165 seconds]
Internet Protocol, Version: 4
    Header length: 20 bytes
    Differentiated Services Field: 0x00 (DSCP 0x00: Default; ECN: 0x00)
    Total Length: 260
    Identification: 0x0000 (0)
    Flags: 0x04 (Don't Fragment)
    Fragment offset: 0
    Time to live: 64
    Protocol: Unknown (0xfd)
    Header checksum: 0x3ebe [correct]
    Source: 193.167.187.132 (193.167.187.132)
    Destination: 193.167.187.107 (193.167.187.107)
Host Identity Protocol
    Payload Protocol: 59
    Header Length: 29
    Packet Type: 16
    Version: 1, Reserved: 1
    HIP Controls: 0x0000
        .... .... .... ..0. = Certificate (One or more CER packets follows)
        .... .... .... ...0 = Anonymous (Sender's HI is anonymous)
    Checksum: 0x8e11 (correct)
    Sender's HIT: 20010014766EFBEEF74DEC73D6C528C0
    Receiver's HIT: 200100150A097C449CA1257176DD0872
    HIP Parameters
        ESP INFO (type=65, length=12)
            Reserved: 0x0000
            Keymat Index: 0x0090
            Old SPI: 0x00000000
            New SPI: 0x00000000
        SEQ (type=385, length=4)
            Update ID: 0x00000005
        ACK (type=449, length=4)
            ACKed Peer Update ID: 0x00000002
        HMAC (type=61505, length=20)
            HMAC: 3C8EB1422F5EA42AB6D42F37FD16035D115A5FFC
        HIP_SIGNATURE (type=61697, length=129)
            5 (RSA)
            Signature: 8BB884A8F7BE1507BDE90A1A2AC3305A2DFF5D047DCD2845...
        ECHO_REQUEST (No sig.) (type=63661, length=4)
            Opaque Data: D52EA94D
```

Figure 5.3 Second UPDATE packet.

5.1.3   Effect of ESP anti-replay window

When several IP addresses are available at a host, one of them is assigned to be a *preferred*
locator. The addresses used in the base exchange become the default preferred locators of
each host. As a rule, a host sends data to the preferred locator of the peer host, and uses own
preferred locator as a source address of transmitted packets.

```
[Time delta from previous packet: 0.129540000 seconds]
Internet Protocol, Version: 4
    Header length: 20 bytes
    Differentiated Services Field: 0x00 (DSCP 0x00: Default; ECN: 0x00)
    Total Length: 148
    Identification: 0x0000 (0)
    Flags: 0x04 (Don't Fragment)
    Fragment offset: 0
    Time to live: 63
    Protocol: Unknown (0xfd)
    Header checksum: 0x402e [correct]
    Source: 193.167.187.107 (193.167.187.107)
    Destination: 193.167.187.132 (193.167.187.132)
Host Identity Protocol
    Payload Protocol: 59
    Header Length: 15
    Packet Type: 16
    Version: 1, Reserved: 1
    HIP Controls: 0x0000
        .... .... .... ..0. = Certificate (One or more CER packets follows)
        .... .... .... ...0 = Anonymous (Sender's HI is anonymous)
    Checksum: 0xb87c (correct)
    Sender's HIT: 200100150A097C449CA1257176DD0872
    Receiver's HIT: 20010014766EFBEEF74DEC73D6C528C0
    HIP Parameters
        ACK (type=449, length=4)
            ACKed Peer Update ID: 0x00000002
        HMAC (type=61505, length=20)
            HMAC: 54207D9753B9A20C3127FBB7614BECEC926721BC
        HIP_SIGNATURE (type=61697, length=42)
            3 (DSA)
            Signature: 083659D85D21D1A3F856922DA59161E3D4B2986E5557B5F1...
        ECHO_RESPONSE (No sig.) (type=63425, length=4)
            Opaque Data: D52EA94D
```

Figure 5.4 Third UPDATE packet.

In the base specification, a HIP association includes two uni-directional ESP security associations, one in each direction. Several IP addresses can be added to an SA and the sender can transmit and receive HIP data packets through any of the addresses. The important property of HIP is that the source address of a packet is not used to lookup a proper SA, but the SPI is used for this purpose. Any number of SAs can be established between two HIP hosts as described below.

The problem with using a single SA for multiple IP addresses is the ESP anti-replay window. Packets sent from separate interfaces are likely to travel via different paths in the network. The anti-replay window limits the sequence numbers of packets acceptable within SA to prevent acceptance of maliciously captured and retransmitted packet duplicates. The ESP anti-replay window is configurable at the host and can be set sufficiently large to permit the use of a single SA if replay attacks are not a perceived issue. In a general case, however,

```
[Time delta from previous packet: 0.399505000 seconds]
Internet Protocol, Version: 4
    Header length: 20 bytes
    Differentiated Services Field: 0x00 (DSCP 0x00: Default; ECN: 0x00)
    Total Length: 136
    Identification: 0x0000 (0)
    Flags: 0x04 (Don't Fragment)
    Fragment offset: 0
    Time to live: 63
    Protocol: ESP (0x32)
    Header checksum: 0x4105 [correct]
    Source: 193.167.187.107 (193.167.187.107)
    Destination: 193.167.187.132 (193.167.187.132)
Encapsulating Security Payload
    SPI: 0x940e5f06
    Sequence: 1

[Time delta from previous packet: 0.000118000 seconds]
Internet Protocol, Version: 4
    Header length: 20 bytes
    Differentiated Services Field: 0x4d (DSCP 0x13: Unknown DSCP; ECN: 0x01)
    Total Length: 136
    Identification: 0x0000 (0)
    Flags: 0x04 (Don't Fragment)
    Fragment offset: 0
    Time to live: 64
    Protocol: ESP (0x32)
    Header checksum: 0x3fb8 [correct]
    Source: 193.167.187.132 (193.167.187.132)
    Destination: 193.167.187.107 (193.167.187.107)
Encapsulating Security Payload
    SPI: 0x545434d4
    Sequence: 1
```

Figure 5.5 ESP packets after handover.

specifications recommend that each physical interface has a separate SA. Furthermore, a host should use the same source address when transmitting to a given host address.

One or more IP addresses can be assigned to an SA. The addresses in the same group should be related – for example, assigned from a single subnet. Packets sent from related addresses are likely to follow the same network path that avoids issues with ESP anti-replay window. Related addresses typically have *fate sharing*. When one address from the group becomes obsolete, other addresses follow.

Figure 5.6 shows an example where two pairs of SAs are created between hosts A and B. Each SA, except SPI AB2, has two destination IP addresses assigned to it. The figure shows a recommended scenario where there is a pair of SAs between groups of addresses. It is also possible to create asymmetric SAs. As an example, in Figure 5.6, it would be possible not to establish SA with SPI_{BA2}. Although such scenarios can be experimented with, it should be noted that not all HIP implementations may support it and interoperability problems may arise when using asymmetric SAs.

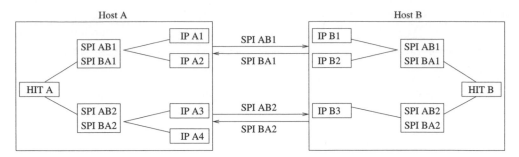

Figure 5.6 Relationship between HITs, SPIs, and IP addresses.

A single address can be assigned to more than one SA. It creates no ambiguity since the packet is assigned to an SA based on its SPI value instead of its source address. However, the current specification does not support such a case.

5.1.4   The LOCATOR parameter

Figure 5.7 shows the format of the LOCATOR parameter. In addition to the usual Type and Length fields, it includes Traffic and Locator Types, Locator Length, the Preferred Locator bit (P), Locator Lifetime, and the encoded Locator. The LOCATOR is a HIP critical parameter; an implementation that receives such a parameter but does not understand it must not process the packet further and it has to notify the peer host. Sending a LOCATOR with zero locator fields has the effect of deprecating all host addresses.

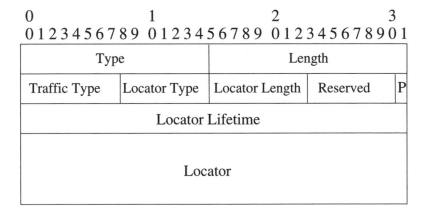

Figure 5.7 Format of LOCATOR parameter.

The Traffic Type parameter specifies if the locator is for HIP data or control packets or both. The value 0 permits a peer host to send both HIP signaling (control) packets and data to this locator. The value 1 permits sending control packets only while value 2 permits sending data packets only. The Preferred bit can be set to indicate that the locator is preferred for a given Traffic Type. By default, the locators from the base exchange are preferred unless overwritten later by the Preferred bit.

The Locator Type field tells the format of the following locator fields. Currently, two formats are defined, an IPv6 or IPv4-in-IPv6 address and the first 32 bits of an SPI concatenated with IPv6 or IPv4-in-IPv6 address. The first type (type 0) is meant for non-ESP use while the second type (type 1) is meant for use with ESP encapsulation.

The Locator Length specifies the length of the locator field in multiples of four bytes. The maximum supported length is then 1020 bytes.

The Locator Lifetime determines how long the locator remains active. The value given is in seconds and counted from receiving the LOCATOR parameter. After lifetime expiration, the locator becomes deprecated and should not be used if another active and reachable locator is available. The lifetime of zero is undefined and must not be used.

The LOCATOR parameter is typically sent in UPDATE messages that are acknowledged. The parameter can be also sent in R1 and I2 packets. Their standard retransmission mechanism ensures robust delivery of the LOCATOR. Experimentally, LOCATOR can be also included in NOTIFY messages. In that case the recipient can only treat the parameter as informational, but should not change its state before verifying new locators.

5.1.5 Locator states

A locator can be in three separate states: ACTIVE, UNVERIFIED, or DEPRECATED. Figure 5.8 shows possible transitions between the states. In ACTIVE state, address is tested to be reachable and has not yet expired. In UNVERIFIED state, address reachability has not yet been tested. The DEPRECATED state indicates an expired locator.

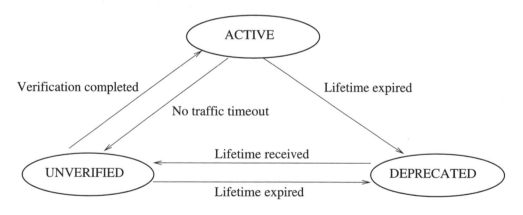

Figure 5.8 Three states of a locator.

To move a locator from UNVERIFIED to ACTIVE state, address verification must complete successfully. After the locator's lifetime expires, it is moved from an UNVERIFIED or ACTIVE to DEPRECATED state. When a host receives new lifetime for a locator, it can move it back from DEPRECATED to UNVERIFIED state. However, a transition from

DEPRECATED to ACTIVE state directly is not possible. The ACTIVE to UNVERIFIED transition can occur if the local implementation requires periodic verification of peer locators after an inactivity timeout.

5.1.6 Credit-based authentication

A malicious host could re-direct traffic to a third party IP address to cause a DoS attack. To limit such a possibility, Credit-Based Authorization (CBA) must be used by a HIP host when sending data to an IP address in UNVERIFIED state. The CBA approach is not HIP-specific but also used by other mobility protocols such as Mobile IP.

The attacker can forge the source IP address of data packets to make them appear as arriving from the third party host. The idea of CBA is to limit the rate of data transmission from a HIP host to the third party host to be no more than potential attacker's traffic. Therefore, the attacker does not gain additional traffic rate to the victim through a HIP host, compared with sending its own traffic directly to the victim.

Using CBA, a host maintains a credit for each peer. The credit contains the number of bytes received from the IP address of the peer. The CBA turns on if there are no ACTIVE addresses for the peer, but an address in UNVERIFIED state is available. Under these conditions, the host can only transmit a packet to the peer if its credit is more than the packet size. The credit is increased with each arriving packet from the peer and decreased by the packet size transmitted to the peer. If the credit is too low to transmit a packet, the host can either drop the packet or queue it until the credit is sufficient or the address verification completes.

To prevent an attacker building up a large credit and purging it to a victim in a large burst, HIP hosts deploy a mechanism called credit aging. Credit aging decreases the credit over time in fixed-length intervals. The specifications propose 5 seconds as the aging period and 0.875 as the aging factor. The algorithm should work well for bulk TCP data transfers with an RTT less than 0.5 seconds.

5.1.7 Interaction with transport protocols

The multihoming support in Henderson (2007) is given for the purpose of *failover*, when a host starts using a spare locator when a current locator fails. A host deploying multihoming for *load balancing* can simultaneously transmit data from several locators to utilize bandwidth over parallel network paths or reduce the latency. Such a scenario creates several issues at the transport layer, related to congestion control and error recovery. In particular, if packets from a single TCP connection are sent over different paths they can experience different propagation delays.

When packets take different times to reach the destination, they can arrive in a different order than transmitted, an effect known as *packet reordering*. Packet reordering easily confuses reliable transport protocols, such as TCP and SCTP, or the application if unreliable UDP transport protocol is used. Below we describe the problem with TCP in detail as an example.

When receiving an out-of-order packet, TCP sends a duplicate acknowledgment to its peer containing the highest sequence number received so far. The sender does not know if duplicate acknowledgments result from packet losses or reordering. As a precaution the sender waits for three duplicate acknowledgments before concluding that a packet was indeed lost and not reordered. However, if characteristics of two paths differ significantly, packet reordering length of more than three packets can easily occur. In that case, packet re-ordering causes a *spurious fast retransmit* in TCP, when a packet which arrived out of order is mistakenly retransmitted by the sender. In addition to wasting network bandwidth, spurious fast retransmits also slow down the transmission. TCP interprets packet loss as a sign of congestion in the network and halves the transmission rate.

As packet reordering is not uncommon in the Internet, researchers proposed several solutions to overcome it. One solution, called the Eifel algorithm, is based on the TCP timestamp option. By comparing the timestamp in the duplicate acknowledgment with a locally stored timestamp the sender can determine if the fast retransmit is spurious and undo it. Another solution is based on Duplicate Selective Acknowledgments (D-SACK). However, all solutions at the transport layer still incur some performance cost and work well only for a moderate rate of reordering. The use of a reordering buffer below the transport layer may be a preferable solution.

The use of paths with different characteristics can also impact the estimate of a retransmission timer at the sender's transport layer. TCP uses a smoothed average of the path's Round Trip Time and its variation as the estimate for a retransmission timeout. After the retransmission timeout expires, the sender retransmits all outstanding packets in go-back-N fashion.

When multihoming is used for simultaneous data transmission from several locators, there can easily be scenarios when the retransmission timeout does not correspond to the actual value. When packets simply experience different RTT, its variation is high, which sets the retransmission timeout value unnecessary high. When packets are lost, the sender waits excessively long before retransmitting. Fortunately, modern TCP implementations deploying Selective Acknowledgments (SACK) and Limited Transmit are not relying on retransmission timeouts except when most outstanding packets are lost.

In another scenario, the sender can use a low-RTT path for a while letting the retransmission timeout become low. Switching to an alternative high-RTT path may trick the TCP sender into thinking that it is not getting acknowledgments longer than the retransmission timer because of packet losses. After a *spurious retransmission timeout* the sender unnecessarily retransmits outstanding packets and reduces the transmission rate. Fortunately, the recommended minimum retransmission timeout value for TCP is 1 second, which should accommodate most paths in the Internet. Furthermore, methods such as the Eifel algorithm and F-RTO can detect and recover from spurious timeouts.

Load balancing among several paths requires some estimate of each path's capacity. The TCP congestion control algorithm assumes that all packets flow along the same path. To perform load balancing, the HIP layer can attempt to estimate parameters such as delay, bandwidth, and loss rate of each path. A HIP scheduler can then distribute packets among the paths according to their capacity and delay, to maximize overall utilization and minimize reordering. The design of the scheduler is a topic of current research work.

Different network paths can have different Maximum Transmission Unit (MTU) sizes. Transport protocols perform MTU size discovery typically only in the beginning of a connection. As HIP hides mobility from the transport layer, it can happen that packets on the new path get fragmented without knowledge of the transport protocol. To solve this problem, the HIP layer could inform the transport layer upper in the protocol stack of mobility events.

This method, known as transport triggers, is still under research although initial specification attempts have been made in the IETF.

5.2   Rendezvous server

A HIP mobile host can change its IP address often. Other hosts cannot initiate a HIP association to a mobile host if its current IP address is unknown. The rendezvous server solves this problem by providing its stable IP address in place of mobile hosts' address. The Initiator obtains the RVS address from DNS as described in Section 5.3 and sends an I1 packet to RVS. The rendezvous server relays the I1 packet during the HIP base exchange (Laganier and Eggert 2008).

5.2.1   Registering with a rendezvous server

The mobile host registers with RVS using the registration protocol described in Section 5.4. Every time the host's address changes, the host updates its registration with RVS. A mobile node can notify its association peers of IP address changes directly. Therefore, the use of RVS is limited for initial contact only for hosts that do not have active HIP associations with a mobile host.

Figure 5.9 illustrates the base exchange through a RVS server. The figure does not show a registration procedure that the mobile node has performed with the RVS server. Only the I1 packet travels through the RVS server, while the following control messages flow directly between hosts.

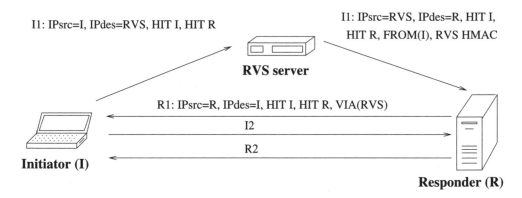

Figure 5.9 HIP base exchange using a RVS server.

To distinguish a relay request from an attempt to establish a HIP association to itself, a RVS compares the Responder's HIT in I1 message with its own. If they mismatch, RVS checks if there are active registrations for such HIT. If found, the RVS rewrites the IP header of I1 packet replacing the destination address with the address of the HIT owner. RVS also replaces the source address in I1 with its own to bypass egress filtering (egress filtering is widely used to prevent IP address spoofing by dropping packets whose source IP subnet prefix does not match the local subnet). Finally, the RVS recomputes the IP checksum of I1 packet and sends it to the HIT owner.

As the RVS is identifying the Responder according to the I1 HIT, the Initiator cannot use opportunistic HIP mode through RVS. If a RVS receives an I1 with NULL Responder's HIT, RVS interprets this as an attempt to establish HIP association to itself.

5.2.2 Rendezvous parameters

The Responder host needs to know the IP address of the Initiator host to reply with an R1 message. Since the RVS can replace the Initiator source address with its own, RVS adds a FROM parameter containing the Initiator's source address. The RVS also adds an RVS_HMAC parameter to I1. Its only difference from the HIP HMAC parameter is a different code number that places RVS_HMAC after the FROM parameter in the HIP header.

The Responder replies with R1 directly to the Initiator's IP address taken out of the FROM parameter. The Responder adds a VIA parameter containing the IP address of the RVS server from where the I1 has arrived. The VIA parameter is meant for diagnostics of HIP operations.

Figure 5.10 shows the format of FROM and VIA parameters. The only difference is that VIA can contain IP addresses of several RVS servers. The IP address is either a pure IPv6 address or IPv4-in-IPv6 format address.

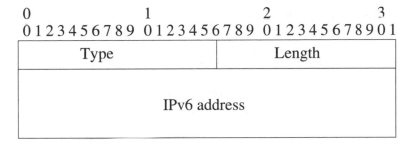

Figure 5.10 Format of FROM and VIA parameters.

5.3 DNS extensions

The use of HIP does not eliminate the need for human-friendly host names. In an initial deployment phase, legacy DNS servers can store HITs in AAAA records meant for host IPv6 addresses. Recall that HITs have the same length as an IPv6 address and can be identified using a special prefix. Therefore, a legacy DNS server can store both HITs and IPv6 addresses for a HIP-aware host.

5.3.1 HIP requirements to DNS

To support other HIP-related parameters, DNS servers need modification to store HIP-specific Resource Records (RR). The DNS extensions are described in a document from the HIP IETF working group (Nikander and Laganier 2008). Currently defined parameters for a HIP RR include host public key, HIT, and the domain name of a rendezvous server.

A legacy application in a HIP host requests a FQDN to IP address lookup typically using the getaddrinfo() call. The resolver also performs a FQDN to HI lookup. The resolver returns a HIT to the application if the request family is INET6 or UNSPEC. An LSI is returned to the application if the request family is INET. Internally, the HIP layer stores IP addresses of a given Host Identity as returned by the DNS query after removing other records. In this way, even if the DNS query returns multiple IP addresses for a host, the application would only see a single HIT or LSI. Selecting an address to contact the Responder host is a task of the HIP layer.

A native HIP application can directly request the HIP RR by passing a special option to the resolver query.

A major benefit of storing HIs to DNS is prevention of Man-In-The-Middle attacks. Although a HIP host can execute a base exchange with Responder host opportunistically, it gives no guarantees on the Responder's host identity. However, if the host identity is recorded to DNS, the Initiator host can verify that the Responder's claimed identity matches its DNS information. To prevent spoofing attacks on DNS resolver queries, the use of DNSSEC is strongly recommended.

5.3.2 Storing a RVS address

The HIP rendezvous server is described in Section 5.2. A host that often changes its IP address cannot always update its DNS information fast enough. Therefore, the host can store the DNS name of its rendezvous server. Alternatively, the host can store the IP address of the rendezvous server in the DNS instead of own IP addresses. In either case, the Initiator host sends its I1 messages directly to the rendezvous server, which forwards them to a current IP address of the Responder host.

Figure 5.11 shows the structure of the HIP Resource Record. RR contains the length of HIT, the public key algorithm, public key length, HIT, public key, and zero or more DNS names of rendezvous servers. Currently, two defined public key algorithms are DSA and RSA. The public key itself is encoded as defined for IPSECKEY RDATA RR in RFC 4025 and can be longer than 512 bytes. The rendezvous server names are encoded as specified in RFC 1035. The specific length information for each RVS name is not needed as the length is evident from the encoding. The RVS names are stored in order of preference.

A HIP host first queries a HIP RR from DNS and then the IP address of the Responder host or its rendezvous server. This process is shown in Figure 5.12 in nine steps. After the first and second steps, the Initiator receives a HIP RR containing an address of the Responder's RVS server. In this example, the Responder's host is likely to be mobile and update its current IP address directly to the rendezvous server. Therefore, the Initiator does not query the DNS server for the address of the Responder, but for the address of its rendezvous server in the third and fourth steps. Afterward, the Initiator forwards the I1 packet through the rendezvous server and completes the base exchange normally. In this picture, the DNS system is shown as a single server, while in practice a local DNS server forwards the query through the DNS hierarchy to the server storing the Responder's RR.

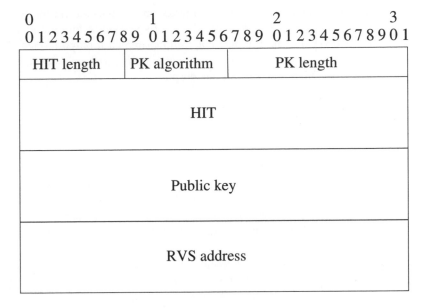

Figure 5.11 DNS Resource Record (RR) for HIP data.

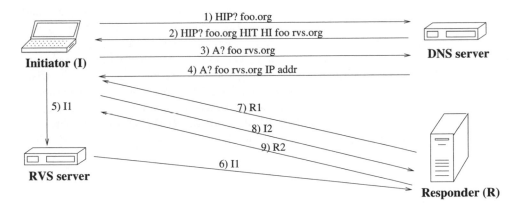

Figure 5.12 DNS query for a HIP mobile host.

5.3.3 DNS security

DNSSEC provides a secure channel from the DNS server to the host issuing a query. DNSSEC is described in RFC4033 and in Section 2.7. In the absence of DNSSEC, an attacker can forge DNS replies using DNS spoofing and fool the Initiator into using a wrong host identity or redirect HIP control packets to itself using the rendezvous server field. The attacker can also redirect the packets to a third party initiating a DoS attack.

There is an increasing number of possible attacks on the one-way property of hash functions such as SHA-1. The attacks decrease the efforts needed to find a collision on a hash output, so that two separate hash inputs map to the same hash output. Since HIT is a hash of the host public key, the Initiator host should not identify the Responder solely based on HIT, but should use the full public key for this purpose.

5.4   Registration protocol

The HIP registration protocol is specified as an extension to the base protocol (Laganier *et al.* 2008). Using the registration protocol, a HIP host can register to a service, such as a rendezvous server or HIP-aware firewall. The registration protocol is implemented as a HIP base exchange with additional parameters. In particular, the registrar announces a list of offered services using REG_INFO parameter in R1 and UPDATE messages. The requester includes the REG_REQUEST parameter to I2 or UPDATE messages to register with a service.

In some cases, the registration can be mandatory for the requester to establish a HIP association to the registrar. In this case, the registrar can terminate a HIP base exchange with a NOTIFY message containing REG_REQUIRED parameter if the requester does not supply the REG_REQUEST parameter in I2.

5.4.1   The process of registration

Upon receiving I2 with REG_REQUEST parameter, the registrar can authenticate and authorize the requester based on its public key. The details of the authorization process depend on the particular service offered. As an example, a firewall can check if requester HIT is included into appropriate Access Control List (ACL). If authorization successfully completes, the registrar creates state corresponding to the service and requester. Otherwise, the registrar returns REG_FAILED parameter indicating the failure type to the requester. If the requester tried to register for several services simultaneously and some of them failed, the R2 or UPDATE message contains both REG_RESPONSE and REG_FAILED parameters.

Figure 5.13 shows a registration process during the base exchange and its refresh when the registration is about to expire. In this example, the registrar announces the existence of three services, S1, S2, and S3. The requester attempts to register to S1 and S2. While registration to S1 succeeds, S2 registration fails. The refresh process is initiated by the registrar with an UPDATE message and completed as a re-registration with two other UPDATE messages.

5.4.2   Packet formats

The format of REG_INFO parameter is shown in Figure 5.14. The lifetime field is encoded in an exponential form in the range from 4 ms up to 178 days. The minimum and the maximum registration lifetimes give the boundaries for registration lifetime for the requester. All services should support a minimum lifetime of 10 seconds and the maximum lifetime of 120 seconds. The REG_INFO also contains one or more Registration Type fields that are offered by the registrar. The REG_INFO parameter is terminated by padding if necessary to end it on 32-bit boundary.

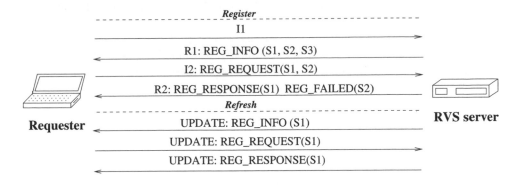

Figure 5.13 Registration with a RVS server and its refresh.

0 1 2 3
0 1 2 3 4 5 6 7 8 9 0 1 2 3 4 5 6 7 8 9 0 1 2 3 4 5 6 7 8 9 0 1

Type		Length	
Min Lifetime	Max Lifetime	Reg Type 1	Reg Type 2

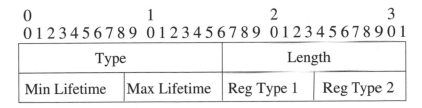

Figure 5.14 Registration Information (REG_INFO) parameter.

REG_REQUEST and REG_RESPONSE share the same format as in Figure 5.15. Both include registration lifetime and a list of one or more Registration Types in the order or preference terminated by padding. For best interoperability, the sender should include more than one parameter into a HIP message while the receiver should be able to process them. If a lifetime less than the minimum or greater than the maximum has been requested, the registrar grants the minimum or the maximum lifetime correspondingly.

0 1 2 3
0 1 2 3 4 5 6 7 8 9 0 1 2 3 4 5 6 7 8 9 0 1 2 3 4 5 6 7 8 9 0 1

Type		Length	
Lifetime	Reg Type 1	Reg Type 2	Reg Type 3

Figure 5.15 Registration Request (REG_REQUEST) and Response (REG_RESPONSE) parameters.

The registration times out after a certain period. Either requester or registrar can cancel the registration before its expiration if the registration is not needed any more.

The registration is canceled by sending REG_REQUEST or REG_RESPONSE with a zero lifetime.

The REG_FAILED parameter has a similar format to REQ_REQUEST, except the lifetime field is replaced by a failure type. Currently, two failure types have been defined. Zero means that additional credentials are required for registration, while one means that the service is not supported.

6

Advanced extensions

This chapter includes a description of HIP opportunistic mode in Section 6.1, piggybacking of data to the HIP base exchange in Section 6.2, HIP service discovery in Section 6.3, simultaneous multiaccess in Section 6.4, and HIT dissemination using a presence protocol in Section 6.5. In Section 6.6, we present Host Identity specific multicast, which resolves IP multicast issues with host authentication and mobility.

6.1 Opportunistic mode

An Initiator can attempt to establish a HIP association without prior knowledge of the Responder's HIT. Such mode is called *opportunistic* and is introduced in the base specifications (Moskowitz *et al.* 2008). The Initiator starts an opportunistic base exchange by sending an I1 with destination HIT set to zero.

6.1.1 Initiating opportunistic base exchange

When a Responder receives an opportunistic I1 with NULL HIT, it can choose any of its Host Identities to set up an association. If the Responder does not allow the opportunistic mode, it can reply with a NOTIFICATION parameter in a NOTIFY packet set to BLOCKED_BY_POLICY error type.

When receiving R1, the Initiator normally ignores its source IP address by locating relevant I1 using HIT. When the opportunistic mode is used, however, the Initiator can only locate the I1 using the source IP address of received R1.

The opportunistic mode creates security vulnerabilities and, as a rule, should be used in a trusted environment. As the Initiator has no knowledge of the Responder's identity, the opportunistic mode is susceptible to Man-In-The-Middle attacks where a malicious third party can place itself between the Initiator and intended Responder to intercept and modify their communication. A similar problem is well-known for SSH users. When first connecting to an unknown host, there is no guarantee that the received public key in fact belongs to that host. Therefore, the user is shown a fingerprint of the received key and requested to type

Host Identity Protocol (HIP): Towards the Secure Mobile Internet   Andrei Gurtov
© 2008 John Wiley & Sons, Ltd

"yes" to accept the key. This procedure it known as a *a leap of faith* as the user has to place certain trust that the key is authentic.

It is best to perform the leap-of-faith in a trusted environment such as a home network or the company's intranet to "pair" the hosts. Attacks in a trusted environment are unlikely and with a high probability the exchanged Host Identities are authentic. Afterward, HIP associations can be established in the public Internet without the use of opportunistic mode since the hosts have learned and stored each other HITs.

6.1.2   Implementation using a TCP option

One possible way to implement the opportunistic mode is to use a special TCP option to carry the Initiator's HIT in a TCP SYN message (Lindqvist 2006a). This approach allows to establish a normal TCP connection without delay if the Responder does not support HIP. If the Responder does support HIP, it can treat the incoming TCP SYN as equivalent to an opportunistic I1. Therefore, the Responder replies with R1 to the Initiator. The Responder at this point can discard the SYN packet and does not create any TCP state to avoid the possibility of Denial-of-Service attacks. The Initiator will retransmit the SYN once the HIP association has been established.

The drawback of using a TCP option to carry HIT is that it limits the operation to only TCP applications. While TCP is by far the most used transport protocol in the Internet, it would be a better design solution to have a generic approach that would work for all transport protocols. Such a solution can be implemented using an IP header option to carry the HIT. Unfortunately, recent Internet measurements show that packets with unknown IP options are often dropped by routers. New TCP options, conversely, are well tolerated (Medina *et al.* 2005).

At the time of writing, the open issues with using the TCP option include compatibility with NAT traversal and rendezvous server HIP extensions.

6.2   Piggybacking transport headers to base exchange

An interested reader may have long wondered why not to combine the HIP base exchange with TCP handshake. In fact, a general solution that allows carrying any transport protocol headers in HIP packets, not just TCP, appears useful. In a normal operation mode, a user TCP application experiences a total delay of 3.5 RTTs before the user data is transmitted. Combining base exchange with TCP handshake can reduce the delay to 2 RTTs.

6.2.1   Piggybacking to I2

A current solution proposes *piggybacking* TCP SYN packet to HIP I2 packet (Lindqvist 2006b). The Responder can first check if the puzzle solution in I2 is correct, and only then process the SYN, create state, and transmit TCP SYN-ACK in R2 packet. With such an approach, Denial-of-Service resistance capabilities of HIP are preserved. The approach works with normal and opportunistic base exchange, as well as with base exchange triggered by a TCP option (see Section 6.1).

Technically, piggybacking TCP SYN to I2 is accomplished by setting the Next Header field in I2 to value 6 to indicate a TCP header. Correspondingly, R2 has the Next Field set to 6 and includes TCP SYN-ACK packet after all I2 fields. Although the current specifications do not support it, setting a "piggybacking compatible" flag in I1 and R1 packets can inform the Initiator that the Responder implements piggybacking. Only in that case would the Initiator transmit SYN with I2, preventing a potential base exchange timeout or failure if the Responder does not understand piggybacking.

6.2.2 Security concerns

Unfortunately, the presented approach saves only one RTT reducing the total delay of establishing a TCP connection over HIP to 2.5 RTTs. Further reduction of delay would require transmitting TCP SYN already in I1, which produces several difficulties. Firstly, processing or storing any additional data at the Responder until completion of puzzle verification in I2 undermines DoS resistance of HIP. However, since DoS attacks are experienced only occasionally, a Responder may allow data in I1 if it is lightly loaded. Secondly, a TCP ACK message sent in response to SYN-ACK can already contain user data according to TCP specifications. The user data needs to be encrypted, which is not possible with the current base exchange specifications. However, earlier versions of the base draft did support including ESP payload into I2 and R2 packets.

Piggybacking TCP SYN and SYN-ACK to I2 and R2 exposes TCP headers to attackers in plain text. Most importantly, TCP port numbers can reveal the type of service the peers use. To address these privacy concerns, the user could be informed of the current HIP operation mode and be able to disable piggybacking for sensitive applications.

6.3 HIP service discovery

The service discovery extensions (Jokela *et al.* 2006) enable a HIP host to find available HIP services on a path towards a given destination or within the local area network. Currently the only service specified for HIP is the rendezvous server, although other services, e.g., a HIP-aware firewall or NAT, can be defined in the future. The registration process to a discovered service follows the registration extensions described in Section 5.4.

During the HIP standardization process in the IETF, the base specification included a Bootstrap (BOS) packet for discovering HIP hosts in the same LAN. The BOS packet was later removed due to insufficient energy in the HIP Working Group to develop it. The BOS packet included the HIT of the host to be broadcast on the LAN. Other hosts can then connect, using HIP, to the host that has announced its HIT. The BOS packet can be seen as a rudimentary service discovery mechanism. In this section, we describe a more flexible protocol to discover HIP services.

6.3.1 Overview of Service Discovery

The experimental specification for HIP Service Discovery specifies two modes of operation: on-the-path and local network discovery. In the first mode, a service provider middlebox is located on the path to the destination peer. A HIP hosts sends an I1, UPDATE or Service Discovery packet towards the destination. When the packet arrives at the Service Provider for relaying, it replies with a Service Announcement packet to the HIP host. In the second mode, a HIP host broadcasts a Service Discovery packet within the local area network or to

IPv6 site-local addresses. The local discovery mode enables location of services lying outside of the path to known HIP peer hosts.

The Service Discovery and Service Announcements packets can include the same fields as in I1 and R1 packets, respectively. This enables a HIP host that had initiated the discovery to reply with I2 to the Service Provider after receiving a Service Announcement packet with a suitable service description. Then, a base exchange can be completed with R2 from the Service Provider avoiding additional delays due to the service discovery process.

Figure 6.1 shows the format of a Service Discovery (SD) packet. It has similar format to I1 but has a different packet type. The SD packet can also include the REG_INFO parameter to specify the type of service being discovered. Without that parameter, any service is of interest to the host that sends the SD packet. The packet does not have any valid HIP control bits.

0	1	2	3
0 1 2 3 4 5 6 7 8 9	0 1 2 3 4 5 6 7 8 9	0 1 2 3 4 5 6 7 8 9 0 1	

Next Header	Header Len	0 Packet Type	VER.	RES	1
Checksum		Controls			
Source HIT (128 bit)					
Destination HIT (128 bit)					
Type		Length			
MinLifetime	MaxLifetime	Reg Type 1	Reg Type 2		

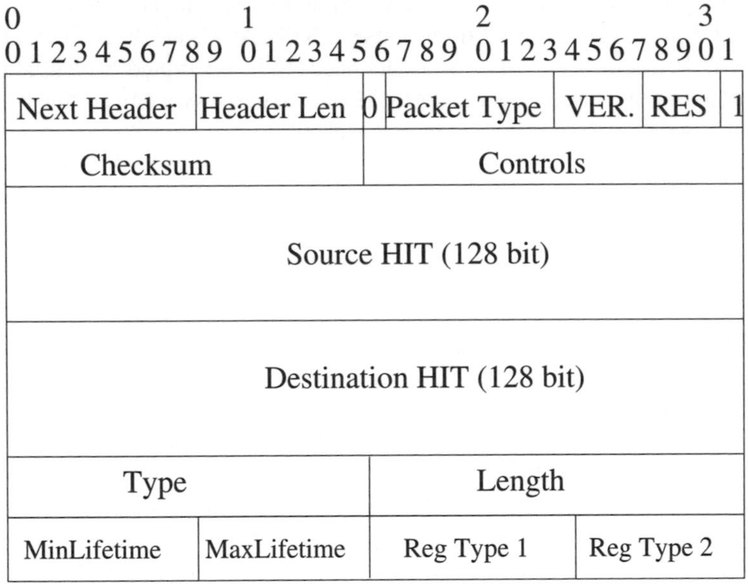

Figure 6.1 HIP Service Discovery packet format.

Figure 6.2 shows the format of the Service Announcement packet. The SAP has a similar format to R1, but a different packet type. It includes the REG_INFO parameter that describes the available services. In active discovery, the SAP packet is sent in response to the Service Discovery packet. In passive discovery, the SAP packet can be sent if an I1 packet or an UPDATE packet passes through a middlebox offering a service. The packet does not have any valid HIP control bits.

6.3.2   On-the-path Service Discovery

Figure 6.3 illustrates the on-the-path discovery process. A HIP host attempts to discover a suitable service (a HIP-aware firewall). It sends a Service Discovery packet setting the

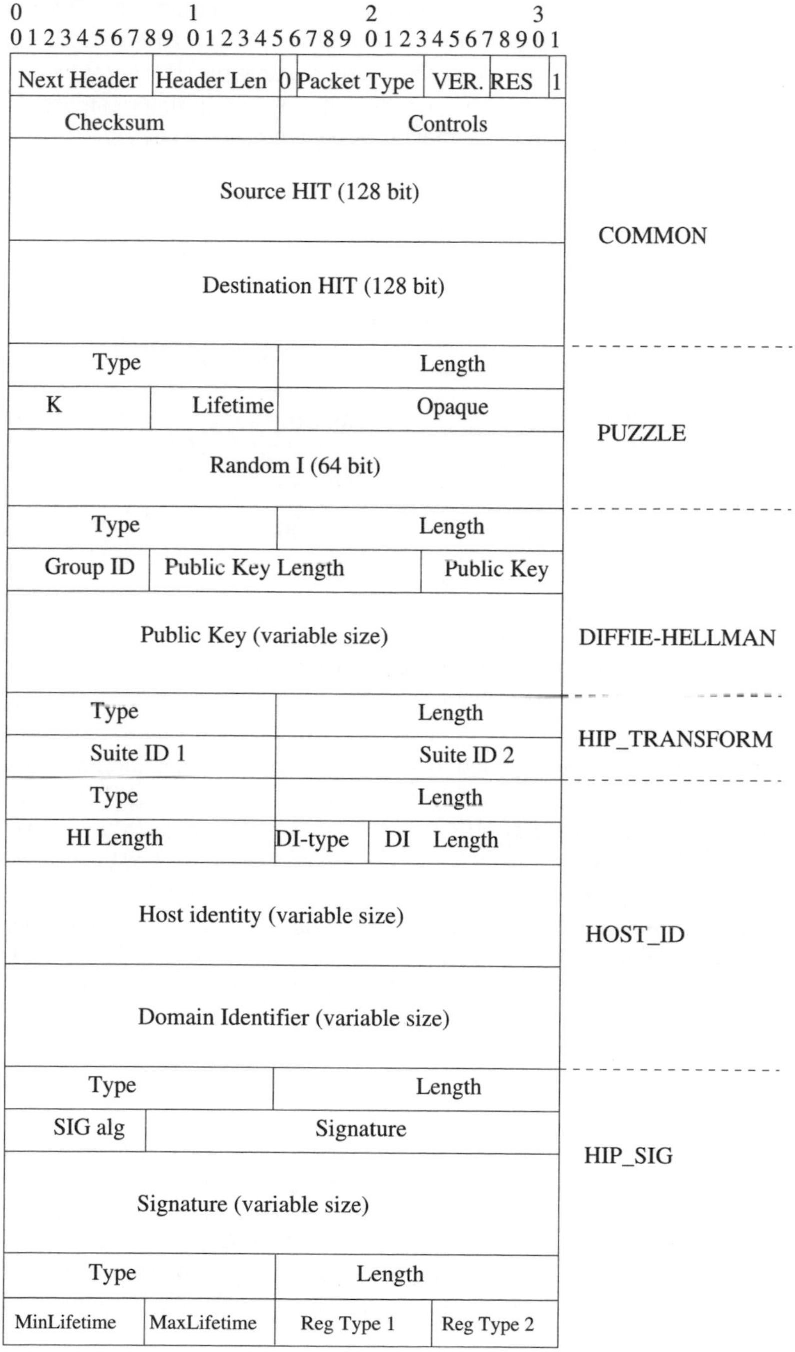

```
0                1                2                3
0 1 2 3 4 5 6 7 8 9 0 1 2 3 4 5 6 7 8 9 0 1 2 3 4 5 6 7 8 9 0 1
```

Next Header	Header Len	0	Packet Type	VER.	RES	1
Checksum			Controls			

Source HIT (128 bit)

COMMON

Destination HIT (128 bit)

Type		Length	
K	Lifetime	Opaque	

Random I (64 bit)

PUZZLE

Type		Length	
Group ID	Public Key Length		Public Key

Public Key (variable size)

DIFFIE-HELLMAN

Type		Length	
Suite ID 1		Suite ID 2	

HIP_TRANSFORM

Type		Length		
HI Length		DI-type	DI Length	

Host identity (variable size)

HOST_ID

Domain Identifier (variable size)

Type		Length	
SIG alg		Signature	

Signature (variable size)

HIP_SIG

Type		Length	
MinLifetime	MaxLifetime	Reg Type 1	Reg Type 2

Figure 6.2 HIP Service Announcement packet format.

destination address to the peer host with which it wants to communicate. The SD packet includes a REG_INFO parameter specifying the desired service as a HIP-aware firewall. The source HIT is set to the HIT of the Initiator host and destination HIT is set to zero. If the Initiator would like to discover services from a known host, the Initiator can use its HIT in the SD packet. Then, other hosts on the path will not reply to such a packet.

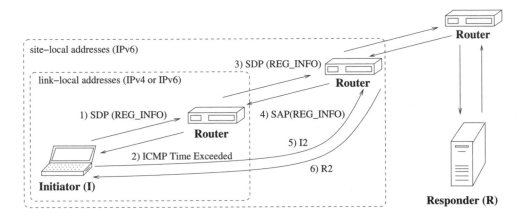

Figure 6.3 HIP Service Discovery process.

To start, the Initiator sets the IP Time-to-Live (TTL) value to one and transmits the SD packet to the network. The first router on the path will reply with an ICMP Time-Exceeded message as it does not implement any HIP service. Before replying, the router must check if the source address of the SD packet is a unicast IP address. If not, the SD packet is silently discarded to avoid security issues with sending a reply, e.g., to a multicast address.

The HIP host will increase the TTL value to two and transmit a new Service Discovery packet to the destination host. During the discovery process the Initiator is in the SD_SENT state, which is similar to I1_SENT state. The Initiator can receive and store a number of Service Announcement packets before deciding with which Service Provider, if any, it wants to register. The Initiator only accepts incoming Service Announcement packets that have the destination HIT equal to the HIT of the Initiator and a unicast source IP address.

The process continues until a pre-defined TTL limit is exceeded, the destination host is reached, or a Service Announcement packet is received. Service Providers only send the SA packet if they provide a service listed in the REG_INFO parameter in the Service Discovery packet. In the example in Figure 6.3, the second router on the path implements a HIP firewall service and replies with a Service Announcement packet. The base exchange and registration from the Initiator host to the firewall service is completed with I2 and R2 messages. Afterward, the Initiator host can make a separate base exchange to the Responder host.

6.3.3 Passive Service Discovery

The service discovery process above is based on an active discovery approach. It may not be suitable in all scenarios, for example, for a lightweight mobile host having battery power constraints. Passive Service Discovery is implemented by a Service Provider replying to the Initiator host after forwarding its I1 or UPDATE packet.

The Service Provider implementing a passive discovery mode should verify for each passing I1 or UPDATE packet if its source HIT is in the list of HITs currently served. To avoid introducing Denial-of-Service vulnerabilities, the Service Provider remains stateless after transmission of the Service Announcement packet. If the Initiator is interested in the offered service, it can complete the registration process by sending I2 to the Service Provider. The HIP implementation at the Initiator must allow for arriving Service Announcement packets in the "I1 sent" or "UPDATE sent" states.

6.3.4   Regional Service Discovery

A HIP host can perform *Regional Service Discovery* by sending a Service Discovery packet to a local multicast or broadcast address. In this case, the destination HIT is set to zero and the destination IP address to a special-purpose address. For IPv4, the host can only use the broadcast address 255.255.255.255, which reaches all hosts on the link. For IPv6, the host can use one of following addresses:

- link-local all multicast address, FF02:0:0:0:0:0:0:1;

- link-local routers multicast address, FF02:0:0:0:0:0:0:2;

- site-local routers multicast address FF05:0:0:0:0:0:0:2.

All Service Providers implementing a service specified in the REG_INFO parameter reply with a Service Announcement packet to the Initiator host after a small random delay to avoid congestion with many SA packets arriving simultaneously. Furthermore, the transmission rate of SA packets is limited to avoid a DoS attack to a spoofed IP address. The SA packet has the source HIT set to the HIT of the Service Provider and the destination HIT to the HIT of the Initiator host. If the Service Announcement packet has the necessary fields as required for an R1 packet, the Initiator can complete the registration procedure with an I2 packet.

Unfortunately, service discovery mechanisms presented in this section are insecure. The Service Discovery packet is not protected by a signature and could be modified in transit or sent from a spoofed IP address. Therefore, the Initiator should take arriving Service Announcement packets only as a hint and use external security mechanisms to verify authenticity of the service provider. Depending on security settings, a HIP host can only implement the active discovery mode. In this case it accepts Service Announcements arriving within a pre-defined time after sending a Service Discovery packet.

6.4   Simultaneous multiaccess

The HIP mobility and multihoming extensions (described in Section 5.1) permit the use of multiple interfaces for communicating with different peer hosts and for failover when the primary path to the peer fails. However, simultaneous use of multiple interfaces towards the same host is not supported because of potential problems with transport-layer protocols (see Section 5.1.7).

6.4.1   Flow binding extension

The experimental HIP extension for simultaneous multiaccess (SIMA) enables a multihomed host to use multiple interfaces in parallel towards the same peer host (Pierrel *et al.* 2006; Ylitalo *et al.* 2003). The task is accomplished by separating transport-layer connections (TCP and UDP) to use different interfaces based on transport port numbers. However, transmission over multiple interfaces by a single transport connection is still a research topic and is not supported by current specifications.

Furthermore, the specification does not include guidelines on how to match different flows to existing interfaces based on Quality-of-Service requirements and path characteristics. Such choices only affect performance, not interoperability, and can be made independently by each implementation. Intuitively, interactive flows such as remote terminal access (SSH or telnet) should be assigned to low-latency paths while bandwidth-demanding flows such as file transfer or media streaming require fast paths with possibly higher latency.

The specification extends the location update procedure and the UPDATE message defined by the HIP mobility and multihoming extensions. The UPDATE message includes a SIMA_FLOW_BINDING parameter that can be used by the peer host to select a Security Association and a destination LOCATOR for outgoing packets. Afterward, the peer host sends an UPDATE message with SIMA_ACK parameter to the multihomed host to confirm SIMA support. It is assumed that the location update is executed before the Flow Binding, although future specification versions can combine location and binding updates.

The current specification supports only flow identification for TCP and UDP transport protocols based on their port numbers. The use of other protocols, such as SCTP and DCCP, requires further extensions. The flow preferences are stored in a flow binding database. For each flow, a list of interfaces in the preference order is supplied. By selecting the Security Association, the flow also selects the interface to use as each SA has a source and destination address.

The multihomed host internally creates the rules for allocating flows to interfaces. It then sends the rules using SIMA_FLOW_BINDING parameter in an UPDATE message to the peer. The peer host follows the rules for selecting the SPI values and the destination address when sending data to the multihomed host. The multihomed host can change its flow binding preferences at any time by re-sending an UPDATE message with a new SIMA_FLOW_BINDING parameter.

Figure 6.4 shows an example scenario illustrating the use of SIMA. In the example, a multihomed host uses a rule binding TCP connections with a destination port 143 (IMAP email) to interfaces if3, if1, and if2. TCP connections to a port 22 (SSH remote shell) are bound to interfaces if2, if1, and if3 in the order of preference. In the beginning, if3 is inactive and the multihomed host informs the peer host to use interfaces if1 and if2 according to binding rules. The multihomed host sets up a Security Association to the peer host from both if1 and if2 by sending an UPDATE message with two SIMA_FLOW_BINDING parameters. Each parameter includes an incoming SPI value, a destination TCP port (22 or 143), and a source TCP port (0 for any).

When if3 becomes available, the multihomed host first sends a normal UPDATE message to the peer hosts with a new LOCATOR. Security Associations are created between if3 and the peer host. Then, the multihomed host sends another UPDATE with the SIMA_FLOW_BINDING parameter with an SPI value of if3 for TCP port 143. This updates the IMAP flow binding to use the new interface if3.

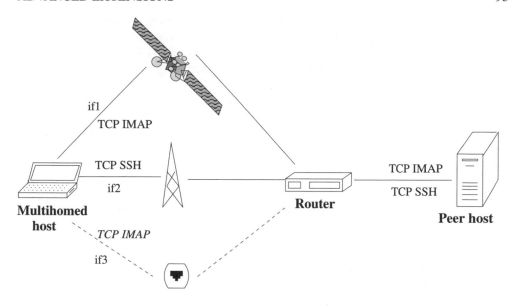

Figure 6.4 Example of simultaneous multiaccess.

6.4.2   Packet formats

Figure 6.5 shows the format of SIMA_FLOW_BINDING parameter having the type value of 664. The parameters bind one or more flow identifiers to a single SPI value. The list of flow identifiers is padded with zeros to the 32-bit boundary. One UPDATE packet can contain multiple SIMA_FLOW_BINDING parameters. The receiver not implementing SIMA can ignore the SIMA_FLOW_BINDING parameter. The lack of an ACK in this case informs the multihomed host that the peer does not support SIMA.

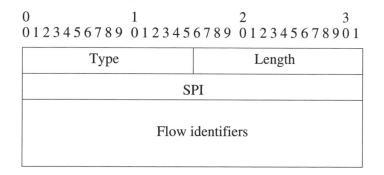

Figure 6.5 The SIMA_FLOW_BINDING parameter.

The format of the flow identifier is shown in Figure 6.6. The protocol number field identifies the transport protocol (such as TCP or UDP). The option length contains the length of the flow identifier without the basic header. The sequence number is incremented on each UPDATE to identify flow bindings. The reserved field is set to zero when sent and is ignored by the receiver. The R flag indicates the use of port ranges for flow identification. The D flag requires the peer to discard all existing flow bindings. The O flag is set to add a flow to the binding database and cleared to remove the flow.

In Figure 6.6, a flow identifier for a TCP connection is given. The protocol number is set to 4 (TCP) and the length is 4 bytes (excluding 4 bytes for a fixed header). The local and remote port number fields identify the flow at the host. A zero value can be used to denote any port. The UDP port flow binding is identical to TCP except the protocol number is 17.

```
0                   1                   2                   3
0 1 2 3 4 5 6 7 8 9 0 1 2 3 4 5 6 7 8 9 0 1 2 3 4 5 6 7 8 9 0 1
```

Protocol #	Option len.	Sequence #	Reserv.	R	D	O
Local port #			Remote port #			

Figure 6.6 TCP or UDP flow identifier.

Figure 6.7 shows the format of the flow binding option with port ranges for TCP or UDP protocols. It is similar to Figure 6.6, except this option can bind multiple flows to use a given interface by specifying a range of local ports and remote ports for either TCP or UDP protocol.

```
0                   1                   2                   3
0 1 2 3 4 5 6 7 8 9 0 1 2 3 4 5 6 7 8 9 0 1 2 3 4 5 6 7 8 9 0 1
```

Protocol #	Option len.	Sequence #	Reserv.	R	D	O
Local port # starts			Local port # ends			
Remote port # starts			Remote port # ends			

Figure 6.7 TCP or UDP flow identifier with port ranges.

Figure 6.8 shows SIMA acknowledgment and negative acknowledgment parameters. The SIMA_ACK is used by a HIP peer host to confirm support of SIMA after receiving a flow binding UPDATE. The SIMA_ACK parameter includes a list of sequence numbers from SIMA_FLOW_BINDING received successfully. The parameter is padded with zeros if the sequence numbers do not end on the 32-bit boundary.

The SIMA_NACK parameter has the same format as SIMA_ACK, but has a different type 662 versus 660. It informs the peer host that the flow binding updates with given sequence numbers were rejected by the host.

```
0                   1                   2                   3
0 1 2 3 4 5 6 7 8 9 0 1 2 3 4 5 6 7 8 9 0 1 2 3 4 5 6 7 8 9 0 1
```

Type		Length	
Sequence #	Sequence #	Sequence #	Padding

Figure 6.8 The SIMA_ACK and SIMA_NACK parameters.

A HIP host sends an UPDATE packet with one or more SIMA_FLOW_BINDING parameters to the peer host when its set of active interfaces changes (e.g., mobility) or when the flow binding rules change (e.g., when a user alters configuration settings). Each SIMA_FLOW_BINDING parameter has a unique sequence number which is incremented by one each time. Upon receiving the UPDATE, the peer host updates its flow binding table with port and protocol information from SIMA_FLOW_BINDING parameters. On each outgoing IPsec packet the peer host performs a table lookup to determine the correct interface for transmission based on packet SPI value.

Incoming ESP packets are handled regularly by a SIMA host. The packet can actually arrive from a different interface than specified by the flow binding table. However, the host should not alter its binding table based on such events. It can send an UPDATE message to the peer host requesting to change its tables. If both hosts are multihomed, there could be a conflict in their flow binding rules. The resolution behavior in this case is not yet defined by specifications.

6.5   Disseminating HITs with a presence service

Establishing a HIP association between two hosts requires knowledge of a peer HIT or the use of opportunistic mode. Peer HITs can be manually configured, distributed using DNS or DHT lookup service. The use of opportunistic mode is convenient but creates possibilities of Man-In-The-Middle attacks. Therefore, we need a mechanism to exchange HITs, especially when HIP is used with SIP.

6.5.1   HITs in the Presence Information Data Format

An experimental extension defines a new information element in the Presence Information Data Format (PIDF) (Papadoglou and Zisimopoulos 2005) developed by the SIP for Instant Messaging and Presence Leveraging Extensions (SIMPLE) working group at IETF. It allows dissemination of Host Identities or HITs using the presence data model in SIP. The element sets up a mapping between Host Identities and SIP URI in hosts. The presence model is an alternative approach to exchanging HITs in a Service Discovery Protocol (SDP) attribute as described in Section 15.2.

The Presence Data Model defines logically separate service (User Agent, UA) and device parts of the architecture. A SIP service can run on multiple devices and be identified by a Globally Routable Unique UA URI (GRUU). Each of the devices can be identified using a HIP host identity. Such an approach benefits from privacy filtering rules defined for the presence service. In particular, the rules limit the set of hosts allowed to retrieve HITs. The SIP UA can indicate, using the presence framework, whether it is available or busy on each physical device.

6.5.2 Disseminating protocol

Figure 6.9 shows HIT dissemination using SIMPLE presence extensions. First, the Responder uses a SIMPLE PUBLISH message to upload its HIT to the presence server. Then, the server sends a SIMPLE NOTIFY message (different than HIP NOTIFY) to distribute the HIT to HIP Initiator. After obtaining the destination HIT, the Initiator can start a SIP session with preconditions, as specified in integration of SIP and resource management mechanisms (RFC 3312). It triggers the HIP base exchange and establishment of ESP security associations. The SIP handshake completes over the HIP association.

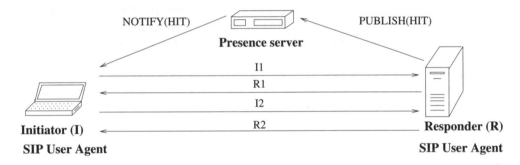

Figure 6.9 HIT dissemination using a SIMPLE extension.

The mechanism in Figure 6.9 also supports the use of a HIP rendezvous server. When publishing its HIT to the presence server, the Responder can supply an IP address of its rendezvous server in the registration message. The rendezvous server will relay the I1 message to the Responder. Unless the RVS server is registered to the presence server, the Initiator may need to perform a separate HIT to IP address mapping to locate a mobile Responder host.

Figure 6.10 shows an extension of XML schema definition to disseminate HITs using PIDF and Presence Data Model. It defines a new namespace "urn:ietf:params:xml:ns:pidf: hitexten" adding the `hitexten` element under *pidf*.

Figure 6.11 shows an example of applying the PIDF extension for disseminating a HIT "2001:0014:766e:fbee:f74d:ec73:d6c5:28c0" on the device with Universally Unique Identifier (UUID) of "992fbae3-f1c9-4223-8ef8-95ef5727d5ce".

6.6 Multicast

The HIP architecture is designed for host-to-host communication, also known as unicast. However, certain applications, such as Internet TV, involve data transmission from one source to multiple destinations. Some applications, such as multi-party video conferencing, involve

```xml
<?xml version="1.0" encoding="UTF-8"?>
   <xs:schema targetNamespace="urn:ietf:params:xml:ns:pidf:hitexten"
                   xmlns:hit="urn:ietf:params:xml:ns:pidf:hitexten"
                   xmlns:xs="http://www.w3.org/2001/XMLSchema"
                   elementFormDefault="qualified"
                   attributeFormDefault="unqualified">

   <!-- This import brings in the XML language attribute xml:lang-->
   <xs:import namespace="http://www.w3.org/XML/1998/namespace"
   schemaLocation="http://www.w3.org/2001/xml.xsd"/>

   <!-- Extension to PDM device element to describe the HIT -->
   <xs:element name="hit-identifier" />
           <xs:simpleType>
             <xs:restriction base="xs:anyURI">

             </xs:restriction>
           </xs:simpleType>
   </xs:element>
```

Figure 6.10 Definition of XML extension schema.

```xml
<?xml version="1.0" encoding="UTF-8"?>
      <presence xmlns="urn:ietf:params:xml:ns:pidf"
       xmlns:dm="urn:ietf:params:xml:ns:pidf:data-model"
       xmlns:hit="urn:ietf:params:xml:ns:pidf:hitexten"
       xmlns:xsi="http://www.w3.org/2001/XMLSchema-instance">
       <tuple id="sg89ae">
        <status>
         <basic>open</basic>
        </status>
        <dm:deviceID>uuid:992fbae3-f1c9-4223-8ef8-95ef5727d5ce
   </dm:deviceID>
        <contact>sip:someone@example.com</contact>
       </tuple>
       <dm:device id="pc122">
        <rp:user-input>idle</rp:user-input>
        <dm:deviceID> uuid:992fbae3-f1c9-4223-8ef8-95ef5727d5ce
   </dm:deviceID>
       <hit:hit-identifier>20010014766efbeef74dec73d6c528c0
       </hit:hit-identifier>
       </dm:device>
      </presence>
```

Figure 6.11 Example of HIT distribution using PIDF.

transmission from several sources to several destinations. Such scenarios are most efficiently handled using multicast data transmission, where the source transmits a single copy of data and routers or hosts in the network multiply packets as needed for delivery to downstream recipients.

Multicast can be implemented on the networking layer as native IP multicast, or as an application service on the overlay network. The network multicast is more efficient than the application multicast, because it can achieve "one link – one packet" principle, whereas application multicast can still transmit multiple copies of the same packet over a link. The Internet Indirection Infrastructure (i3) is one possible system for implementing application-layer multicast. Application multicast is easily deployable while several issues hindered deployment of IP multicast. One issue is lack of access control in the native IPv4 multicast model known as Any Source Multicast (ASM). Any host can join the multicast tree as a receiver by informing its router. A more restrictive version of multicast called Source Specific Multicast (SSM) limits who can transmit to the multicast tree; a complete access control for receivers is still missing.

In standard HIP, a single public–private key pair identifies a single host. Some researchers perceive HIP multicast as using a public key to identify a group of hosts. Therefore, the group must generate and share a private key between members of the group. In this section, we focus on the case where each host has own public–private key pair and is willing to join an IP multicast tree.

6.6.1   Challenges for IP multicast

Mobility of hosts participating in multicast is a largely unsolved problem. Particularly, if the multicast source changes the IP address, the whole multicast tree needs to be reconstructed. Although some solutions to the mobility problem based on bi-directional tunneling are proposed, fundamentally the host identity is coupled with the current IP address. Furthermore, current multicast solutions do not allow construction of native dual-stack IPv4/v6 multicast trees and do not support multihoming.

Two common approaches for multicast receiver mobility are Bidirectional Tunneling and Remote Subscription. With Bidirectional Tunneling, the receiver subscribes to the multicast stream via its home agent located in the user's home network. When the user moves to a foreign network, it creates a tunnel to the home agent that relays the user's multicast signaling and stream data. Therefore, the user can move between networks without affecting the multicast tree. With Remote Subscription, the user asks the local multicast router in a visiting network to join the multicast tree. The old branch of the multicast tree from the previous user's network location eventually times out and the data starts arriving to the new user's location.

Few solutions exist to the mobility of the multicast sender. In SSM, the entire multicast tree is constructed using the IP address of the source as the root. If the sender moves, the entire tree needs to be rebuilt using the new IP address as the root. During the rebuilding process the multicast stream is interrupted and listeners do not receive any data. Bi-directional tunneling can be applied to avoid the tree reconstruction for the mobile multicast sender.

Some solutions to provide authentication to multicast receivers were proposed, including Multicast Control Protocol (MCOP). However, MCOP is only able to authenticate the subnetwork where the user is located, not the host of the user. Therefore, other hosts from the same subnetwork can receive the multicast stream without authentication.

The last problem with native IP multicast is the difficulty of constructing a multicast tree combining IPv4 and IPv6 hosts. Although several solutions for IP version interoperability do exist, they are aimed at unicast communication and do not yet support construction of dual-stack multicast trees. A host that wants to join a multicast tree with a source located in a different IP version network is unable to create a valid join message. The receiver is unable to join the multicast tree even if the source has the same IP version but there is a transit network on the path to the source that has a different IP version.

6.6.2 Host Identity Specific multicast

In traditional ASM multicast, participating hosts are identified with an IP address of the multicast group G. In SSM, the unicast IP address of the multicast source is necessary to construct the multicast tree (S,G) in addition to the multicast address. In this section, we describe a new HIP-based multicast architecture called Host Identity Specific Multicast (HISM) (Kovacshazi and Vida 2007). In HISM, HITs are used instead of IP addresses to identify hosts, HIT_S for multicast source and HIT_R for multicast receivers. The use of HITs enables a host to change its IP address without the need to reconstruct the multicast tree.

In addition to HITs, another identifier is needed to replace the IP-version specific multicast address. The Session ID (SID) is version-independent and has a length of 26 bits. Applications can create IPv4 or IPv6 multicast addresses from SID if necessary. For IPv4, the multicast address is constructed by appending the prefix "111011" before 26-bit SID. For IPv6, the prefix "FFFF::" is used. The prefixes include the address ranges meant for multicast further reserving a part of address space for HISM. The HISM channel is now identified by a pair (HIT_S, SID) preserving the benefits of source specification of SSM while removing dependency of fixed IP addresses.

In traditional multicast, a host can join the multicast channel by following the subscription procedure. The host application triggers subscription by giving the group multicast address G and IP address of the multicast source (for SSM only) to the host operating system. The host sends out join messages to the multicast group using Internet Group Management Protocol (IGMP) for IPv4 and Multicast Listener Discovery (MLD) reports for IPv6. When the join message reaches the first multicast router, called the designated router (DR), it sends out the Protocol Independent Multicast (PIM) join message to attach to the multicast tree. When the PIM join message reaches a multicast router that is already part of the tree, multicast stream data starts to flow to the receiver host.

With HISM, the join procedure is modified as shown in Figure 6.12. The application specifies the pair (HIT_S, SID) that the user wants to listen (1). The operating system constructs IPv4 or IPv6 multicast addresses from SID based on current host connectivity (2). The group management join message is sent containing the HIT_R for user authentication and address control (3). The DR starts authenticating the subscription request based on (HIT_S, SID) and HIT_R (4). The authentication server evaluates whether the receiver host is authorized to view the multicast stream based on HIT_R and resolves the current IP address of the source based on HIT_S (5). The authentication server also maintains a list of dual-stack border routers for IPv4/IPv6 interoperation and gives the address of the dual-stack router closest to the source. If authentication is successful, the DR joins the multicast tree (6).

The normal PIM join is sufficient if the source and receiver use the same IP version. Otherwise, the join request is constructed using the dual-stack router as a multicast source (7). The dual-stack router performs address conversion and continues construction of the tree. The join procedure completes when the join request reaches the DR router of the source or another multicast router that is already a part of the tree (8).

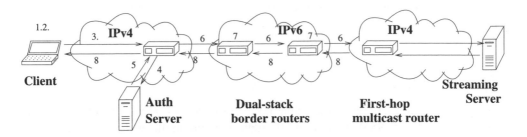

Figure 6.12 Architecture of Host Identity Specific Multicast.

The use of HISM removes the need to reconstruct the multicast tree when the source changes its IP address. It is sufficient that the resolution mechanism returns a correct IP address based on the HIT_S. The deployment of HISM would generate a software update in the multicast routers but does not require support of HIP in the routers. The use of public-key cryptography in the routers would significantly reduce their routing speed.

To support HISM, applications should give the pair (HIT_S, SID) of the multicast stream the user is interested in. The legacy IPv6 applications can easily pass HIT_S instead of source IP address in SSM as HITs have the same length as IPv6 addresses. The HIP layer converts the multicast group address to SID for legacy applications. New applications should be written to use the HIT_S and SID explicitly.

To join a multicast group, hosts use a group management protocol, such as IGMPv3 for IPv4 and MLDv2 for IPv6. A new Version Independent Group Management Protocol (VIGMP) is proposed for HISM hosts that carries host and session identifiers but is backward compatible with IPv4/v6 ASM and SSM multicasting. The advantage of the version-independent approach is that all hosts need just one multicast implementation, independent of the current type of network connectivity.

VIGMP messages include the identifier of the receiver HIT_R that is used for authentication. For backward compatibility, receivers can still join the multicast tree by specifying IP addresses of the group and the source. The authentication server then maps addresses to source and session identifiers if needed. The router periodically broadcasts the multicast query to locate potential listeners. Interested and already subscribed hosts reply with a report containing identifiers of the session, own host, and multicast source. Legacy hosts can still understand the VIGMP query message and reply with an MLD or IGMP report.

An VIGMP-capable router can inform receivers of changes in the multicast parameters with a router session report message. When the multicast source moves, its new unicast IP address is announced to the listeners through this message.

The HISM authentication mechanism is an extension of Multicast Control Protocol (MCOP) enabling authentication of hosts based on HIT_R, not just network subnets.

The receiver host can perform the HIP base exchange to the multicast source if mutual authentication is required. For larger multicast groups, this could overload the sender and additional authentication infrastructure is needed as we describe in Section 6.6.3.

When the first-hop multicast router (Designated Router, DR) receives a join request from the user host, it is forwarded to the authentication server. The policy database at the server contains active multicast session identifiers together with source HIT and IP address information, as well as a list of authorized receiver HITs. Multicast sources need to register their host and stream identifiers with the multicast server when creating a new multicast stream. The mapping between the source HIT and its current IP address is maintained in DNS. The HIP-specific DNS resource record already includes the name of the rendezvous server and can be extended to include HISM-specific fields. For a quickly moving source, the use of DHT for storing the mapping information might be a faster solution.

The authentication server also maintains a list of dual-stack border routers. These routers are able to convert between IPv4 and IPv6 multicast packets. The authentication server adds dual-stack routers to an anycast group identified by an anycast IP address. When sending a message to the group, only the closest router will receive it. After a successful authorization, the authentication server sends the IP address of the source and the address of the dual-stack router to the user's designated router, which forwards the message to the user's host.

During an active session, the authentication server periodically checks if the IP address of the multicast source has changed. If a change is detected, the server informs the Designated Router, which forwards the new source IP address to all receivers using an VIGMP message. The authentication state at the Designated Router and the server eventually times out, requiring the multicast receiver to re authenticate periodically. This soft-state approach accommodates the presence of mobile hosts that can leave the network without removing the multicast subscription.

HISM routing is based on Protocol Independent Multicast, Sparse Mode (PIM-SM). The Designated Router (DR) plays a significant role in construction of the tree by authenticating the user. If several border multicast routers are present, only one is selected as DR to avoid multiple versions of the multicast tree being constructed. After receiving a join message from the user host and authenticating it, the DR sends the PIM join message upstream toward the source. The message includes two new flags, "C" for requiring IP version conversion and "M" for indicating a change of the source IP address for mobility. The "C" flag tells that the current destination of the join message is not the actual source but an intermediate dual-stack router.

The core multicast routers store the forwarding state based on source HIT, not the IP address. The routing record has the form $HIT_S|SID|Input\ interface|Output\ interface$. The routers do not have the capability to resolve HITs to IP addresses. However, using HITs prevents tree reconstruction in the case of source mobility and enables construction of dual-version multicast trees. The IP addresses are only used during the tree construction phase. Figure 6.13 shows the content of encoded source format that includes HIT_S, IP address of the source, and IP address of dual-stack router for tree building. The new HISM-specific fields are shown in italics.

Construction of native dual-stack multicast trees is not possible in the current IP multicast model. If the transit network has a different IP version than the source and receivers, unicast tunnels can be constructed to pass through the transit network as shown in Figure 6.14. However, this is a suboptimal solution that can create multiple packet copies going over

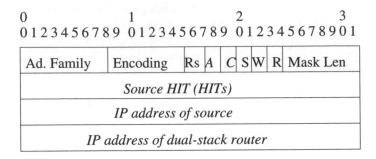

Figure 6.13 Encoded source format for HISM.

the same link and prevents receivers within the transit network from joining the multicast tree. With HISM, the need for tunnels disappears and a native multicast tree is constructed between networks with different IP versions, using dual-stack border routers for address conversions.

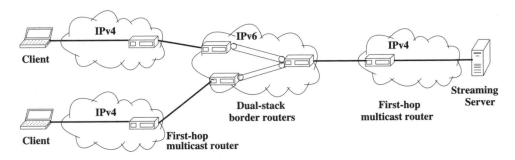

Figure 6.14 Tunneling solution for traversal of IPv6 domains for IPv4 multicast. HISM enables native multicast trees without tunneling.

Legacy ASM and SSM hosts can possibly join the HISM tree as shown in Figure 6.15. Border routers indicate their HISM capability by setting a flag or an option in Hello messages that routers exchange with their link peers. The border routers convert between legacy ASM/SSM join messages and HISM messages by inserting or removing HIT_S and other fields. To find out the necessary information, the border router may need to contact the authentication server. If the overhead of contacting the authentication server for each multicast packet is too high, the border router can create a tunnel through the legacy network to the next HISM-capable border router. The use of tunnels is a must if dual-stack border routers perform IP version conversion. Then, removal of HIT_S from the multicast header can create collisions at the border router as different multicast sources can use the same SID to identify their multicast streams.

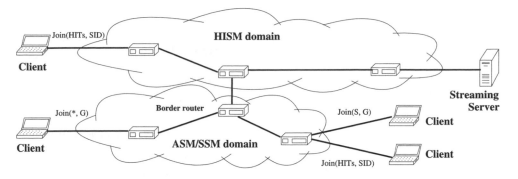

Figure 6.15 Mixed multicast tree with HISM and legacy SSM/ASM parts.

6.6.3   Authenticating multicast receivers

In IP multicast, the source host does not know how many receivers it currently has and therefore cannot easily offer commercial services requiring accounting and payment. Furthermore, deploying IP multicast is a challenge for Internet Service Providers due to current business models. In particular, a provider that forwards more traffic outwards than inwards has to compensate the difference with payments to its peer providers. The IPv6 multicast is still under development with trends towards Source Specific Multicast (SSM), which identifies the sender and disallows data distribution by arbitrary hosts. However, IPv6 multicast requires deployment of IPv6 that is not global yet.

HIP can provide a solution to user authentication and accounting in IP multicast, using certificates to control user access to multicast data. This section describes an architecture proposed by Jokela where the multicast source encrypts the data and retailers authenticate multicast receivers with HIP. The architecture uses Simple Public Key Infrastructure (SPKI) certificates including the following fields. The *issuer* contains the public key of the stream provider who creates the certificate. The *subject* is the public key of the certificate user. *Delegation* is a Boolean flag that allows or disallows further use of the certificate beyond the subject. *Authorization* describes the actions that the subject is entitled to perform, i.e., receive a multicast stream. *Validity* sets the earliest and latest dates when the certificate is valid. *Signature* is created with the private key of the issuer.

The general architecture follows the TINA-C business model for networking applications. Figure 6.16 illustrates components of the system. *Stream Provider* transmits an encrypted multicast stream and creates certificates authorizing its access. *Stream Access Retailer* assists the Stream Provider in creating certificates for receivers. Due to a possibly large number of receivers, it is not feasible for the Stream Provider to handle them all. Instead, it creates certificates for Stream Access Retailers enabling further delegation of certificates. *Stream Access Point* is a multicast router in the network that verifies the certificate of the receiver and adds it to the multicast distribution tree. *End user* obtains a certificate from the Stream Access Retailer after a payment and presents it to the Stream Access Point for receiving the multicast transmission. This architecture assumes that the Stream Provider trusts Stream Access Retailers and Access Points to handle the authorization and accounting of users correctly.

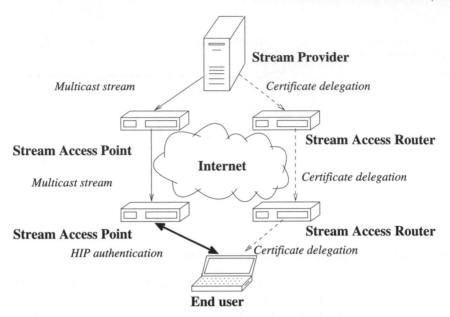

Figure 6.16 Authenticating multicast receivers with HIP.

After obtaining the certificate, the end user executes the HIP base exchange to the local Stream Access Point. The end user supplies the certificate using HIP CER packet and the Stream Access Point checks that the user's public key is the same as in the certificate. The certificate also includes the stream group ID that determines the stream that the user is subscribing to. After authenticating the user, the Stream Access Point can join the multicast tree for the stream and start delivering it to the user. To prevent the user from sharing its identity with other users, the Stream Access Point contacts the Stream Access Retailer to check if the certificate is still unused and mark it as used afterward.

Data is sent encrypted from the Stream Provider, which periodically changes the decryption key. If certificates expire at random different times, the Stream Provider would have to change the key frequently. Instead, certificates should expire at certain intervals, e.g., one hour. The Stream Provider would deliver a new decryption key to the Stream Access Points at this interval, which would further distribute the key to users with valid certificates.

7

Performance measurements*

The current trend of moving mobile telecommunication systems to IP technology is well-recognized. However, the security aspect of using IP protocol stack on lightweight devices, such as PDAs or mobile phones, is not sufficiently explored. In particular, encryption and public key signatures implemented in software are computationally expensive operations that could stress CPU and battery resources of mobile devices. The data throughput and latency can be negatively affected as well.

This chapter is structured as follows. Section 7.1 briefly describes the Nokia Tablet hardware and our port of HIPL implementation. Section 7.2 contains measurement results of the basic HIP characteristics and their in-depth analysis. Section 7.3 concludes the chapter with a summary of results and plans for future work.

7.1 HIP on Nokia Internet Tablet

In this chapter, we describe the performance measurements of our port of HIP for Linux (HIPL) implementation to the Nokia 770 Internet Tablet, Linux-based PDA. Although several previous projects evaluated HIP on standard Internet hosts (Henderson 2003; Heer *et al.* 2008; Jokela *et al.* 2004; Nikander *et al.* 2003), none has targeted a HIP assessment on a mobile device with restricted resources. To check whether running IP-based security on lightweight hardware is feasible, we performed HIP measurements over WLAN with Nokia Tablet acting as a mobile client, thus mainly being the Initiator of a HIP association. In particular, we measured data throughput, latency, and power consumption of the HIP base exchange and mobility update. We then analyzed the results and suggested conditions where unmodified HIP would be suitable for use on lightweight hardware.

The choice of Nokia 770[1] as a target device for our measurements has been supported by several factors. First of all, it is a PDA with a set of limited resources and provides a good example of lightweight hardware to test HIP on. Secondly, such a handheld device

---

*Source: A. Khurri, E. Vorobyeva and A. Gurtov, Performance of Host Identity Protocol on Lightweight Hardware, *Proc. of the 2nd ACM/IEEE Conference on Mobility in the Evolving Internet Architecture 2007 (MobiArch)*, pp. 27–34, DOI: 10.1145/1366919.1366925. © 2006 ACM, Inc. Used with permission.
[1]Afterward Nokia released Internet Tablets N800 and N810, which have better CPU and memory characteristics.

ideally represents the notion of a mobile client constantly moving across the Internet. In this approach, the tablet would be a desired device for HIP to be deployed on to deal with mobility issues. Next, Nokia 770 is becoming more and more attractive for both users and developers resulting in a number of applications (VoIP, Audio, Video on Demand, etc.) that might utilize the benefits of HIP. Finally, since Nokia 770 is a Linux-based PDA it is much easier for any software (including HIP) to be ported to the open source platform.

Nokia 770 Internet Tablet is a Linux based handheld device with a high-resolution touch screen display, built-in WLAN, and Bluetooth support. Mainly designed for easy web browsing, the tablet is also convenient for Internet telephony and instant messaging, reading emails and documents, playing media content. As its core, Nokia 770 has a Texas Instruments (TI) OMAP 1710 CPU running at 220 MHz. The device comes with 64 MB DDR RAM. The source of power for the tablet is a 1500 mAh Li-Polymer battery. The operating system is a modified version of Debian/GNU Linux. For our experiments we used the latest release known as the Internet Tablet OS 2006 edition. It has a GNOME-based graphical user interface and runs the 2.6.16 Linux kernel.

Porting HIP to the Nokia 770 Internet Tablet consisted of a few steps. Since the handheld is running embedded Linux, we used an existing Linux implementation of the protocol, namely HIP for Linux (HIPL) implementation developed at Helsinki Institute for Information Technology. While the HIP daemon and the other utility programs of HIPL are userspace applications, a few modifications to the Linux kernel are necessary in order to support HIP. More specifically, three patches have to be applied to the Nokia kernel, which must also have IPv6, IPsec, AES, 3DES, and SHA1 support.

Low computational power makes it impossible to compile big software projects as well as Linux kernel directly on the PDA. For building both HIPL userspace applications and the Nokia 770 Linux kernel we used a Scratchbox cross-compilation environment. Located on a PC, Scratchbox toolkit emulates the ARM environment and allows building the software for use on the real devices. We then flashed our custom kernel image onto the device and packaged the userspace applications into a Debian binary file (*.deb) to be installed on the tablet.

7.2   Experimental results

This section presents the results of our experiments with the Host Identity Protocol on the Nokia 770 Internet Tablet. First we introduce the platforms and the network environment we used. Then, in the following subsections we report measurement results and their interpretation.

7.2.1   Test environment

We performed our measurements on Nokia 770 Internet Tablet (Tablet) and Intel Pentium 4 CPU 3.00 GHz machine with 1 GB of RAM (PC) connected to each other via a switch and a WLAN access point (AP) in our test network. The network provided both IPv4 and IPv6 addresses. The wireless AP supported IEEE 802.11 g standard and WPA (Wi-Fi Protected Access) encryption. All communicating parties used the same implementation of HIP. To better indicate the Tablet's performance level we repeated our measurement scenarios with a more powerful, 1.6 GHz, IBM laptop (Laptop) connected to the PC over the same wireless link as the Tablet. Through a comparison we then evaluated the impact of the Tablet's lightweight hardware on the maximum achievable data throughput, latency, duration of the base exchange and mobility update.

7.2.2 Basic HIP characteristics

Duration of HIP base exchange

A HIP association is set up by exchanging four control packets between communicating hosts. The purpose of measuring the HIP base exchange time was to determine the duration of various stages of BE such as generating and processing of HIP messages by the Tablet. The measurements were performed using a script that established a HIP association 50 times in a number of varying scenarios. Since we did not find significant differences between IPv4 and IPv6 performance we present only results with the RSA HITs mapped to IPv6 addresses of the hosts.

Figure 7.1 depicts the times that were measured in our experiments. We leave out I1 packet generation time due to its insignificance. T1 represents the time for the Responder to process an I1 packet and generate an R1. According to HIP implementation, Responder does not spend a lot of time for this phase since it chooses pre-created and signed R1 messages and adds a puzzle to them just before sending the packet to a network. The next time, T2, contains a number of CPU-intensive cryptographic operations such as generating and verifying signatures, calculating Diffie–Hellman (DH) session key. During this stage the Initiator must also solve the challenge it received from the Responder. T3 indicates the time needed by the Responder to process an I2 packet, which involves puzzle solution check, Initiator's public key verification, and computation of the DH shared secret. If the puzzle was solved correctly Responder generates an R2 message and signs it. Finally, during T4 the Initiator processes the R2 packet and completes the base exchange. At this point, HIP association is established.

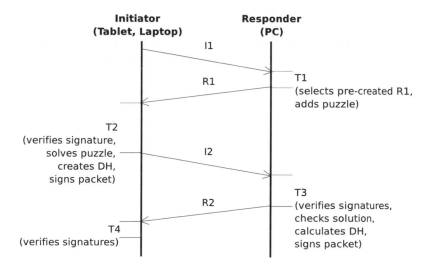

Figure 7.1 The times measured on the Initiator and the Responder.

Figure 7.2 illustrates T1, T2, T3 and T4 times as well as the total duration of the HIP base exchange. We compare the results for two different HIP associations where the Initiators are Tablet and Laptop with the PC acting as the Responder. Thus, T1 and T3 times are measured for the PC whereas T2 and T4 times correspond to both Tablet and Laptop. As the figure indicates the Laptop greatly outperforms the Tablet for all operations involved with BE. T2 time for the Tablet is nearly 1.2 seconds, which is significantly longer than the respective time of the Laptop (0.14 seconds). The majority of T2 is spent by the Tablet on operations with public key signatures. This time heavily depends on the length of a public key. For our tests on both Tablet and Laptop we used the RSA key size of 1024 bits.

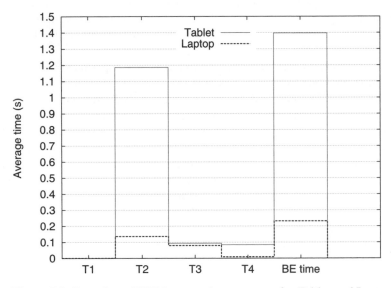

Figure 7.2 Duration of HIP base exchange stages for Tablet and Laptop.

The next test established a HIP association initiated by the PC while the Tablet acted as the Responder. Results suggest that the base exchange time is independent of whether PC or Tablet initiates the handshake. In both cases the base exchange lasts around 1.4 seconds.

While 1.4 seconds to perform a HIP handshake between the Tablet and the correspondent PC might be acceptable for users, HIP communication of two such lightweight devices produces higher delay. We measured the BE duration for a Tablet-to-Tablet scenario as over 2.6 seconds. Tablet spends a similar amount of time for T2 and T3 phases. Since puzzle solution check in T3 is not computationally expensive we assume that the major influence on the BE time is exerted by cryptographic operations costly for Tablet's CPU. Such operations include signature verification and generation, as well as computation of the Diffie–Hellman shared key.

Puzzle difficulty

Upon receiving an R1 packet, the Initiator is expected to solve a cookie challenge (puzzle) it gets from the Responder. This is done to protect the Responder against possible Denial-of-Service attacks by compelling the Initiator to spend a certain amount of CPU cycles to find a right answer. Depending on the conditions, i.e. on the trust level between the communicating endpoints, Responder has an opportunity to adjust the puzzle difficulty to be solved by the Initiator. The difficulty (K) is represented by a number of bits that must match in a hash output sent back to the Responder. In the above presented scenarios the default puzzle difficulty of ten was used.

To see how the duration of the base exchange is affected by the puzzle difficulty we measured the time T2 with varying value of K. Figure 7.3 illustrates this dependency for the Tablet and the Laptop and shows that the time needed to solve the puzzle grows exponentially with increasing its difficulty. There is a time limit during which the Initiator must find a solution to the challenge. With Nokia 770, setting a high value of K by Responder would not be possible since Tablet's CPU will take a long time to solve such puzzle. For example, a puzzle difficulty of 20 would keep Tablet's CPU busy for over 10 seconds, which is unacceptable for most applications and users. The Laptop, in contrast, would solve a similar challenge in 1.3 seconds. Balancing between the puzzle difficulty and the time limit during which a correct solution is valid for the Responder might be an issue when using the lightweight hardware in a hostile environment with a low level of trust.

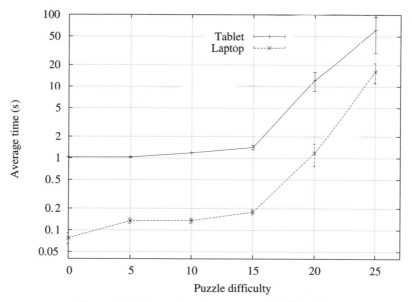

Figure 7.3 Processing time vs. puzzle difficulty.

RTT

RTT (Round Trip Time) equals the time for a packet to travel from a node across a network to another node and back. Our tests evaluate the effect of using HIP on RTT for applications. Tests used the ping6 tool for sending 100 ICMP messages over HIP and over plain IP. We measured RTT for a number of scenarios including Tablet, Laptop and PC as HIP communicating entities.

Table 7.1 contains mean values of the RTT as well as standard deviations measured over IPv6 and over HIP. The first RTT value in each test appears to be high because of the HIP base exchange and an ARP query performed upon the first connection. It is excluded from the average RTT calculations presented in the table. Figure 7.4 shows the whole set of RTT values with the PC acting as the Initiator. On average, HIP raises the latency by 35–45%.

Table 7.1   Average Round Trip Times for Tablet and Laptop.

RTT	Mean (ms)			Standard deviation (ms)		
	IPv6 (64B)	IPv6 (116B)	IPv6/HIP (ESP)	IPv6 (64B)	IPv6 (116B)	IPv6/HIP (ESP)
PC→Tablet	2.223	2.358	2.936	0.470	0.425	0.931
Tablet→PC	1.901	1.900	2.748	0.332	1.235	1.347
PC→Laptop	1.026	1.049	1.177	0.340	0.312	0.243
Laptop→PC	1.065	1.070	1.207	0.338	0.427	0.502

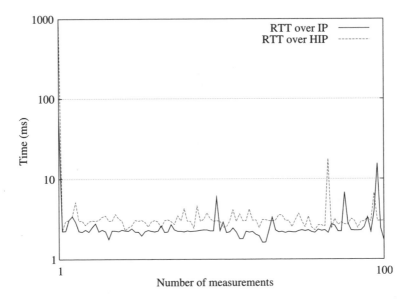

Figure 7.4 Round Trip Time, PC as the Initiator.

The RTT time that we measured includes the transmission time of an ICMP ECHO_REQUEST message from the PC to the Tablet, processing time on both hosts and the latency of delivering an ICMP ECHO_RESPONSE back to the PC. The default size for an ICMP message equals 64 bytes (56 data bytes and 8 bytes of ICMP header). When used with HIP, the size of an ICMP message is augmented by ESP headers and amounts to 116 bytes. We measured the RTT time for a plain ICMP message of the size 64 bytes and 116 bytes as well as for an ESP encapsulated ICMP packet (IPv6 over HIP case). The results indicate that increasing the size of the ICMP packets affects the transmission latency by a small factor (6%). The major impact on the RTT over HIP (2.936 ms) is therefore made by a slow processing of the ICMP messages encapsulated with ESP. HIP increases the RTT value for the PC-to-Tablet connections on average by 37%. In contrast, the same proportion for the PC-to-Laptop scenario is around 15%. According to this comparison ESP encapsulation of the data involved by the Host Identity Protocol affects more seriously the lightweight devices than ordinary PCs or laptops.

Throughput

IPSec ESP data encryption performed by the Tablet can also reduce the maximum achievable data rate of the wireless link. We measured TCP throughput by an iperf tool generating TCP packets to a correspondent node. It is necessary to mention that the WLAN Access Point introduces its own data encryption by means of WPA protocol. Different tests have been performed to evaluate the overhead of ESP and WPA data encryption. The average values of the throughput are presented in Table 7.2. An average value of 4.86 Mbit/s represents an upper bound of the throughput achievable by the Tablet acting as the Initiator (see Tablet-to-PC scenario). This value was measured with plain TCP/IP traffic in a totally open network with no encryption algorithms employed. Although the Tablet's specification claims supporting IEEE 802.11 b/g standard with a maximum data rate of 54 Mbit/s, Tablet's CPU imposes its own constraints. Further analyzing the results, we might conclude that WPA encryption makes a minor impact on the throughput. Enabling the WPA access control on the WLAN AP reduces the data rate only by 0.4% (4.84 Mbit/s vs. 4.86 Mbit/s). In contrast, the ESP influence is much stronger and reduces the throughput by 32% (3.27 Mbit/s vs. 4.86 Mbit/s) in the same network. Mutual impact of WPA and ESP is even bigger as double encryption is used.

Table 7.2   TCP throughput in different scenarios.

| Throughput | Mean/standard deviation (Mbit/s) | | | |
	TCP	TCP/HIP	TCP+WPA	TCP/HIP+WPA
Tablet→PC	4.86/0.28	3.27/0.08	4.84/0.05	3.14/0.03
Laptop→PC	21.77/0.23	21.16/0.18	—	—

In comparison with the Tablet, the Laptop achieves as much as 21.77 Mbit/s of the TCP data rate over the same open wireless link (see Laptop-to-PC scenario). A valuable fact is that with the Laptop the impact of ESP encryption involved by HIP is tiny as compared with the Tablet and equals 3% of the decrease in throughput.

Figure 7.5 graphically depicts the results explained above and shows the distribution of TCP and TCP/HIP throughput values over a WPA-free wireless link. The graph illustrates HIP influence on the TCP throughput as well as a big difference in values achieved by the lightweight Tablet and much more powerful Laptop.

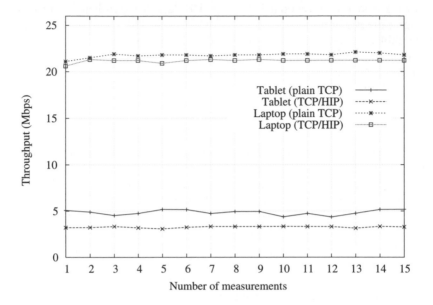

Figure 7.5 TCP throughput in an open wireless network.

End-to-end security provided by HIP might be used not only for data protection itself but also for authentication purposes as an alternative to WPA access control in wireless networks. However, as the results above indicate for the devices with limited computational power the data throughput and latency are significantly affected and might become a concern.

Duration of Mobility Update

HIP sends Mobility Update packets when the IP address of a HIP mobile terminal changes. We measured the time to exchange Mobility Update packets by manually changing the IP address of the network interface to simulate a simple mobility case. We repeated our tests 50 times and calculated the average. The average duration of Mobility Update between the Tablet and the PC is 287 ms (see Figure 7.6). However, in reality this delay can be lower for applications. Once the correspondent node receives the first UPDATE packet it knows the Tablet's new location and can transmit data to the new address using Credit-Based Authorization (CBA). CBA limits the transmission rate to a new IP address until it is verified to be reachable by the last two UPDATE packets. Such practice prevents hijacking of arbitrary IP addresses. The average time for generating, sending and processing the first UPDATE packet is around 20 ms. Comparing to the Tablet, our 1.6 GHz laptop is capable of completing the three-way Mobility Update with its correspondent node in 100 ms.

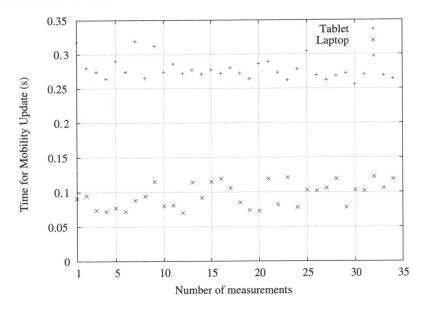

Figure 7.6 Duration of Mobility Update.

Battery lifetime

Power consumption is a crucial issue for any portable device. The capacity of the Nokia 770's battery keeps the device in standby mode for a few days. However, the battery resources are exhausted quickly by applications requiring data transmission over WLAN. The objective of measuring battery lifetime on the Tablet was to assess how expensive the Host Identity Protocol operations might be in terms of power consumption. We used an external multimeter to measure the consumption of the battery's current while the device was busy with various applications (see Table 7.3). Given the capacity of the battery and the current consumed by an application we were able to compute a theoretical time to deplete the battery. Alternatively, we ran the same application on the Tablet with a fully charged battery until its depletion to verify our empirical assumption about the lifetime. With HIP, the average current measured by the multimeter was 0.38 A. A fully charged 1500-mAh battery kept the Nokia Tablet working for about three and a half hours.

In fact, we found almost no difference in power consumption between the HIP-enabled and non-HIP applications. Establishing a HIP association, Mobility Update as well as ESP encrypted traffic all consume a similar amount of the current (0.36–0.38 A) equivalent to a plain TCP/IP data connection. We interpret these results as caused by the low computational power of the Nokia's CPU which is kept busy all the time upon transmitting data over WLAN regardless of the protocol and the application being used. We also believe that HIP does consume more power than the non-HIP applications if the data throughput is compared. In other words, due to a lower bitrate caused by ESP data encryption a HIP application would require a notably longer time for a similar task to be completed. For instance, the Tablet is able to transmit 100 Mbytes of data in 170 seconds over plain TCP/IP while HIP would spend an additional 98 seconds (and total time of 268 seconds) for the same piece of work.

In terms of power consumption the use of HIP would therefore involve longer CPU utilization and consequently more energy consumed for a task.

Table 7.3   Power consumption by applications.

Application/Mode	Current (A)
HIP Base Exchange	0.36
ESP traffic (iperf with HIP)	0.38
Plain TCP (iperf without HIP)	0.38
Video stream from a server	>0.50
Local video	0.27
Audio stream from a server	0.40–0.50
Local audio	0.20
Browsing (active WLAN)	0.35–0.50
Passive WLAN	0.12
Activating screen	0.12–0.14
Standby mode	<0.01

7.3   Summary

This chapter presented measurements and performance evaluation of Host Identity Protocol on the Nokia 770 Internet Tablet. We found several interesting results on the use of asymmetric cryptography on lightweight devices.

- The unmodified HIP may be used in scenarios where a lightweight device communicates through a single proxy server in the Internet. A HIP association establishment in such a case is 1.4 sec and Mobility Update is 287 ms.

- For scenarios involving two mobile hosts or multiple parallel HIP associations, unmodified HIP is too heavy for lightweight devices. For two Tablets, the HIP association establishment is already 2.6 sec.

- Surprisingly, the Tablet only achieves 4.86 Mbps in a WLAN capable of 22 Mbps even without HIP. The use of WPA encryption has negligible effect on throughput, while ESP encryption with HIP reduces the throughput to 3.27 Mbps, which is still sufficient for most Tablet applications.

- The RTT over WLAN is only several milliseconds. HIP increases the RTT by a few milliseconds, which does not noticeably affect the applications.

- The use of ESP encryption with HIP does not affect the battery consumption in the Tablet, although the energy cost per byte is higher with HIP due to reduced throughput. We noticed that the Tablet CPU is always fully utilized when an application transmits data over WLAN that depletes the battery in 3–4 hours.

- We believe that the measurement results are applicable to a wide range of mobility and security protocols in addition to HIP. Most such protocols rely on similar public-key and IPsec ESP operations like HIP.

Our measurements served as a motivation for proposing Lightweight HIP that uses hash chains instead of asymmetric cryptography (Heer 2007). Lightweight HIP achieves up to two orders of magnitude reduction of HIP computational cost at the expense of public key authentication. In future work, we plan to compare LHIP and HIP on newer Nokia N800 Tablet with video capability, as well as evaluate HIP implementation on the Symbian OS platform.

8

Lightweight HIP

Tobias Heer[1]

Apart from mobility and multihoming support, HIP also supports host authentication, payload encryption, and a cryptographic namespace without requiring changes to the network infrastructure or network applications. However, in particular, mobile devices with few CPU resources are slowed down by HIP. The poor performance of HIP on mobile devices results from the extensive use of public-key cryptography for securing the main functions of HIP. These public-key operations are complex and CPU intensive.

In this chapter, we present Lightweight HIP (LHIP), a HIP extension that enables HIP to offer mobility and multihoming support without using public key cryptography. The goal of LHIP is to speed up HIP to make it usable for weak mobile devices.

8.1   Security functionality of HIP

HIP uses several cryptographic algorithms and protocols to provide secure communication and tamper-proof protocol functionality. Some of these technologies involve CPU-intensive calculations. Especially, public-key algorithms are known to be costly in terms of CPU cycles. In this chapter, we present our performance analysis of cryptographic functions in HIP. We state the need for a lightweight version of HIP for CPU-poor devices and we define the scope of lightweight HIP. We first analyze how the cryptographic components of HIP interact and discuss the simplest but insufficient solutions for a lightweight HIP before we discuss an alternative authentication scheme that allows to securely operate HIP with reduced use of public-key cryptography.

---

[1]This chapter is contributed by Tobias Heer (RWTH Aachen University, Distributed Systems Group).

Host Identity Protocol (HIP): Towards the Secure Mobile Internet   Andrei Gurtov
© 2008 John Wiley & Sons, Ltd

8.1.1   Performance limitations of HIP

The performance measurement results in Chapter 7 show that the cryptographic components of HIP significantly slow down the base exchange and the update process on the Nokia N770 Internet Tablet, depending on the public keys in use. The Initiator as well as the Responder have to calculate one signature with their own private key and one signature verification with the public key of their peer during the base exchange. Furthermore, they have to compute the Diffie–Hellman shared secret. No payload traffic between the two hosts can flow before the base exchange is completed. Moreover, the CPU is busy during the RSA, DSA, and Diffie–Hellman computation. Real-time applications, like streaming audio and video, might experience interruptions during these computations.

RSA and DSA signatures are also used during the update process. The Initiator of an update must sign two packets with RSA or DSA while the Responder must sign one message. Furthermore, the hosts may verify the messages of the other peer[2]. Finally, a host must update all its HIP associations after its set of IP addresses has changed. Signing several packets increases the delay even more. Therefore, the update process will take several seconds on weak hosts communicating with multiple peers.

Depending on the application scenario, the computational overhead is appropriate or not. Applications that exchange sensible or valuable data require appropriate measures of protection. The use of expensive public-key cryptography is adequate for such cases. However, HIP slows down unnecessarily applications that do not require such protection.

8.1.2   Problem statement

HIP aims at being a general purpose solution. The new namespace and the separation of the IP address and the host identity provide a convenient and efficient way to address hosts regardless of their location. However, using HIP is only possible if both hosts support the protocol. To achieve a widespread deployment, HIP should be applicable to many applications, environments, and platforms.

Using HIP for all communication, however, introduces the performance penalties for all communication flows. Especially on weak devices, these performance issues limit the applicability of HIP in general. For instance, web browsing on mobile devices, with a delay of seconds, can discourage users from using HIP. Web pages with contents from several servers even aggravate the situation because clients must establish several HIP associations at the same time to display the web page. Mobile hosts that establish several HIP associations with different hosts can also experience delays of several seconds whenever they change their point of network attachment. Only few users will find such long delays acceptable when it comes to audio streaming or voice telecommunication. Moreover, users will experience CPU load peaks after mobility events. In the worst case, these peaks cause the freezing of all applications for several seconds. Even non-networking applications, such as audio and video players, are affected. The question arises if the gain of usability due to support of mobility really outweighs the reduced usability due to the HIP performance issues. The proliferation of *Peer-to-Peer* (P2P) applications even aggravates the situation. Hosts within P2P networks open and maintain dozens or hundreds of long-lasting connections to distinct hosts at the

---

[2]They can also use the HMAC signature instead of verifying RSA or DSA signatures. The RSA and DSA signatures are included in the UPDATE packet to allow verification by middleboxes.

same time. Using HIP in these very popular networks leads to extremely long delays for location updates as a large number of associations must be updated concurrently. However, P2P telecommunication applications such as *Skype* benefit from the HIP mobility support because, e.g., Internet phone calls are not interrupted after mobility events.

Many application cases require neither encryption nor authentication on the network layer. For instance, many web sites do not provide data that requires protection. Moreover, data that requires protection is typically encrypted on higher network layers. Many applications use their own authentication mechanisms and do not rely on lower layers to provide authenticated communication. Consequently, many applications do not necessarily require HIP-based authentication and encryption services. Nevertheless, these applications benefit from HIP-based mobility and multihoming support. Applications with long-lasting connections sustain mobility events and are applicable to mobile hosts without restrictions.

One might argue that insufficient CPU resources is a temporary problem for mobile devices and that future generations of mobile devices will have sufficient CPU resources to run HIP without delays. However, increasing the CPU frequency means using more energy. Thus, devices that do not require faster CPUs to perform their basic duties are unlikely to be equipped with a faster CPU just to support HIP because this would require more expensive hardware and more battery resources to keep the device running for the same time. Moreover, the additional CPU resources are only required during the load peaks caused by the HIP base exchange and the update process. Therefore, the CPU would run idle for most of the time to provide enough computational power during these peaks.

Currently, users of mobile devices have to decide whether they are willing to accept the performance of HIP to have mobility support. However, the long delays, which might even surpass the time required to reestablish HIP-less connections after a mobility event, might influence the users to favor HIP-less solutions. Thus, limiting the scope of HIP to applications that require both security plus mobility or multihoming will hinder the deployment for small mobile devices, which would benefit most from it. Moreover, it might hinder the deployment of HIP in general since a large group of devices that actually require the features of HIP will not be able to use it without compromises. Manufacturers and developers might, therefore, favor HIP-less solutions as HIP-enabled devices require more processing power, leading to higher hardware costs. However, wide support from developers and manufacturers is essential to elevate HIP from an isolated solution for special problems to an omnipresent general purpose solution. We argue that a lightweight version, which uses the host identity namespace and provides HIP-like mobility support without CPU-intensive operations, is required to make HIP attractive for users of weak mobile devices.

8.1.3 Scope of LHIP

Performance tests have shown that especially the public-key cryptographic operations consume a considerable amount of time during the crucial HIP actions. Consequently, the first goal for LHIP is to replace the public-key algorithms in HIP. It is clear that, without public-key cryptography, LHIP cannot provide the same security nor the same functionality as HIP. Therefore, it is important to define the scope of LHIP before continuing with the protocol details.

The primary goal of LHIP is to increase the performance of HIP on CPU-poor devices. Mobile devices, such as mobile phones and PDAs with CPU frequencies below 200 MHz, should be able to use LHIP without serious limitation of usability.

Without public-key cryptography, host authentication, such as provided by HIP, is clearly out of scope for LHIP. Using the same namespace with authentication support requires the same authentication measures as for HIP. We sacrifice host authentication for the sake of performance and assume that applications will use their own authentication mechanisms. Moreover, applications are free to use HIP instead of LHIP in cases that require authentication. Thus, an attacker can use LHIP to impersonate a host but it cannot cause harm because of the properties of the data or due to higher level authentication mechanisms. Nevertheless, it must be possible neither for an attacker to steal an established HIP or LHIP association, nor to impersonate one of the communicating peers during an established HIP association.

Protection against MITM attackers during the base exchange is difficult to achieve without verifying the cryptographic properties of the HIs. A basic assumption of LHIP is that no such attacker is present during the base exchange and that it is unmodified. Furthermore, one can assume that applications implement measures to identify these attacks if such protection is necessary. Alternatively, HIP-aware applications can decide to use HIP instead of LHIP. Payload encryption and authentication is also out of scope for LHIP. LHIP trusts that the transport layer and the upper layers provide adequate measures to authenticate and encrypt data whenever necessary.

However, the LHIP should provide protection against MITM attacks that take place after the base exchange. We assume that a MITM can be on the communication path after location changes or after multihoming events. This assumption is important to support mobile devices. Frequent subnetwork changes should not endanger the security of the protocol.

The way HIP works should not change in general, which means that the concept of decentralized end-host mobility and the way HIP provides its services should not be altered. LHIP should provide the same functionality as HIP does but without authentication and encryption. However, we need to replace certain mechanisms in HIP and to introduce new techniques to gain performance. LHIP should use the same namespace and the same name resolution infrastructures as HIP does. It should not be necessary to deploy a second namespace infrastructure for LHIP.

LHIP should also support middleboxes and network infrastructure elements in a similar way as HIP but host authentication for middleboxes is also out of its scope. Above all, LHIP must provide basic protection and security. In other words, LHIP must not allow attacks that are worse or easier to mount than attacks without LHIP.

8.1.4 Threat model

It is necessary to define a threat model that allows the identification of which kinds of attackers and attacks have to be expected for a lightweight version of HIP. The following assumptions characterize a typical attacker for LHIP.

- An attacker can receive all packets, including the base exchange and UPDATE packets.

- The attacker can modify, drop, resend, duplicate, and delay packets. However, we assume that the attacker does not take active measures during the base exchange.

- The attacker has considerably more CPU resources than the victim. However, the attacker cannot compute calculations that are considered to be computationally infeasible. In particular, an attacker cannot reverse hash functions nor can it forge RSA or DSA signatures.

The LHIP design copes with such an attacker and provides adequate measures against it.

8.2 HIP high-level goals

HIP uses a wide range of different cryptographic mechanisms to achieve security, confidentiality and authentication. To determine the security requirements for HIP and a lightweight version, we discuss how these security mechanisms are used. HIP achieves three high-level goals in terms of security: *payload security*, *protocol security*, and *secure host authentication*, which we refer to as *namespace security*.

H1: Payload security ensures that the access to payload, transferred by HIP, is restricted to the peers that share the same HIP context. Two hosts can send data over insecure communication channels in a secure way. It is computationally infeasible for an attacker to decrypt the transferred payload. Furthermore, an attacker cannot manipulate the payload unnoticeably.

H2: Protocol security ensures that the communication protocol is not susceptible to malicious interactions with attackers or misbehaving peers and network components. An attacker cannot influence the protocol behavior in a way that causes harm to one of the communicating peers or any other third party. A secure protocol must be able to prevent and discover accidental or intended misuse that leads to limited functionality. Protocol security is not restricted to the communicating peers only, but also takes the communication channel and intermediate nodes such as gateways, directly or indirectly participating in the communication process, into account.

H3: Namespace security enables trusted communication between hosts and provides a way to securely address hosts, using HIP. Namespace security mechanisms must protect the namespace from intended or accidental misuse. Above all, HIP must provide protection against identity theft and impersonation.

A combination of different techniques and cryptographic protocols ensures each of these goals. Figure 8.1 depicts the high-level goals and the techniques used to achieve them. Some of these techniques depend on others, leading to transitive dependencies. The illustration depicts these dependencies. It shows a directed dependency graph between high-level goals and cryptographic techniques, protocols, and algorithms.

For achieving these security goals, HIP strongly relies on public-key cryptography. The protocol enforces protocol security by utilizing RSA, and HMAC signatures. These signatures ensure that packets have been sent by the legitimate peer. The HMAC algorithm requires a shared secret that both peers must possess. This shared secret is provided by the Diffie–Hellman key exchange. The key exchange is protected with RSA signatures to ensure that the Diffie–Hellman parameters have not been modified by a Man-In-The-Middle attacker. RSA signatures are also used in UPDATE packets, since middleboxes, such as firewalls, cannot use the HMAC signatures to verify packet authenticity and integrity because they are not in possession of the shared secret. The high-level goal of *protocol security*

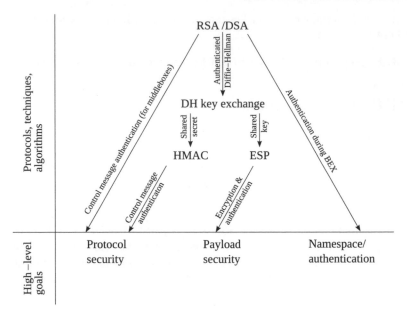

Figure 8.1   HIP high-level goals and dependencies between techniques, protocols, and algorithms that are employed to achieve the goals.

is, therefore, reached by utilizing RSA, HMAC signatures, and the Diffie–Hellman key exchange.

Payload security for HIP is provided using ESP encapsulation, utilizing symmetric encryption algorithms to protect the payload. These encryption algorithms require both hosts to share secret symmetric keys, generated during the authenticated Diffie–Hellman key exchange. Payload security, therefore, depends on the ESP protocol, the Diffie–Hellman key exchange, and RSA signatures.

Namespace security and host authentication depend on RSA signatures. Signing R1, I2, and R2 packets with the private key belonging to a host identity guarantees that the host is using its own identifier. The RSA signature reveals spoofed identifiers.

As shown in Figure 8.1, removing a single public-key component from HIP directly affects the high-level goals. For instance, the HMAC algorithm can only provide protocol security if a shared secret was provided with the Diffie–Hellman key generation. Furthermore, statements about the identity of a host are only possible if the Diffie–Hellman key exchange was authenticated. Thus, removing the public-key algorithms for the sake of performance leaves HIP without appropriate measures to provide protocol, payload, and namespace security.

8.2.1   LHIP high-level goals

The dedicated goal of LHIP is to reduce the computational cost of HIP while retaining the high-level goals *H2 – protocol* and *H3 – namespace security*. However, LHIP relaxes the goal *H3*. It is only required that the namespace cannot be used for attacks worse than attacks on plain IP. Host authentication is not required as long as identity theft during the lifetime of an LHIP association is not possible. Hence, the high-level goal *H1 – payload security* is sacrificed for the sake of performance. However, the level of payload security should still be comparable to plain IP.

Compatibility with the HIP and upgradability from LHIP to HIP are further goals. Upgradability refers to the ability of hosts to transform an LHIP association into a HIP association without tearing down and re-opening the association. These additional goals are summarized in a new high-level goal *compatibility*. The high-level goals of LHIP are defined as follows:

LH1: Increased performance for weak devices. The performance penalties of public-key cryptography should be reduced to amounts acceptable for use on mobile devices with little CPU power. Especially, the delays during the base exchange and the connection disruption during updates are noticed by the users.

LH2: Protocol security must be provided in LHIP. The protocol should not allow attacks worse than today's attack on TCP/IP. This includes attacks on the communicating peers, third parties, and infrastructure elements such as middleboxes.

LH3: Namespace security should be provided to a certain extent. Host authentication is not required but attackers must not be able to impersonate a communication peer during the lifetime of an LHIP association. Furthermore, LHIP must be able to deal with namespace conflicts. The relaxed goal of host authentication must not yield new ways of attacking HIP or LHIP hosts.

LH4: Compatibility with HIP should be provided to a certain extent. Infrastructure elements such as name resolution services should be used by HIP and LHIP in the same way. LHIP should resemble HIP and provide the same functionality regarding mobility and multihoming support. LHIP should extend HIP rather than defining a new HIP-like protocol. The message format and the packet contents should not differ much from HIP packets to reduce the implementation overhead for hybrid HIP and LHIP implementations. Furthermore, middleboxes should be able to learn about new IP addresses of hosts and new SPI values in a secure way. LHIP associations should be upgradable to HIP associations during their lifetime without the need to close and reestablish the association. Additionally, compatibility with legacy applications should also be provided the same way as in HIP.

The requirements stated above can be summarized as: *Modifying HIP to achieve secure mobility and multihoming for TCP/IP without payload encryption and host authentication.*

The HIP dependency graph in Figure 8.1 shows that ESP encryption and, therefore, the high-level goal *H1 – payload security*, can be removed from HIP without affecting the other high-level goals. However, the vector for performance improvements reached by removing payload encryption is limited. Furthermore, the ESP protocol does not influence the long connection establishment delay on weak devices. A much greater gain in performance can be achieved by reducing the use of public-key cryptography during the base exchange and the update process. We focus on this challenge to make HIP applicable to weak devices.

8.2.2 Possible approaches

We first discuss three simple approaches to reduce the number of public-key operations in HIP. These approaches show that it is not trivial to speed up HIP. They either violate one of the desired high-level goals or achieve insufficient performance improvements.

Alternative 1: remove Diffie–Hellman

The first trivial solution is to remove the Diffie–Hellman key exchange while keeping all other protocol functions as they are. This solution obviously reduces the cost of establishing a new HIP association since the CPU-intensive Diffie–Hellman shared key computations are omitted. However, it also affects the symmetric cryptographic elements and the hashed message authentication codes because no shared keys are provided anymore. A shared key for ESP is not needed in LHIP as payload encryption is not necessary. ESP can operate in *NULL mode*[3], which does not require a shared key. This mode does not encrypt the payload and, therefore, can be used without shared keys. Still, a shared secret is required for calculating the HMAC signatures.

One can also remove the HMAC signatures since they are useless without a shared secret. HIP uses HMACs and RSA to authenticate HIP control messages. The security properties of RSA are sufficient to provide adequate authentication. Moreover, the RSA signatures have to be used in any case to enable middleboxes to authenticate UPDATE and CLOSE messages. So far, this approach seems to achieve the desired high-level goals.

However, RSA signature verification is much more CPU-intensive than HMAC verification. The absence of the HMAC signature means that the peers cannot check the authenticity of UPDATE messages in a cost-efficient way before calculating expensive public-key operations. This introduces a new weakness, which can be exploited for DoS attacks. Both LHIP peers, the weak client as well as its peer, possibly a commercial server, become susceptible to attacks aiming at CPU-cycle consumption. The RSA signatures would always have to be checked without any possibility to filter out forged messages from an attacker.

An attacker that is aware of HIP associations of a victim can send invalid UPDATE messages with spoofed host identifiers. The attacker can use the HIs of the victim's peers. The victim can only filter out the forged messages by verifying the RSA or DSA signatures while the attacker can send random bits as forged RSA and DSA signatures. This gives the attacker an advantage over the victim and provides an efficient way to consume large amounts of CPU cycles on the target host. Due to this weakness, the first approach cannot achieve the second high-level goal *LH2 – protocol security*.

Alternative 2: remove DSA/RSA

The first approach illustrated that an attacker can use unprotected RSA signatures for attacks. Furthermore, these signatures cause an unacceptable amount of computational overhead during the base exchange and during the update process. As stated above, host authentication is not a necessary requirement for LHIP. Therefore, removing RSA from HIP could be considered an option. The Diffie–Hellman key exchange generates a shared secret that can

---

[3] Note that using ESP and AH in NULL mode violates the specifications in Manral (2007). Thus, another payload transport format might be necessary.

be used for host authentication during the lifetime of a HIP association. The HMAC in the UPDATE messages provides authentication for the communication peers. This is sufficient to provide weak authentication with *temporal separation*. Provided that the base exchange is authentic, an attacker cannot impersonate one of the communication peers during the lifetime of the association.

However, the computationally expensive Diffie–Hellman key exchange is still required for creating the shared secret. Furthermore, middleboxes have no means of verifying the authenticity of UPDATE messages. The missing RSA signatures leave them no secure option to learn of updated IP addresses or new SPIs. This limits the usability of LHIP with packet inspecting middleboxes since they will not support HIP mobility without authenticated updates. Alternatively, it will make new attacks, targeted at middleboxes, possible if these react to unauthenticated HIP UPDATE messages. Again, the second high-level goal cannot be achieved by removing RSA from HIP.

Moreover, completely removing RSA from HIP enables attackers to use spoofed HIs. The legitimate owner of a host identity has no means of proving its identity to a remote host. This raises severe security issues, especially when HIP and an RSA-less version of HIP are used in parallel. The high-level goal *LH3 – namespace security*, therefore, cannot be achieved without RSA.

Alternative 3: reduction of the key size

One could argue that HIP, as it is, offers enough speed-up potential to support weak devices. RSA as well as the Diffie–Hellman key exchange can use keys of different length. Short RSA keys could be used to generate weaker identifiers and short Diffie–Hellman keys could be used to generate weak session keys. Although these session keys do not offer sufficient security for long-term data security, they provide short-term session keys for the HMAC authentication. However, the potential for improvements is limited to medium-sized keys as very short RSA and Diffie–Hellman public keys are easy to compromise and should not be used.

A host could choose medium-sized keys to lower its processing cost of generating signatures and calculating a shared secret. However, it still needs to verify the HIs of the peer. Especially, the verification of long DSA HIs is CPU-intensive. The weak device cannot lower the verification cost for these keys, since it has no influence on the key length selection of its peer. Consequently, a server that wants to support weak clients as well as clients with sufficient CPU power either has to lower its provided level of security for all hosts or it must use different identifiers for weak hosts. Using two identifiers leads to practical problems when the identifiers are passed from one host to another. It also complicates the mapping between a host and its identifiers. In particular, legacy applications have no means of expressing which keys should be used and what level of identifier hardness is required. This conflicts with goal four, *LH4 – compatibility*.

One solution to this dilemma could be to compile several keys into one identifier. The HIT would be the hash over all these keys. This would tie these keys together. However, this makes key revocation difficult. Identifier components cannot be changed without changing the complete identifier and the HITs generated from them. This would make compound identifiers short lived and would seriously reduce the usability of the host identity namespace.

Hence, decreasing the key length would violate goal *LH3 – namespace security* or *LH4 – compatibility*.

LHIP approach

The previous paragraphs pointed out that a non-trivial solution is necessary to provide efficient mobility and multihoming in a secure way. The following sections will introduce our LHIP approach, discuss its advantages and disadvantages, and demonstrate its effectiveness and practical usefulness.

8.3   LHIP design

In this section, we present the design and the rationale of LHIP. Section 8.3.1 discusses hash chains as an alternative to public-key-based cryptography. In Sections 8.3.4 and 8.3.6, we introduce a lightweight authentication layer based on hash chains, enabling message authentication with little computational overhead. Section 8.3.7 identifies weaknesses that arise if the host identity namespace is not protected by asymmetric cryptography. Section 8.3.8 shows how an LHIP association can be transformed into a HIP association and Section 8.4 evaluates the performance of LHIP.

The previous sections pointed out that the high-level goals for a lightweight version of HIP cannot be reached by removing the public-key operations alone. Nevertheless, hosts with little processing power require alternative authentication mechanisms that reduce the computational cost of HIP. However, there are no computationally inexpensive algorithms with identical properties. Hash chains, however, offer an efficient way to authenticate messages without a shared secret but require different application and offer different security properties. Despite the differences they are an alternative to Diffie–Hellman and RSA in the context of HIP. The following will introduce the Interactive Hash Chain authentication and the variant that is used by LHIP.

According to our threat model, we assume that no active MITM attack is performed during the base exchange. This assumption and the relaxed high-level goal *LH3 – namespace security* allow us to use RSA signatures only in cases of conflicts and namespace violations. Hash chain signatures provide authentication for LHIP control messages. Since middleboxes can also verify the hash chain signatures, there is no need for RSA signatures in these packets.

Provided that no namespace violations occur, LHIP can operate completely without computing public-key operations. However, LHIP relies on RSA and DSA signatures in the case of namespace conflicts. Figure 8.2 illustrates the dependencies between the employed cryptographic techniques and the LHIP high-level goals. The abstract high-level goals of performance and compatibility are not depicted as they depend on the proposed design rather than on concrete mechanisms. Optional processes are indicated by dashed lines.

8.3.1   Hash chains for HIP authentication

The basic idea behind hash chain (cf. Section 2.4.3 for details on hash chains) based packet authentication is to let the sender of a message sign the message with an HMAC created with a secret value as the key. In the absence of an encrypted channel to the receiver, the sender must transmit the message, the signature, and the key in cleartext to the receiver to

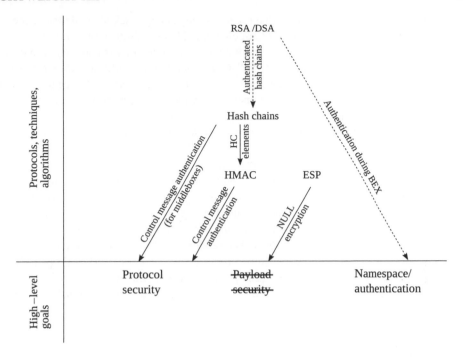

Figure 8.2 LHIP high-level goals and dependencies between techniques, protocols, and algorithms being employed to achieve them. The abstract high-level goals of performance and compatibility are not depicted. Optional processes are indicated by dashed lines.

allow signature verification. To secure the signature process, a host only accepts messages as authentic if the signature has been created before the key was disclosed. This ensures that only the host that was in possession of the secret before it had been disclosed was able to sign the message. Using hash chains (cf. Section 2.4.3) as source for the keys allows to relate several consequent signatures to the owner of a hash chain.

8.3.2 Time-based signatures

Before going into the details of interactive hash chain signatures we illustrate the concept of time based signatures as the concept of interactive hash chain signatures stems from the same principles. The *Timed Efficient Stream Loss-tolerant Authentication* (TESLA) (Perrig *et al.* 2005) protocol uses a time-based approach to sign messages. The approach requires loose time synchronization between the communicating peers. The sender sends a hash anchor to a group of receivers. Each message is protected by an HMAC, which is created with an unrevealed element of the hash chain.

The sender uses each unrevealed element of the hash chain to sign messages for a certain time span. The element is disclosed after this time span. Receivers can authenticate the messages as soon as the corresponding element is disclosed. Messages signed with an element that has already been disclosed are dropped by the receiver to prevent attacks.

The time synchronization ensures that receivers only accept packets while the sender has not disclosed the corresponding hash chain element. This guarantees that only the sender, who is in possession of the hash chain, can sign messages. An attacker can only sign packets after the disclosure of the element. At that time, other hosts would not accept any more packets signed with the hash chain element. When disclosing an element, the next unrevealed element of the hash chain becomes the element with which future packets are signed. Again, it is only valid for a certain fixed interval before it is disclosed.

Using a time-based instead of a message-based authentication method enables TESLA to deliver authenticated messages with little additional overhead. Signing messages requires the use of one hash chain element per time interval and the usual cost for generating and verifying an HMAC. TESLA introduces an additional delay. A receiver cannot verify the message before the sender discloses the corresponding hash chain element. The receiver has to buffer the messages until then. Another disadvantage is that hash elements are revealed even when no signed payload was sent during the last time interval.

Figure 8.3 illustrates the message flow. The sender uses h_i to generate the HMAC of the messages. The receiver accepts and buffers all packets while h_i is valid. It drops messages signed with h_i if they arrive in the next time period. This happens to *message4* in the figure. The receiver can start to verify the buffered messages after receiving the disclosed element of the hash chain h_i.

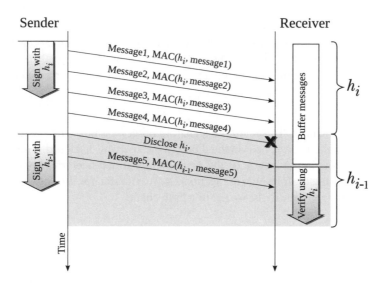

Figure 8.3 Signatures with the TESLA protocol. The sender signs and sends multiple messages. The receiver verifies the messages after the sender has disclosed h_i.

8.3.3   Interactive signatures based on hash chains

Torvinen and Ylitalo (Torvinen and Ylitalo 2004) use an approach that is similar to TESLA to sign messages with hash chains and hash functions. As in TESLA, undisclosed hash chain elements are used to generate message authentication codes. Like in TESLA, a mechanism is required to separate the time span when a hash chain element is used for signing messages from the time for message verification. In contrast to TESLA, which uses loose time synchronization, the interactive approach relies on a strictly sequential message exchange between the communicating peers.

Figure 8.4 illustrates the signature process. A signer sends a packet containing a *message* and an HMAC generated with an unrevealed hash chain element to the verifier in the first signature packet (S1). The verifier acknowledges the receipt by sending an acknowledgment packet (A1) and buffers the *message*. The signer discloses the corresponding hash chain element in a second signature packet (S2). Like in TESLA, the receiver can verify the buffered message as soon as it receives the disclosed hash chain element.

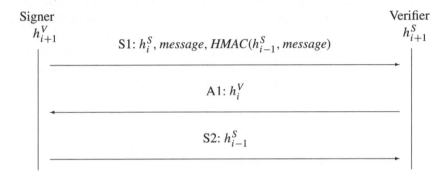

Figure 8.4 Interactive hash chain signature. (S1) The signer signs and sends a message. (A1) The verifier acknowledges the receipt. (S2) The signer discloses the secret hash chain element.

The protocol must fulfil three basic requirements to be secure.

1. The verifier must receive the message (S1), signed with a hash chain element h_i, before the signer discloses h_i.

2. The verifier must not accept any messages signed with h_i after the signer has disclosed h_i.

3. The signer must not disclose h_i before it receives the acknowledgment (A1) from the verifier.

Requirement number one ensures that only the host, which possesses a hash chain, can sign messages. An attacker cannot sign messages with the signer's hash chain as it is not in possession of the required undisclosed hash chain element. The signer does not disclose its next hash chain element before it has received the signed packet. It waits for the acknowledgment from the verifier before it sends the next hash chain element. Requirement number one demands that the signer can distinguish valid acknowledgments from acknowledgments an attacker could have forged.

Requirement number two ensures that an attacker cannot sign any messages with a disclosed element of a hash chain. The verifier would not accept any packets after it has received the signature from the signer. Providing this property demands that the verifier must be able to distinguish signature packets from bogus packets that an attacker could send to disturb the communication. Otherwise, an attacker could constantly send forged packets to force the verifier to deny valid signature packets. The interactive approach utilizes hash chains to meet these requirements. Not only the signer but also the verifier uses a hash chain. Every packet contains a hash chain element as clear text. This element ensures that the packet was sent by the legitimate signer.

The receiver of an S1 message verifies that the message was sent by the legitimate signer by hashing the contained cleartext hash value and comparing it with the last hash value revealed by the peer. This check indicates that the signer has sent *some message* but it cannot verify the message contents. Receiving the signer's hash chain element signals the verifier that it can not accept any more packets containing this hash chain element. This guarantees property *two*.

The verifier adds an element of its own hash chain to the A1. This hash chain element provides the signer with the knowledge that the verifier has received the signed message. At this point, it is safe to disclose the next hash chain element in the S2 message. This enables the verifier to verify the message.

An attacker cannot forge any packets without interaction with the legitimate hosts as it cannot calculate undisclosed elements of the hosts hash chains. Even a Man-In-The-Middle attack can be prevented as an attacker cannot sign messages before the signer has disclosed the next hash chain element and it will not disclose this element before the verifier has acknowledged the receipt of the signed packet. This acknowledgment cannot be forged by an attacker because it contains an undisclosed hash chain element.

Two hash chains are an unforgeable chain of triggers with each element of the one chain triggering an element of the other chain to be published. The cleartext hash chain elements in the packets are, therefore, called *trigger values*.

Interactive hash chain signatures do not require any form of time synchronization nor do they introduce a fixed delay until the verifier can verify the packets. A delay of one additional *Round-Trip-Time* (RTT) is induced for every signed packet. The approach has more overhead than TESLA due to the additional interaction. However, the packets containing only *trigger values* are very small and require a negligible amount of network bandwidth.

8.3.4   LHIP authentication layer

Hash chains cannot provide the same level of security nor can they be used in the same way as shared secrets because they can only provide temporary secrets. Interactive Hash Chain (IHC) signatures differ in three ways from regular HMAC signatures that have been created with a shared secret. First, messages cannot be verified instantly. Hence, the receiver must buffer each message until it receives the hash chain element required for the verification. Second, two hash chain elements of an identical hash chain must never be on the wire at the same time. This requires queueing messages and sending signed messages one by one. Third, the interaction necessary to delivering a signed message is more complex than for shared secret HMACs. Packet loss, retransmissions, duplicate packets, invalid messages, and undeliverable packets must be handled with respect to the security properties of hash chains.

Integrating IHC signatures tightly into HIP would complicate the protocol and require complex modifications to HIP. Instead, LHIP creates a new authentication layer, which extends HIP. The authentication layer is transparent for most of the functionality of HIP. This allows us to reuse nearly all of the mechanisms that HIP utilizes to provide secure mobility and multihoming. The isolated authentication layer ensures message authenticity with hash chains while HIP can create and process packets as if a shared secret were present.

Figure 8.5 illustrates the LHIP authentication layer and the way it interacts with HIP. The authentication layer accepts unsigned control messages from HIP and determines whether it must apply a signature or not. This is determined by the HIP header type and the message contents. The authentication layer enqueues all messages that require protection and sends them one by one. The authentication layer takes care of the necessary interaction messages required for IHC signatures. These packets are not passed to the HIP layer but are handled by the authentication layer. The authentication layer verifies that all signed messages are valid. It handles tasks such as packet retransmissions and duplicate packets, and provides support for undeliverable packets.

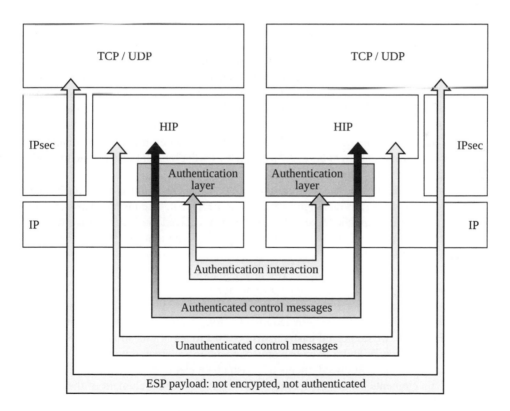

Figure 8.5 The LHIP authentication layer handles the authentication of messages requiring this service. It handles the interaction necessary for IHC signatures. Payload is sent encapsulated into unencrypted ESP packets.

During the following sections the term *signer* (S) denotes the host that initiates the signature process while the term *verifier* (V) denotes its peer.

Improved IHC signatures

The interactive hash chain approach as presented in Section 8.3.3 leaves some space for improvements that provide better resilience against Man-In-The-Middle attacks, simpler packet handling, and increased efficiency for acknowledged messages. We present three improvements: *dedicated hash chains*, *pre-signatures*, and *pre-acknowledgments*. These improvements are based on the *second preimage resistance* and *collision resistance* of cryptographic hash functions. They simplify the state machine of the authentication layer and avoid some security issues. The complete signature and verification process is depicted in Figure 8.7. The differences between the original IHC-based signatures and the improved IHC-based signatures are illustrated in the following sections. Throughout these sections, we assume that both hosts use the same cryptographic hash function H. The output length of H is l bits. We use a 160-bit output length for practical examples. The popular hash function SHA-1, for example, generates such 160-bit output.

Dedicated hash chains   The IHC signature approach uses one hash chain on the signer and one hash chain on the verifier side. Both hosts can exchange authenticated messages over insecure channels. However, this scheme is not secure when it is used for mobility updates as they are performed by HIP. Hosts may change their location at any time and hosts may even change their location concurrently. Therefore, each host must be able to send signed messages at any time. An MITM attacker can mount an attack whenever both hosts send their next hash chain element at the same time due to simultaneous location updates. Figure 8.6 illustrates the attack. Assume hosts A and B have established a communication context and have exchanged their hash chain anchors h_{i+1}^A and h_{i+1}^B earlier[4]. Both hosts decide to send a signed message at the same time.

A Man-In-The-Middle attacker M can delay both S1 packets and use the hash chain element h_i^B to acknowledge A's message before B has received it. It sends h_i^B to host A. This is interpreted as a valid acknowledgment that host B has received the first message from A. A would react to the acknowledgment by disclosing h_{i-1}^A. The attacker would now be in possession of h_i^A and h_{i-1}^A though B has never acknowledged the receipt of h_i^A. Therefore, it would accept messages signed with h_{i-1}^A. The attacker could now generate a valid signature for an arbitrary message, which would appear to be signed by A.

The main problem of this IHC signature scheme is that A cannot distinguish hash values of B that are meant for acknowledgments from hash values that are meant for signatures. LHIP uses two dedicated hash chains per host to avoid this situation. Every host generates a *signature chain* and a *trigger chain*. Signatures are created with elements from the signature chain while acknowledgments use hash chain elements from the trigger chain. The trigger chain elements trigger the release of the next hash chain element of the signer's signature chain.

Both hosts exchange the two anchors during the base exchange. After the exchange, a host can send updates whenever it is necessary. Even simultaneous updates are no threat to

---

[4]The superscript to h indicates which host the hash chain element belongs to.

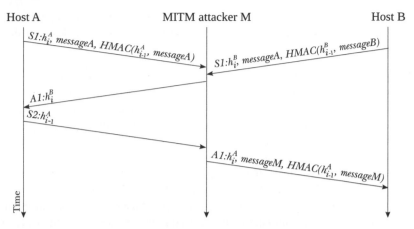

Figure 8.6 MITM attack on an IHC signature scheme that uses only one hash chain per host. An attacker can exploit simultaneous messages.

this construction because an attacker cannot sign messages with hash chain elements from a trigger chain nor can it acknowledge messages with elements from a signature chain.

Using two hash chains per host does not increase the overhead because IHC signatures, with and without dedicated hash chains, use one hash chain element on each side. The only additional effort comes from exchanging two hash chain anchors per host instead of one.

Although both hosts can initiate the signature process whenever necessary, it is not possible to send any further signed messages before the process is completed. Thus, using one pair of hash chains per host creates one duplex channel. Consequently, multiple channels can be created by utilizing multiple pairs of hash chains. Several channels enable parallel signature interaction handling and, therefore, reduce delays. However, the sequential way in which HIP handles UPDATE messages does not require such parallel channels.

Pre-signatures The original IHC signature approach consists of three packets for each signed message. The S1 packet contains a hash chain element h_i, the message m and the HMAC that is generated from m and the next undisclosed hash chain element h_{i-1}. This packet must be buffered by the verifier and cannot be verified before h_{i-1} has been disclosed by the signer. This scheme bears some problems. For example, it requires strict serialization. Multiple messages cannot be sent at once as only one hash chain element may be on the wire at any time. Packing several signed messages into one signed packet is not possible in general as the number of bytes per packet is limited by the maximal packet size defined by HIP. Sending several messages signed with one hash chain element collides with the interactive scheme. This considerably reduces the maximum throughput.

We propose not to send the actual message in the S1 packet but to send it in the S2 packet instead. The first packet only contains the signer's next hash value h_i^S and the signature $HMAC(h_{i-1}^S, m)$. The verifier acknowledges this packet with a hash chain element of its trigger chain. Consequently, the signer discloses h_{i-1}^S and the message m in a second packet. Figure 8.7 illustrates this process. We call the signature without the message *pre-signature* as it signs a message that is sent later.

Pre-signatures offer four appealing advantages. Not all of these advantages are necessary to protect HIP UPDATE messages. We state these advantages, nevertheless, for the sake of completeness.

Small size The Responder must buffer the first message of the IHC signatures to verify it by using the hash chain element from the second message later on. Using pre-signatures instead of messages plus signatures reduces the amount of data that needs to be buffered. The benefit for end-hosts is marginal as they typically have enough memory to buffer the signatures plus the messages. In contrast, memory is a scarce resource in middleboxes, such as firewalls and NAT boxes. Moreover, middleboxes need fast access to the memory to achieve high throughput. These packet-forwarding middleboxes must buffer the first of the IHC signature messages to verify the signatures. Using pre-signatures drastically reduces the amount of memory that is required for message authentication in middleboxes. Only the small pre-signature must be buffered. The middlebox buffers the small pre-signature instead of the larger cleartext message. Moreover, pre-signatures reduce the amount of data written and read from the memory of the middlebox. Details on how middleboxes verify IHC signatures are provided in Section 8.3.6.

Cumulative transmission Many pre-signatures fit into one HIP control packet because of their small size. Therefore, many pre-signatures can be placed at the verifier side with one message. All of these pre-signatures are generated with the same undisclosed hash chain element. The verifier acknowledges the receipt of a cumulative pre-signature just as it would acknowledge a single pre-signature. The signer can now send all pre-signed messages in parallel. This increases the performance of the system.

Efficient retransmission A signature can only be verified if both packets from the signer to the verifier are transmitted without any errors. In the case of the first packet being erroneous, it must be retransmitted, while both packets must be retransmitted in the case of the second packet being erroneous. This means that the first packet requires to be retransmitted in every error case while the second packet only needs to be retransmitted when an error occurs in the second packet. Sending the actual message m in the first packet means that the large message needs to be retransmitted in both error cases. Placing m in the second packet reduces the probability of retransmissions of the potentially large message m.

Confidentiality The original IHC signature scheme places the cleartext message m in the first packet. However, the verifier cannot trust this message until it is verified by using the hash chain element disclosed in the second packet. An attacker, however, can read and use the message from the packet under the assumption that it is authentic. This assumption gives the attacker an advantage because it gains knowledge of the message and can react to the message before the verifier can use the message. The attacker can use this knowledge to foresee and disturb the actions of the verifier. Pre-signatures do not reveal the contents of the *message* until it can be verified. Hence, an attacker cannot use the interval before the message is verifiable to react to it. This is important whenever two or more parties compete for a limited resource. Prior knowledge of the message gives an attacker an unfair advantage. LHIP does not necessarily require confidentiality. However, it might turn out beneficial for interaction with certain kinds of LHIP-aware middleboxes.

Pre-acknowledgments Each signed message from a signer S to a verifier V requires a three-way IHC signature process. Sending a signed acknowledgment for a signed message requires another set of three packets. We reduce this overhead by introducing *pre-acknowledgments* (PACK) and *pre-negative acknowledgments* (PNACK). The pre-acknowledgments contain pre-signed acknowledgments likewise the pre-signatures contain pre-signed messages. They contain hashed information, enabling S to verify whether V has sent an acknowledgment or a negative acknowledgment.

Figure 8.7 depicts the signature process with pre-acknowledgments; V generates the PACK and PNACK value after it receives the pre-signature. At this point it is unclear whether the signature process will be successful or not. Therefore, it creates pre-signatures of acknowledgments for both cases: PACKs and PNACKs. It uses a constant number $const_{ack}$ and $const_{nack}$ as messages and creates pre-signatures with the next undisclosed hash chain element h_{i-1}^V. However, using ordinary pre-signatures for acknowledgments would mean that the acknowledgment as well as the negative acknowledgment would be signed. S could verify the PACK as well as the PNACK value. To avoid this, V includes a secret number $secret_{ack}$ in the PACK value. V includes a different secret number $secret_{nack}$ in the PNACK value. An attacker cannot forge these values as they contain an undisclosed element of the hash chain of S. The PACK and PNACK values are created as follows:

$$PACK = H(const_{ack} \parallel secret_{ack} \parallel h_{i-1}^V) \tag{8.1}$$

$$PNACK = H(const_{nack} \parallel secret_{nack} \parallel h_{i-1}^V) \tag{8.2}$$

V sends back both values in the second packet of the IHC signature process. S stores both values and waits for V to disclose its next hash chain element and the $secret_{ack}$ or the $secret_{nack}$ value. After receiving the message and verifying it, V sends back either an acknowledgment or an negative acknowledgment. It does so by V sending a packet containing h_{i-1}^V as the trigger value and $secret_{ack}$ or $secret_{nack}$, respectively.

S verifies the acknowledgment by using Formula 8.1 with the disclosed values. The constant values $const_{ack}$ and $const_{nack}$ are publicly known. The acknowledgment is valid if the result matches the PACK value sent in the second packet of the IHC signature process. The corresponding check is done for negative acknowledgments with Formula 8.2 in case $secret_{nack}$ is contained in the fourth packet.

All of the contents of the PACK and PNACK values ($secret_{ack}$ / $secret_{nack}$, $const_{ack}$ / $const_{nack}$, and h_{i+1}^V) have a fixed length to avoid length extension attacks on the hash function (cf. Section 2.4.3).

V must choose new values for $secret_{ack}$ and $secret_{nack}$ for every acknowledgment. Otherwise an attacker can use previously disclosed secrets to forge acknowledgments and negative acknowledgments. It could, for example, replace the acknowledgment secret in the fourth packet by a negative acknowledgment secret that was used and disclosed earlier. This secret would pass the verification test because V uses the same secrets for all PNACK values. Therefore, unique secrets are essential for the protocol to function properly.

Using PACK and PNACK values enables acknowledgments for signed messages with only one additional packet. They are efficient to compute and to verify as only few hash

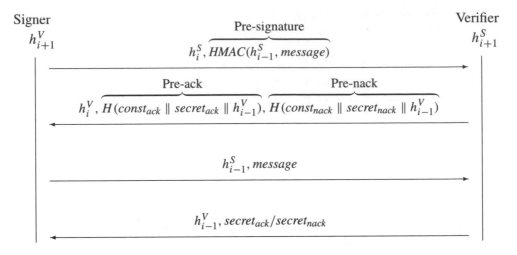

Figure 8.7 Interactive hash chain signatures with pre-signatures and pre-acknowledgments. The message is sent in the third packet instead of in the first packet.

operations are necessary. Their usage saves two packets and one RTT respectively, compared with signed acknowledgments without PACK and PNACK.

Retransmissions

Packet retransmissions are necessary for two reasons. First, IP packets can get lost on their path or arrive corrupted. This corruption is revealed by the checksum field, contained in the IP header. Erroneous packets are discarded and not processed. These packets must be retransmitted. Second, syntactically correct packets with valid checksum fields can carry invalid signatures. These invalid signatures are possibly caused by an attack. These errors require to start the signature process over.

The LHIP authentication layer must react differently to both situations. Simple packet loss requires to retransmit the last packet without using a new hash chain element. These retransmissions are always initiated by the signer. These retransmissions do not require using new hash chain elements because no further hash chain elements have been disclosed after the packet was lost.

Invalid signatures require restarting the signature process with a new PSIG packet that uses new hash chain elements. The invalidity of the signature is detected when the A2 packet is received. Either the S1 or the S2 packet is malicious. Therefore, both packets must be retransmitted. The old hash chain value of the first S1 packet cannot be reused as its preceding hash chain element was already disclosed in the S2 packet. As a consequence, the signer must use a new undisclosed hash chain element to re-start the signature process. This means that each failed signature requires the use of two new hash chain elements on the signer side (for S1 and S2) and the use of two new hash chain elements on the verifier side (for the A1 and A2). The number of retries for each communication interface must be limited to avoid hash chain exhaustion attacks. A sender should use another interface for resending the S1 packet if the limit is reached.

8.3.5   LHIP integration

The LHIP authentication layer is conceptually located below HIP. All HIP control packets are passed through the authentication layer. This layer determines whether a packet requires protection or not. This can be done by examining the parameters contained in the packets. Update packets containing new location information, new IP addresses, and new hash chain anchors need to be protected to avoid impersonation and connection disruption. Packets that require authentication are buffered and sent with the IHC signature scheme proposed earlier. The authentication layer on the remote peer verifies the hash chain signature and delivers the packet to the HIP instance. In contrast to protected packets, unprotected packets are sent directly. They are not processed by the authentication layer.

The authentication layer reuses the HIP packet format and parameter format to attach control information to HIP packets. This section describes the packets and parameters used by the authentication layer and the interaction with HIP HMAC signatures.

Pre-signature creation

The LHIP authentication layer mimics the properties of the HMAC signature in the HIP packets and, thus, provides a similar kind of authentication that protects the same part of the HIP packet. The HIP layer (the unmodified part of HIP) signs messages with the HMAC algorithm as it would without hash chain signatures. The authentication layer signs the contents of the HMAC parameter with IHC signatures. In the absence of a shared key, the HIP layer uses a fixed zero key for the HMAC creation. The HMAC is, therefore, not more than a message digest. Using this message digest as basis for the LHIP authentication has the advantage that the HIP layer can still use all mechanisms that rely on the presence of the HMAC signature. Signing the message digest instead of the message does not reduce the security of the IHC signatures. Using message digests is very common for most signature algorithms (cf. Section 2.4.3).

The authentication layer of the signer generates the pre-signature from the contents of the HMAC parameter and the next undisclosed element of its signature chain. It then uses the HMAC algorithm to generate the pre-signature using the following formula:

$$psig = HMAC\left(h_{i-1}^S, HMAC(0\ldots0, m)\right) \qquad (8.3)$$

The LHIP authentication layer on the receiving host verifies the IHC signature and passes the verified message to HIP. HIP can verify that the right portion of the message is covered by the HMAC signature and that the message digest is valid. It can verify the HMAC with the zero key.

Parameters

The LHIP authentication layer must exchange authentication data to provide its services. This data is carried by HIP packets and parameters. We define several new parameters for use with the authentication layer. These parameters carry the trigger values, the hash chain anchors during the base exchange, the pre-signatures, and the acknowledgments. All of these parameters use the TLV structure described in Section 4.1.2. The selection of the parameter type numbers reflects the position of the parameter in the HIP packet: all parameters in the packet must be in ascending order.

Hash Chain Anchor Parameter   The *Hash Chain Anchor Parameter* (HCAP) is a critical parameter that carries the anchor of a hash chain. These anchors are exchanged during the base exchange. A message may carry several HCAPs. Besides the HIP parameter header, the HCAP contains an 8-bit anchor type field and a variable bit field for the anchor value. The 8-bit type field determines whether the variable bit field is an anchor for the *signature chain*, *trigger chain*, or some other hash chain. The HCAP is located in the parameter type range that is signed with RSA and HMAC signatures. It is used during the base exchange and within UPDATE messages.

Hash Chain Value Parameter   The *Hash Chain Value Parameter* (HCVP) is also critical and carries the trigger values necessary for the IHC signature interactions. It is appended to all packets of the authentication layer. The parameter type number must be higher than any parameter type number used by the HIP peers because it is added after the HIP processing of the outgoing control messages.

Pre-Signature Parameter   The *Pre-Signature Parameter* (PSP) is used in the PSIG packet and contains the pre-signature of the message. It just consists of the HIP parameter header and a variable bit field for the hash function output. The PSP is appended behind the HCVP. This parameter is non-critical because it may be contained within R1 and I1 messages (cf. chained bootstrapping in Section 8.3.6).

PACK, PNACK, HACK and HNACK parameter   The PACK and the PNACK parameters carry the PACK and the PNACK values. The *Hash chain ACK* (HACK) and the *Hash chain NACK* (HNACK) parameter carry the positive and negative acknowledgments in the ACK packet. These parameters just consist of variable length field that holds the $secret_{ack}$ or $secret_{nack}$ hash value. All of these parameters are critical.

Further parameters   LHIP introduces some more parameters that are not directly used by the authentication layer. We will describe these parameters when they are used in this document.

Packet types

The authentication layer uses three new kinds of packets: the PSIG, TRIG and ACK packets. The MSG packet is the actual message with IHC signature parameters appended. Therefore, no special packet number is assigned for the MSG packet. The packet numbers are strictly preliminary.

PSIG packet   The PSIG packet contains the HIP header, the hash chain trigger encoded as HCVP, and the pre-signature encoded as PSP.

TRIG packet   The TRIG packet contains the HIP header, a hash chain trigger in an HCVP, and the PACK and PNACK parameters.

MSG packet The MSG packet is the HIP UPDATE packet requiring protection. The authentication layer adds a trigger encoded as HCVP to it. This trigger indicates that the packet is part of the signature process.

ACK packet The ACK packet also contains the HIP header and a hash chain trigger as HCVP. It contains either the HACK or the HNACK. The presence of a valid HACK indicates that the signature process was successful, while the presence of a valid HNACK indicates that the signature was invalid.

8.3.6   LHIP associations

The LHIP authentication layer provides protocol security for HIP. This section illustrates how both hosts establish an LHIP communication context and how LHIP ensures secure mobility updates. Other HIP specific protocol functionalities such as closing an LHIP association require protection as well. We show how LHIP secures this functionality.

LHIP base exchange

HIP uses the base exchange to create the HIP communication context, the so-called HIP association, on both hosts. Likewise, LHIP creates a communication context at the beginning of a communication. LHIP extends the HIP base exchange to establish the LHIP communication context. LHIP hosts generate the hash chains and exchange the hash chain anchors with the remote peer during the base exchange. Unlike the HIP base exchange, no public-key operations are performed during the LHIP base exchange.

The LHIP base exchange is kept similar to the HIP base exchange to maintain compatibility with HIP and to enable hybrid LHIP and HIP implementations. It consists of the same four-way handshake as the HIP base exchange. All HIP specific fields except the RSA and HMAC signatures are present in the LHIP base exchange.

LHIP does not change the semantics nor the functionality of HIP parameters such as the *puzzle*, or the *R1 generation counter*. However, LHIP extends the functionality of the *transform parameter* and uses the new LHIP parameters during the base exchange.

HIP transforms The fourth LHIP high-level goal is compatibility with HIP. Two LHIP-enabled HIP hosts need a way to agree on a HIP or LHIP association. This way should allow LHIP-enabled HIP hosts to communicate with pure HIP hosts by using HIP. A host that is not LHIP-enabled should always establish a HIP association with LHIP hosts. The choice of whether to use LHIP or HIP has to be negotiated during the HIP base exchange because the LHIP parameters must be exchanged during this stage.

The first base exchange packet, the I1 packet, is identical for both HIP and LHIP. The I1 only indicates that the Initiator requests to establish either a HIP or an LHIP association. The type of the association is not determined yet.

LHIP defines an LHIP transform suite ID. An LHIP-enabled Responder can express its preference by adding the LHIP transform suite ID to the HIP transform parameter in the R1 message (cf. Section 4.1.2). This LHIP transform suite ID indicates that the Responder supports LHIP for message authentication. The order of HIP transforms indicates whether the Responder prefers LHIP or HIP. The absence of the LHIP transform suite indicates that the host does not support LHIP or that it is not willing to establish an LHIP association for the selected HI.

The Initiator selects a HIP transform suite from the given transform suites and sends it back to the Responder. LHIP will be used if the Initiator selects the LHIP transform suite. An Initiator that does not support LHIP will not select the LHIP transform suite and ignore it. Such an Initiator will always use plain HIP associations.

This way of agreeing on a HIP or LHIP selection maintains full compatibility between LHIP-enabled and normal HIP hosts. No additional packets or parameters are needed to agree on an LHIP association.

Hash chain anchors   LHIP exchanges all hash chain anchors during the base exchange. These anchors are added to the I2 message and the R2 message. This enables the Responder to stay stateless until it receives the I2 message. Furthermore, this approach enables the Responder to use pre-created R1 messages.

Adding the Initiator's hash chain anchors to the I1 message would require the Responder to establish state for storing the anchors after receiving the I1 message. This is not desirable as the HIP I1-R1 mechanism enables the Responder to verify the routability of the Initiator's IP address before establishing state. Adding the hash chain anchors to the I1 message would make this stateless locator verification impossible. Consequently, we add the hash chain anchors to the I2 message.

Adding the hash chain anchors of the Responder to the R1 message would also mean that the Responder cannot pre-create the R1 message. Instead, it would be required to generate distinct hash chains for every R1 message. Using one pre-created message with only one set of hash chains is impossible because a hash chain must only be used for one LHIP association. Reusing a hash chain would render the IHC signature system insecure. Although, the hash chain creation is computationally inexpensive, computing new hash chains for every message would enable attacks with a storm of I1 messages. Therefore, we decided to add the hash chain anchors to the R2 message.

RSA and DSA signatures   The R1 message is the only message of the LHIP base exchange that must be signed with RSA or DSA for compatibility reasons with HIP. All other messages are not necessarily signed with public-key signatures. As a consequence, hosts cannot authenticate their peers. However, this is not a design goal for LHIP. This means that weak hosts do not need to calculate public-key signatures in most cases. Nevertheless, public-key signatures are necessary in case of namespace conflicts. Section 8.3.7 describes the possible attacks on LHIP and how public-key signatures are used to prevent these.

Hash chain signed base exchange   The four-way base exchange cannot be protected by the authentication layer because the hash anchors are exchanged in the last Initiator and in the last Responder messages. Using the authentication layer to sign these messages would require additional packets. Furthermore, it would not significantly increase the security because an MITM attacker could still replace the hash anchors transmitted during the base exchange.

A way to gain increased security is to use chained bootstrapping in a similar way to how WIMP does. Figure 8.8 illustrates the process. The hash chain anchors are denoted a_{sig} for the signature chain, a_{trig} for the trigger chain and a_x for further chains. The corresponding seed values are denoted s_{sig}, s_{trig} and sx for further hash chain seeds. Recall that the hash chain is calculated from the seed values.

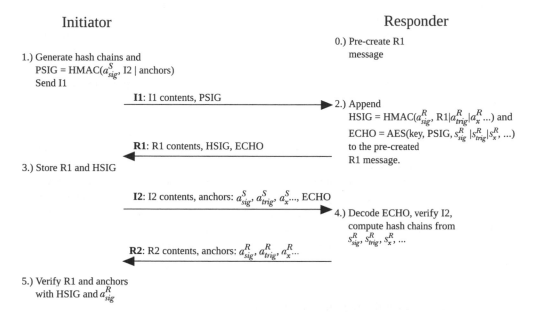

Figure 8.8 Chained bootstrapping for LHIP. The LHIP base exchange is depicted in a simplified form. Both peers sign their base exchange messages with IHC signatures.

The I1 can contain the pre-signature of the I2 packet. Placing the signature of the I2 packet in the I1 packet requires that the Initiator creates both packets at the same time. Fields in the I2 packet that depend on values contained in the R1 message must be left out of the signature. These values are not known at this point. The Responder buffers the pre-signature in the I1 packet and verifies the I2 message as soon as it arrives. The pre-signature of the I2 packet is created with the anchor value of the signature chain in the I2 message. The Responder can verify the I2 message with the pre-signature from the I1 message and the anchors in the I2 message. The chained bootstrapping mechanism ties together the I1 and I2 message and protects the hash chain anchors in the I2 message from manipulations. However, an MITM attacker can still modify the pre-signature and the hash chain anchor to forge packets.

Chained bootstrapping is also applicable to the R1 and the R2 message. The Responder can add an HMAC signature, created with the anchor value of its signature chain, to the R1 message. This signature can contain the contents of the R1 message and the hash chain anchors in the R2 message. The Initiator can verify the I1 message and the hash chain anchors after it has received the R2 message. This means that the Initiator must react to the unverified R1 message to continue the base exchange. Chaining the R1 and R2 message ensures that both messages have been sent by the same host and that neither the anchor values nor the contents of the R1 message have been modified. However, it is possible that an MITM attacker modifies both messages and forges R1 and R2 messages with different contents and anchor values.

Chained bootstrapping requires the Initiator and the Responder to compute their hash chain anchors in the first step of the base exchange. The Responder must store the pre-signature of the I1 packet until the I2 message arrives. Furthermore, it must create and store the hash chains used for the association before it sends the R1 message. It can either establish state or it can use the echo request field to store the values. To avoid state establishment after receiving the I1, the Responder can create an encrypted envelope with a symmetric encryption algorithm, such as AES. This envelope contains the seed values of the hash chains and the pre-signature from the I1 packet. The Responder sends this encrypted envelope to the Initiator but keeps the encryption key secret. The Initiator sends back the echo response in the I2 message. The Responder can open the envelope and use the pre-signature to verify the I2 message. It uses the seed values to re-compute the hash chains that were used for creating the signature in the R1 packet. This enables the verifier to stay stateless and still use IHC signatures for the base exchange.

Encrypting the seed values instead of the complete hash chains keeps the echo request parameter small. It enables the Responder to encrypt only single hash values instead of whole hash chains. However, this technique requires the Responder to calculate the hash chains twice: during the R1 and the R2 creation. Nevertheless, this is preferable to establishing state because the computation of short hash chains is inexpensive. However, attackers can force the Responder to generate hash chains for forged I1 packets

Diffie–Hellman and public-key signatures   The Diffie–Hellman and the RSA or DSA HIs are exchanged during the LHIP base exchange. However, the Diffie–Hellman shared secret is not calculated and the RSA or DSA identifier of the peer is not verified. These values are stored for later use during the upgrade process from LHIP to HIP.

Hybrid LHIP- and HIP-enabled Responders must sign the pre-created R1 message to maintain full HIP compatibility. However, this overhead is acceptable because only one signature is required for potentially many associations.

Pre-created hash chains   Hosts can speed up the connection establishment by pre-creating hash chains. Unlike the public-key signatures and the Diffie–Hellman key exchange, there is no need to calculate the hash chains during the base exchange. To further reduce the computation effort during the base exchange, a host can pre-calculate a pool of hash chains. The host takes hash chains from the pool whenever it establishes a new association. The hosts refill their hash chain pools whenever they have CPU cycles to spare.

HIP payload   Like HIP, LHIP requires to utilize a tunneling protocol to transfer its payload. Currently, the only transfer protocol being specified is IPsec ESP. LHIP also uses ESP to sustain the compatibility with HIP and HIP-aware middleboxes although the ESP traffic cannot be encrypted securely due to the missing shared secret. Thus, LHIP hosts use the ESP NULL mode for packet encapsulation and a fixed zero-string for packet authentication[5]. This way of operating IPsec neither encrypts nor authenticates the ESP packet but still performs all operations required to tunnel the payload to the destination host.

---

[5]NULL mode for authentication and ESP encapsulation conflicts with Manral (2007).

Concluding the LHIP base exchange At the end of the LHIP base exchange, both hosts have agreed to use LHIP and both hosts have exchanged their hash chain anchors. An unencrypted ESP BEET tunnel is established and the hosts can exchange payloads via this tunnel.

Both hosts are able to send authenticated UPDATE messages that an attacker cannot unnoticeably modify. Both hosts have stored the Diffie–Hellman keys and HIs of their peer for later use.

Neither of the hosts has computed RSA signatures, nor done RSA signature verification, nor the Diffie–Hellman shared secret computation during the base exchange. This means that both hosts save one RSA signature, one RSA signature verifications, and one Diffie–Hellman shared key computation.

Mobility and multihoming signaling

LHIP uses the new authentication layer to protect mobility and multihoming updates. LHIP does not modify the basic update process nor does it replace the update messages with other packets. This maintains compatibility with HIP-aware middleboxes and allows us to reuse the HIP update functionality, including present and future extensions without modifications.

The LHIP authentication layer protects the first pair of HIP UPDATE messages. The U1 and U2 messages need protection because they contain the new location information and the SPI values of the peers. Otherwise, an attacker can disrupt an LHIP association or reroute traffic if these messages are unprotected. Such an attacker does not necessarily have to be located on the communication path. Guessing the HIs, SPIs, and IP addresses of the peers would be sufficient to perform an attack. Therefore, these messages require authentication.

Unprotected third update message The third UPDATE message serves as a response to the return routability test. It contains an *ack* parameter and the *echo response* parameter. This message does not need protection. To show that this decision does not influence the security of LHIP we assume two kinds of attackers: eavesdroppers who are not on the communication path and MITM attackers on the communication path.

We first analyze the case of an MITM attack. An MITM attacker can either drop or modify the message. An attacker cannot modify the message without invalidating it because it only contains an acknowledgment and the echo response. The receiver can easily discover modifications of these parameters.

Both actions, modification, and destruction, therefore, lead to the same result: the path cannot be verified. A signature cannot protect against this attack because it can only identify the manipulation, not prevent it.

An eavesdropper can send a forged packet with invalid contents but it cannot drop the actual UPDATE packet. This means that an eavesdropper can possibly deliver a forged UPDATE message before the legitimate UPDATE packet reaches the host. However, the receiver can ignore invalid messages and wait for a packet with a valid echo response. It can use timeouts to mark a locator as unroutable if it receives no valid response within a certain time span.

We decided not to protect the third message since a host will notice manipulations even without signatures. This saves three IHC signature packets that would be required to protect the message. Note that further HIP extensions might need to send protected data in the third UPDATE packet. Such extensions would also need to redefine the LHIP behavior.

LHIP update process   The three-way HIP update process is extended by the packets for IHC signature authentication. Two full IHC signature authentication cycles are added for the packets carrying new locator information and ESP parameters. This makes the update process a nine-way update process. However, the ACK packet and the following packet are sent in parallel. This reduces the transmission time required for the update process to 3.5 RTTs.

Figure 8.9 illustrates the process. Both hosts exchange UPDATE packets. These MSG packets, specifically the U1 and U2 packets in the HIP update process, are preceded and trailed by IHC signature packets. The last UPDATE packet is unauthenticated.

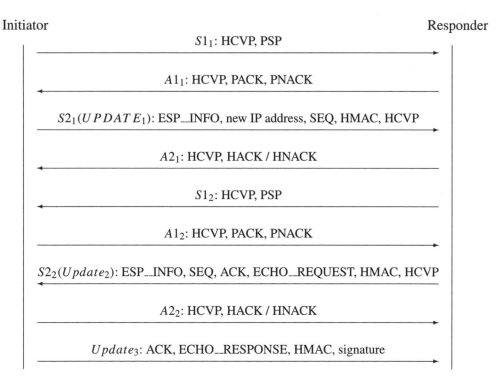

Figure 8.9   The LHIP update process is protected by IHC signatures. The original update packets are marked with U1, U2, and U3. The U1 and U2 packets are the MSG packets of our IHC signature scheme.

In contrast to HIP, neither RSA nor DSA signatures need to be calculated because the middleboxes can use the IHC signatures to authenticate the packets. Therefore, the LHIP update process reduces the computational overhead of HIP updates although it increases the communication overhead.

Updating hash chain anchors   Hash chains are finite sequences. This means that a host might run out of hash chain elements after some time. In this case, a new hash chain anchor must be transmitted to its peer. This hash chain anchor must be signed to avoid identity theft during the lifetime of an LHIP association.

LHIP uses UPDATE messages to exchange new hash chain anchors. The anchors can either be piggy-backed on UPDATE packets used to propagate new IP addresses or be sent in dedicated UPDATE packets. LHIP uses the IHC signature scheme to authenticate the packets that carry the anchors. A host that receives a new anchor from its peer stores the anchor and waits until its peer decides to use the new hash chain.

A host does not explicitly announce that it begins using the new hash chain. Instead it simply uses the first undisclosed hash chain element of the new chain in a new authentication cycle. Its peer will notice that the HCVP parameter in the packet is invalid when it verifies it with the last known hash value of the peer's old hash chain. In this case, the receiver of the packet attempts to verify the HCVP parameter with the newly announced anchor value. It continues to use the new hash chain values if the verification succeeds.

This implicit way of hash chain activation requires no additional packets or parameters. The receiver calculates two hashes instead of one hash if the HCVP verification fails. This allows the hosts to send hash chain anchors to the peers at any time and to activate them any time it becomes necessary.

LHIP interaction with middleboxes

HIP UPDATE messages are processed not only by the communicating peers but also by middleboxes on the communication path. These middleboxes learn about new IP addresses and SPI values by inspecting the UPDATE messages. HIP uses RSA signatures to sign these messages. This enables middleboxes to filter out forged messages that aim at disrupting a HIP association by modifying the state of the middlebox.

The middleboxes cannot use the HMAC signatures for message verification because they are not in possession of the shared secret. IHC signatures do not rely on a shared secret. Thus, the middleboxes can use these signatures to verify HIP UPDATE messages without using costly public-key operations.

Update verification   A middlebox learns about the update packet by observing the trigger values in the packets. Assume host A and host B have established an LHIP association. A middlebox M learns the hash chain anchors of both hosts during the base exchange or during the update process. It learns of the anchor value h_{i+1}^A of the signature chain of host A and the anchor value h_{i+1}^B of the trigger chain of host B. Host A sends an IHC signature protected UPDATE packet when it changes its position. M has to learn about the new IP address of host A during the update process. M observes the first PSIG packet from A to B. It contains the next undisclosed element h_i^A of the signature chain of A. It can verify this PSIG packet by comparing $H(h_i^A)$ with h_{i+1}^A. It buffers the pre-signature and waits for the TRIG packet from host B.

The TRIG packet from host B to host A indicates that host B has received the pre-signature. M can verify the origin of the TRIG packet with the HCVP parameter contained in it. Eventually, host A sends the MSG packet. This packet contains the new locator of host

A and h_{i-1}^A. The middlebox can verify the MSG packet and therefore the authenticity of the new IP address and update its mapping between HITs and IP addresses accordingly.

Middlebox state update   Middleboxes learn of the hash chain elements during the base exchange and during updates. Figure 8.10 illustrates the process. Host A and host B have established an LHIP association and exchanged their anchor values. Middlebox 1 stores these values to verify packets later on. Host A changes its location and sends an update from a new IP address. Middlebox 1 can verify the hash chain value in the PSIG packet. It stores the new hash chain values during the update process. Host A has moved behind Middlebox 1. This middlebox allows outgoing traffic from A and stores the hash chain values during the update process. It can verify PSIG messages from B whenever B moves to another location.

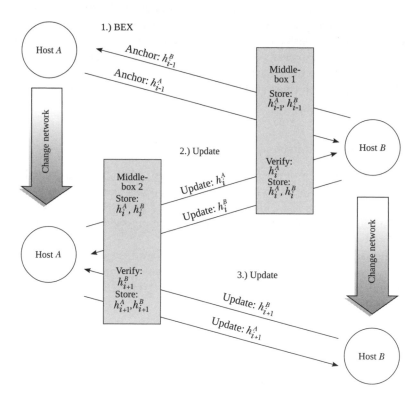

Figure 8.10 Middleboxes learn the hash chain elements during the base exchange and during updates.

Middleboxes must buffer the first message of the IHC signature until the message with the corresponding hash chain element arrives. The pre-signatures, proposed in Section 8.3.4, keep this first message small, making it easier for middleboxes to handle IHC signatures.

The middlebox must let PSIG and TRIG packets pass through without verification of the source IP addresses contained in them. However, it should check that only a limited number of PSIG and TRIG packets are sent to the same IP address to avoid DoS attacks with these packets. It should also check that these packets contain no payload besides the HCVP, the PSP, the PACK, and the PNACK parameters and should drop packets that contain further parameters to prevent attacks.

Closing an LHIP association

HIP uses CLOSE messages to indicate that a HIP association should be closed. Both hosts delete the state bound to the communication context and tear down the BEET tunnel. HIP CLOSE messages are protected by HMACs and public-key signatures.

LHIP must naturally protect this message to prevent attackers from closing an LHIP association with forged CLOSE messages. A naive approach would be to protect the closing procedure with IHC signatures. This would make the two-way close process an eight-way IHC-signed process. However, this is not necessary for CLOSE messages.

CLOSE messages only transfer one piece of binary information. Either a host requires to close a HIP association or not. We can use an approach which is similar to Lamport's one-time signatures to secure these messages. We exchange a short hash chain that only consists of the random seed value r and the anchor value h_1. This hash chain is called a *close chain*. We call hash chains that only consist of two elements *binary hash chains*.

Each host publishes the anchor of this binary chain during the LHIP base exchange. The host discloses the seed value r whenever it decides to close the LHIP connection. Its peer can verify the CLOSE message by comparing $H(r)$ with the anchor of the close chain h_1. A host can be sure that its peer requests to close the HIP association if those values match.

It sends back a CLOSE message containing an *ack* parameter and the seed of its own close chain. Thus, both hosts can verify that their peer is closing the LHIP association.

This procedure does not require additional packets nor the use of public-key signatures. Middleboxes that have observed the base exchange are also able to authenticate the close messages. However, middleboxes that have not observed the base exchange cannot authenticate CLOSE messages. They have not learned the anchor of the close chain because the base exchange packets were routed over a different communication path. They must use conventional approaches, e.g., timeouts for inactive associations, to determine when to delete the state allocated for the communicating peers. An alternative way to inform middleboxes about the anchor of the close chain is to include this anchor in the signed part of every UPDATE message. Accordingly, even middleboxes that have not observed the base exchange are able to verify CLOSE messages.

8.3.7 Security considerations

The third high-level goal of LHIP is namespace security. We stated that it is not necessary for LHIP to authenticate hosts. The namespace must, nevertheless, be protected from malicious use. HIP verifies the authenticity of the HIs and, therefore, the public-key signatures during the base exchange. LHIP reduces the computational overhead by omitting these checks. This section illustrates attacks on the host identity namespace that are possible when host identifiers are not verified and it concludes with defenses against these attacks.

Attacks without namespace protection

The attack vectors opened by the LHIP design presented so far can be used by an attacker for impersonation and DoS attacks. The attacks build on the fact that the HIs of LHIP hosts are not verified during the base exchange. For the attack to work, the attacker needs *a priori knowledge* of HIP associations before they are established. The attacker must learn with which host the victim will establish an LHIP communication context before the base exchange takes place. This knowledge can be gained by communication pattern analysis of victim hosts. E-mail communication, automated update services, instant messaging, or shared folders and printers are typical examples of frequent communication between the same hosts and can, therefore, be subject to the attack.

HI blocking attack   The identity blocking attack exploits the fact that HIP and LHIP use a pair of HIs to identify a HIP context. This means that only one HIP or LHIP association can exist for every pair of HIs. This situation does not lead to problems for plain HIP because only a host that is in possession of the private keys is able to establish a HIP association. The host identity is statistically unique and collisions between HIs are expected to be extremely unlikely. In contrast to HIP, LHIP does not use RSA signatures to authenticate the identities of the hosts during the base exchange. Therefore, every host can impersonate any other LHIP host. However, host authentication is not a goal of LHIP. An attacker exploits the fact that only one LHIP association can be assigned to a pair of host identifiers. The attacker blocks connection attempts of a victim by establishing a HIP association with the victim's host identity before the legitimate host does.

This example shows that attackers can misuse the unprotected host identity namespace for attacks that surpass simple impersonation. These attacks are worse than today's attacks on the IP or TCP protocol and must be dealt with.

The use of public-key signatures in case of conflicts prevents the attack discussed above. For every I1 message in the base exchange, the Responder checks whether a HIP association exists that uses the HIs contained in the I1. If that is the case, the Responder requires the Initiator to sign its I2 message with its HI. If the signature is valid, the Responder discards the existing LHIP or HIP association and establishes a new association with the Initiator.

Accordingly, legitimate hosts are protected from namespace attacks. Moreover, legitimate hosts can prove their identity to servers that have already established an association with an attacker. This policy does not introduce a performance overhead in the common case of legitimate communication because the use of public-key signatures is only necessary in the case of conflicts.

The presented approach opens the following attack vector: attackers can try to provoke host identity collisions and force a server to verify the HIs of spoofed I2 packets. However, when a server suspects that it is under attack, it is able to force the client to solve a cryptographic puzzle by utilizing the HIP puzzle mechanism.

HI stealing attack   An attacker with a priori knowledge of the peers which an LHIP host will connect to can mount a second kind of attack. The attacker exploits the fact that a host that has already established a HIP association with another host would reuse this association for all further communication between the hosts.

An attacker A can impersonate a server S by using the server's host identity HI_S. The attacker cannot impersonate the server in a global scope without influence on the name resolution infrastructure but it can spoof the HIT of the server and establish an LHIP association with a victim V. This circumstance is problematic when a process on the victim tries to open an LHIP association with S. The LHIP layer reuses the LHIP association with HI_S and sends the packets to the attacker. This allows the attacker to circumvent the name resolution infrastructure and to impersonate other hosts. This is problematic whenever the Initiator assumes that the name resolution infrastructure is trustworthy.

This attack is possible because the host neither checks HIs for incoming nor for outgoing traffic and one HIP association is used for all traffic between two hosts. Consequently, an attacker can open an LHIP association without proving its identity and circumvent the host identity resolution infrastructure. Authenticating either the HIs of peers for outgoing or incoming requests for LHIP associations prevents this attack. A host that typically acts as a client should authenticate the HIs of its peer whenever it acts as a Responder to prevent the HIT stealing attack. Consequently, a server should authenticate the HIs of its peers whenever it acts as Initiator. This increases the computational overhead of LHIP. However, in most cases clients do not act as Responders while servers typically do not act as Initiators. Thus, clients use unauthenticated LHIP associations for outgoing traffic and servers use unauthenticated LHIP associations for incoming traffic.

Defense mechanisms

The beginning of this section shows that the host identity namespace is vulnerable to attacks if host identifiers are used without verification. Conditional use of RSA for host identity verification prevents these attacks. LHIP must provide a way for hosts that discover namespace conflicts to inform their peers that they must authenticate. We introduce mechanisms for conditional authentication and discuss security implications in the remainder of this section.

LHIP flags   We introduce the *LHIP flags parameter* (LFP), which holds information about an LHIP association. It expresses whether a host requires authentication from its peer and whether it is willing to authenticate itself. During the base exchange, the Responder adds the LFP to the R1 message. It sets the *authentication required* flag when the Initiator must authenticate. It sets the *authentication granted* flag when it is willing to authenticate to the other host.

If requested, the Initiator uses RSA to sign its I2 message to continue the LHIP base exchange. The Initiator adds an LFP, indicating whether the Responder should authenticate or not, to the I2 message. The Responder must prove its identity by using RSA to sign the R2 message if requested by the Initiator.

The base exchange fails if a host that is required to authenticate is not willing to authenticate. In this case, opening an LHIP association is not possible. The Initiator can determine this situation on receipt of the R1 and proceed with a regular HIP base exchange alternatively. Attackers can exploit the conditional signatures to mount DoS attacks targeted at the CPU resources of a host. Therefore, a host may deny to use RSA according to its policies. It expresses this denial by setting the according bit in the LHIP flags parameter to 0.

The LHIP flags are added to the R1 message after all other parameters because the parameter is appended to the pre-created R1 packet. We, therefore, assign 63404 as the parameter type number. The even number indicates that the parameter is not critical. We decided to use a non-critical parameter that is ignored by non LHIP implementations.

Prevention of replay attacks   RSA signatures are susceptible to replay attacks if the signed part of the message does not contain data that is unique for every message. An attacker can exploit this weakness by replaying such old LHIP I2 and R2 messages. The Diffie–Hellman key exchange prevents such replay attacks for HIP. However, the Diffie–Hellman shared key is not generated by LHIP hosts. The Responder cannot distinguish replayed LHIP I2 and R2 messages from legitimate messages if these messages do not contain unique data. Therefore it will discard the old HIP association and establish a new HIP association with the attacker.

Including a unique echo request parameter in the R1 and I2 messages prevents this attack. On receipt of the signed echo response, the Responder checks that no other LHIP association has been established with the same echo response before. The echo request parameter allows the Responder to hide information in it. This supports the Responder in verifying the uniqueness of the echo response.

Authenticated hash chains   The RSA signatures being used to authenticate a host also cover the hash chain anchors of the host. Consequently, the hash chain is tied to the identity of a host. The hash chain can be used to authenticate the host later on. The peer and the middleboxes on the path can verify that updates and close messages were sent by a legitimate owner of an HI. This authenticated LHIP mode has the same namespace properties as HIP. The identity of the host can be authenticated unambiguously. It is not possible to impersonate the host, neither during the base exchange nor during the lifetime of an LHIP association.

8.3.8   Association upgrades: from LHIP to HIP

One design goal of LHIP is upgradability to HIP. An LHIP association should be transformable to a HIP association without the need to close and reestablish the association. To upgrade from LHIP to HIP, both hosts must calculate the shared secret, generate the necessary keys from it, verify their HIs, and replace the NULL mode encrypted BEET tunnel with an encrypted BEET tunnel. Nearly all information that is needed to complete the upgrade is already exchanged during the base exchange. This section describes which modifications to the base exchange are necessary, how the upgrade is signalled, and how it is performed.

Second HIP transform

The HIP transforms indicate algorithms used during a HIP association. The selected HIP transform defines the hash functions used for the HMAC algorithm as well as the symmetric key algorithms used for ESP encapsulation. The LHIP transform suite indicates that a host will use the lightweight authentication layer. The suite also determines which hash function should be used by the authentication layer. In HIP, the Initiator of the base exchange selects one HIP transform suite from the selection of transform suites offered by the Responder.

In order to upgrade an LHIP association, both hosts must agree on a transform suite to be used after the upgrade. LHIP hosts use the transform parameter to agree on two choices:

an LHIP transform suite and a HIP transform suite. An LHIP-enabled host that wants to use upgradable LHIP sends back the LHIP transform as the first choice and a HIP transform as the second choice. Both hosts store the selected HIP transform and use it during the upgrade process.

Upgradable associations

An LHIP association is upgradable only if both hosts have exchanged their Diffie–Hellman parameters, the second HIP transform, and an echo request parameter during the base exchange. Furthermore, both hosts must be willing to upgrade the association depending on local policies. This willingness is expressed by setting the *upgradable flag* in the LHIP flags parameter (cf. Section 8.3.7). The connection is upgradable when all necessary parameters have been exchanged and both hosts express their willingness.

Upgrade process

Upgrades are necessary if an application opens a socket, for example, and requires authentication and encryption for this socket. An existing LHIP association with the socket's destination host can be updated in that case. Another case is that local policies or a change of location might cause an application to require increased security.

The upgrade process is a two-way message exchange. We use HIP UPDATE messages to carry the upgrade information. We refer to these UPDATE messages as UPGRADE messages though we do not define a new message type for these. We refer to the host that initiates the upgrade process as *Initiator* and its peer as *Responder*. Figure 8.11 illustrates the upgrade process.

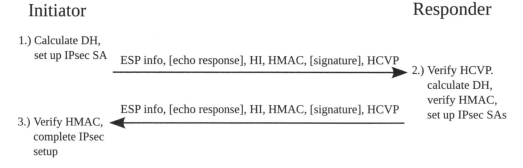

Figure 8.11 The LHIP upgrade process. A host has to add the echo response parameter and the RSA signature if it has not already authenticated itself during the LHIP base exchange.

LHIP uses RSA and DSA signatures during the UPGRADE process. The two-way upgrade process is not protected by client puzzles. Attackers could open numerous LHIP associations and upgrade them at the same time to put a heavy load on the Responder. It is an open question how such DoS attacks can be prevented. Exchanging a fresh puzzle before the upgrade would require two additional messages, which would slow down the upgrade process. Using the puzzles from the base exchange gives the attacker enough time to compute the puzzles before performing the attack. However, hosts can use the LHIP flags to state that they are not supporting the LHIP upgrade process. In this case, the upgrade process consists of establishing a new HIP association through a standard HIP base exchange.

First upgrade message   Before initiating an upgrade, the Initiator calculates the Diffie–Hellman shared secret based solely on the Diffie–Hellman keys exchanged during the base exchange. It generates the necessary symmetric keys and the keys for the HMAC creation from this shared secret. It sets up new IPsec security policies, using the algorithms selected in the second HIP transform. It also sets up a new outgoing IPsec security association with the symmetric keys.

The Initiator prepares an UPGRADE message containing the SPI value of the new IPsec SA and adds an HMAC to protect the SPI values by using the freshly created key. The host must use its private RSA key to prove its identity by including the echo response from the base exchange before the HMAC and the RSA signatures are applied. A host can skip this authentication if it has authenticated during the base exchange.

Role of the echo response   The echo response protects the UPGRADE progress from replay attacks (cf. Section 8.3.7). It ensures that the RSA signature was created for the current LHIP association and that the UPGRADE message is not a replay of a former LHIP UPGRADE exchange. Accordingly, the Responder can verify that the Initiator is in possession of the secret corresponding to the host identity.

Role of the HMAC, RSA signature ordering   The fact that the HMAC signature is included in the RSA signature prevents the following kind of attack. We assume that the HMAC was added after the RSA signature and it would not be part of the signature to illustrate this attack.

An MITM attacker replaces the Diffie–Hellman parameter during the base exchange. Neither the Initiator nor the Responder notices this modification because the Diffie–Hellman parameter is not protected by signatures.

During the LHIP upgrade, the Initiator signs the upgrade message and appends the HMAC signature. The MITM attacker then replaces this HMAC signature with the HMAC signature generated with its own Diffie–Hellman shared key. The Responder would use the attacker's Diffie–Hellman shared key to calculate the shared secret and would verify the RSA signature and the HMAC. Consequently, the Responder accepts the packets of the attacker instead of the Initiator as authentic.

The order of the HMAC signature and the RSA signature prevents this kind of identity theft. The HMAC signature is included in the RSA signature and cannot be replaced by an attacker. This ties the HMAC key and, consequently, the Diffie–Hellman public key of the Initiator to its HI.

Protection of the first upgrade message

The UPGRADE message triggers the Diffie–Hellman shared key calculation at the Responder. This operation is CPU-intensive and can be used in an attack to provoke the Diffie–Hellman calculations with a forged message. Similar to the HIP CLOSE message, the UPGRADE message only triggers the update and does not contain any explicit information not being protected by the HMAC. Therefore, it is sufficient to use the same protection mechanism as for CLOSE messages (see Section 8.3.6). Both peers exchange the hash chain anchor of a binary hash chain, the *upgrade chain*, during the base exchange. The disclosure of the next hash chain element of this hash chain indicates that a host has initiated the LHIP upgrade process. The disclosed upgrade chain element is attached as HCVP.

The Responder first checks the HCVP parameter in the message. It hashes it and compares it with the anchor value of the upgrade chain. This comparison is inexpensive and proves that the legitimate peer initiated the upgrade process. It calculates the Diffie–Hellman shared secret if the hash chain verification is successful. The Responder generates the symmetric keys and the key for the HMAC signatures from the shared secrets.

The Responder can verify the HMAC signature in the UPGRADE message by using the freshly calculated keys. The Responder modifies the IPsec security policies to use the encryption algorithms selected in the second HIP transform. It then sets up the outgoing and incoming IPsec SAs corresponding to the remote peer's SPI and the symmetric keys.

It is necessary to include the HIs of the peers in the UPGRADE messages to support HIP-aware middleboxes. The HIs in the packets enable these middleboxes to learn the public keys of the hosts, in case the middlebox has not observed the base exchange. This is the case if one of the hosts has moved behind the middlebox during the LHIP association.

Concluding the upgrade   The Responder sends back an HMAC-protected UPDATE message containing its SPI value for the incoming IPsec SA. The Responder must authenticate, using its RSA or DSA private key, if it has not authenticated during the LHIP base exchange. It must also include the echo response from the LHIP base exchange to avoid replay attacks. The Initiator can use the HMAC to verify the UPDATE message. It sets up its outgoing SA with the SPI given by the remote peer. At this point, both peers have upgraded to HIP and established an encrypted BEET tunnel. They do not use the LHIP authentication layer any more.

The LHIP upgrade only requires two messages. The Diffie–Hellman shared secret and the RSA signatures can be pre-calculated by a host to speed up the upgrade process. In this case, the upgrade is inexpensive and can be executed with little computational overhead.

The first and the second UPGRADE packets are symmetric. This prevents problems when both hosts decide to send an UPGRADE packet at the same time. The fact that both UPGRADE packets contain identical parameters solves this problem. Each host will take the UPGRADE packet of its peer as the reply to its own packet. Therefore, no additional measures are necessary to cope with simultaneously sent UPGRADE packets.

HIP downgrade   A downgrade from HIP to LHIP is not desirable because both hosts are already in possession of the shared secret that enables efficient message authentication and symmetric encryption.

The computational cost of an established HIP association could, nevertheless, be lowered by using NULL encryption instead of symmetric encryption for HIP. The runtime modification of a HIP association is out of scope for LHIP.

However, the hash-chain-based update process might be of interest for middleboxes and weak hosts because of its low computational overhead. Such a mixed HIP and LHIP mode is future work.

8.4   LHIP performance

In this section, we briefly summarize the performance properties of LHIP. We used the HIP for Linux implementation and our LHIP-enabled implementation to compare the performance of HIP with the performance of LHIP. We measured the time consumed by relevant processes in the hipd. We mainly focused on the packet handling and cryptographic functions. The measurements show how much time is spent on processing for the base exchange and the update process.

We measured on the same systems that have been used for the HIP performance evaluation in Chapter 7. In practice, the N770 typically acts as a client and, therefore, we used the N770 as Initiator in our tests. We repeated the tests with the 3.2 GHz Xeon to provide a basis for comparisons. A 3.0 GHz Xeon server acted as the Responder in all measurements.

The N770 was connected to the Xeon server via a 802.11/b Wi-Fi network. Due to the high variance of wireless channels, measurements that include packet transmission over the wireless link are less precise and show a higher variance than the measurements that only include operations on the hosts.

8.4.1   LHIP base exchange

We first present our time measurements for the LHIP base exchange and its individual operations. We measure three different categories of processes:

Cryptographic computations performed on the hosts to generate keys or to sign and verify messages.

Protocol processing that is performed to process packet parameters, send packets, and to modify the state of a HIP or LHIP association.

Packet processing that contains all operations required to process a packet including cryptographic computations and other protocol processing.

Total processing time contains packet processing, packet transmission time and other delays caused by external influences.

Figure 8.12 depicts the time measurements for the LHIP base exchange. The total duration of the base exchange is 14.56 ms. This low delay is achieved by removing all asymmetric cryptography from the base exchange. The hash chains for the IHC signatures have been pre-calculated before the base exchange. On-line computation of the hash chains delays the LHIP base exchange by 1.17 ms (8.0%) on the N770 Initiator and by 0.07 ms (0.5%) on the Responder (3.0 GHz Xeon). The computational cost of the LHIP base exchange is basically reduced to the time required for packet transmission and protocol processing.

The impact of LHIP processing on the initial delay of a connection is low for typical Internet RTTs. For instance, with RTTs of 50 ms, the duration for computations on the hosts is 6% (7 ms) of the total time of the base exchange for LHIP (109 ms).

LHIP cuts down the time for packet processing during the base exchange to 6.1 ms processing time on the N770. This time contains the operations absolutely necessary to establish the HIP state on both hosts and to set up the IPsec tunnel. LHIP leaves little potential for further optimization as decreasing the time for packet processing from 6 to 3 ms

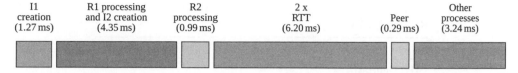

| I1 creation (1.27 ms) | R1 processing and I2 creation (4.35 ms) | R2 processing (0.99 ms) | 2 x RTT (6.20 ms) | Peer (0.29 ms) | Other processes (3.24 ms) |

Figure 8.12 The LHIP base exchange between the N770 and a 3.0 GHz Xeon server.

would only lead to marginal improvements, even on weak devices such as the N770. A main design goal of LHIP was to reduce the overhead of the public-key operations during the base exchange. Our LHIP design has succeeded in this point as it enables LHIP to completely remove all public-key operations from the base exchange without introducing other CPU-intensive processing steps.

The base exchange performance improvement in LHIP is also important because a HIP association is established with every peer a HIP host communicates with. Introducing a long delay for every association is problematic especially when it is unclear whether the services of HIP, namely authentication, encryption, and mobility support, are required or not. The demand for these services is typically not clear before the association is established. Developing reliable heuristics to determine the need for these services before establishing an association is difficult. Requiring user interaction for each association reduces usability of HIP. Therefore, the most practical solutions are to use HIP for all associations or for none. The LHIP base exchange allows letting the applications determine whether encryption and authentication is required after the base exchange and provides mobility and multihoming support with low computational overhead.

Note that LHIP provides less security than HIP. The given values assume that none of the hosts has to authenticate during the LHIP handshake. Host authentication delays the I2 or R2 generation by the time that is required to compute the RSA signature. The I2 and R2 processing are delayed by the time to verify the RSA signature.

8.4.2 LHIP update

The LHIP authentication layer adds preceding and trailing IHC signature messages to every HIP UPDATE message to authenticate these messages. This means that more packets must be processed and transmitted. This section will determine the overhead of the improved IHC signature scheme and discuss how the IHC signatures perform in different scenarios.

Each HIP UPDATE message is accompanied by three IHC signature packets: the S1, A1, and A2 packets. The IHC signatures replace the RSA and DSA signatures in the update packets. Therefore, we compare IHC signatures with RSA signatures. We measured the processing time of IHC signature packets on the Nokia N770 and a 3.2 GHz server. We used the *Internet Table OS 2005* on the N770 for our measurements. Table 8.1 summarizes the results. The results are the mean durations derived from 150 IHC signatures during LHIP updates. The measurements for RSA 1024, RSA 1536, DSA 1024, and DSA 1536 are also depicted to allow comparisons with RSA and DSA.

The IHC signatures replace the RSA and DSA signatures in the update packets. Therefore, we compare the efficiency of IHC signatures with the efficiency of these public-key signatures. The IHC signatures only consume 1.2% of the processing time of a 1024-bit

Table 8.1   Performance of IHC signatures compared with RSA and DSA.

	N770	Xeon 3.2 GHz
IHC signatures		
Send PSIG	0.33 ms	0.03 ms
Process PSIG, send TRIG	1.47 ms	0.05 ms
Process TRIG, send MSG	1.52 ms	0.05 ms
Verify MSG, send ACK	1.60 ms	0.05 ms
Process ACK	0.49 ms	0.05 ms
Sender (total)	2.34 ms	0.13 ms
Receiver (total)	3.07 ms	0.10 ms
RSA and DSA signatures		
RSA 1024 sign	181.32 ms	9.09 ms
RSA 1536 sign	578.70 ms	27.03 ms
DSA 1024 sign	96.71 ms	1.34 ms
DSA 1536 sign	229.27 ms	2.18 ms
RSA 1024 verify	10.53 ms	0.15 ms
RSA 1536 verify	23.47 ms	0.23 ms
DSA 1024 verify	118.73 ms	1.61 ms
DSA 1536 verify	287.76 ms	2.81 ms

RSA signature and 2.5% of the processing time of a 1024-bit DSA signature on the tablet. Even the N770 can perform all computations necessary to send an IHC signed message in less than 2.5 ms. It can receive IHC signed messages with a computational overhead of 3.07 ms.

Depending on the application scenario, hosts may open HIP or LHIP associations to several peers. Mobile hosts have to update all of these associations whenever they move. Assuming that the N770 uses a 1024-bit RSA host identity and has 10 open associations to other HIP hosts, it must compute 20 RSA signatures to send the first UPDATE messages and the second UPDATE messages to all peers. This results in a computational cost of 3626.4 ms for processing the RSA signatures for 10 updates. In contrast to RSA signed updates, transmitting 10 IHC signed updates consisting of an IHC signed message from the Initiator to the Responder and one IHC signed message from the Responder to the Initiator[6] causes a computational cost of 55 ms for packet processing on the N770 and 2.3 ms on the Xeon host. This reduces the processing time for signing LHIP update messages to 1.5% of the computational cost of HIP with 1024-bit RSA HIs. Therefore LHIP achieves the high level goal *LH1 – performance* for the update process.

IHC signatures significantly reduce the time for signing UPDATE messages on weak hosts. However, they introduce an RTT dependent delay due to the necessary interaction. As RSA, DSA, and HMAC signatures do not require interaction, these signatures can be delivered with a single transmission. IHC signatures delay the message delivery by one additional RTT because the PSIG and TRIG packets need to be exchanged before the actual message is sent. The efficiency of IHC signatures, therefore, mainly depends on the latency

---

[6]The third LHIP UPDATE packet is not protected because modifications would result in an obvious mismatch between ECHO_REQUEST and ECHO_RESPONSE.

of a link between two peers. The IHC signature process is faster than the RSA and DSA protected UPDATE messages if the time for processing the IHC signature on both peers plus one RTT is lower than the time for the RSA or DSA signature creation[7]. The PSIG creation and the TRIG handling on the N770 consumes 1.85 ms processing time. The PSIG handling on the 3.0 GHz server took 0.05 ms. Therefore, sending IHC signed signatures from the N770 to the 3.0 GHz server is less time consuming than 1536-bit RSA signatures as long as the RTT between both peers is not higher 578 ms. The N770 calculates 1024-bit RSA signatures in 181.32 ms. This means that IHC signatures from the N770 to the Xeon host are faster if the RTT is below 178.88 ms. The IHC signatures delay the message delivery for higher RTTs. As a rule of thumb, IHC signatures require less time if the RTT of the network is smaller than the time for the public-key signature generation.

LHIP-aware middleboxes can use the IHC signature packets to verify messages. This means that these middleboxes do not have to compute RSA or DSA verifications. The improved IHC signature scheme reduces the amount of data that must be buffered by the middlebox. Verifying an IHC signature only requires to buffer the small pre-signature of the message to verify the message in the MSG packet. Nevertheless, middleboxes must provide buffer space to store pre-signatures. However, HIP-aware middleboxes must provide buffer space for HIP associations, too. They must store the HIs of the communication peers to verify the RSA and DSA signatures.

8.5 Discussion

We defined four high-level goals for LHIP in Section 8.2. These high-level goals determine how LHIP should perform and what security properties it must provide. In this section, we discuss whether LHIP achieves its high-level goals.

8.5.1 LH1 – performance

Our performance measurements show that LHIP can be used on hosts with low processing power, such as mobile devices, without imposing long delays for establishing and maintaining an LHIP association. IHC signatures drastically reduce the computational overhead of UPDATE messages. Even devices with little CPU resources, such as the N770, are capable of performing multiple location updates with different peers at the same time. However, the delays caused by long RTTs affect the LHIP update process more strongly than does the HIP update process.

LHIP CLOSE and UPGRADE messages are protected by single hash chain elements. The generation and verification of these elements is cheap in terms of CPU cycles. LHIP speeds up the closing procedure as it makes RSA and DSA signatures in the close messages unnecessary. The LHIP to HIP upgrade process requires both hosts to compute the shared secret and the RSA signatures. The computational overhead is, therefore, similar to the overhead of the base exchange.

---

[7]The RSA or DSA signature verification is optional for HIP hosts because RSA and DSA signatures in HIP update packets are mainly used by middleboxes, while HIP hosts can use the HMAC signatures for message authentication.

LHIP reduces the computational cost of HIP during the base exchange and the update process to less than 2.5% of the cost of HIP with 1024-bit RSA or DSA HIs. Therefore LHIP achieves its first high-level goal *LH1 – performance*.

8.5.2   LH2 – protocol security

LHIP provides considerably less security than HIP. It does not authenticate the HIs, nor does it provide payload encryption. However, it decreases the computational overhead of HIP in scenarios that do not require these security properties.

It signs UPDATE messages with IHC signatures and, therefore, guarantees that attackers cannot forge important HIP control messages. The improved IHC signatures are no longer susceptible to MITM attacks, in the case of simultaneous signature packets from both peers. LHIP also protects the CLOSE messages and the UPGRADE messages to prevent attacks aiming at state manipulation and connection disruption.

However, the first contact between two hosts must be authentic. LHIP cannot protect against MITM attacks without authenticated I2 and R2 messages. LHIP allows to use RSA and DSA signatures for these packets, depending on whether signatures are necessary or not. Hosts can request strong authentication from their peer if required. Nevertheless, the basic functionality of LHIP does not require public-key authentication. LHIP provides secure mobility updates and protects all important protocol processes with hash chains and IHC signatures. Therefore, LHIP achieves the second high-level goal *LH2 – protocol security*. However, the question of how to protect the LHIP upgrade process with client puzzles remains.

8.5.3   LH3 – namespace security

LHIP uses RSA and DSA signatures in case of name space conflicts and to prevent harmful impersonation attacks. The optional authentication of the I2 and R2 messages and the authentication of UPGRADE packets ensures that a host cannot block HIP or LHIP nodes. Furthermore, these signatures prevent attacks aiming at impersonating a HIP node by first establishing an LHIP association and then upgrading to HIP. We are not aware of further attacks on the HIP namespace not prevented by these measures. LHIP, therefore, achieves its high-level goal *LH3 – namespace security*.

8.5.4   LH4 – compatibility

LHIP was designed to interoperate with pure HIP implementations. The decision whether an LHIP association is possible or not is taken during the base exchange. LHIP hosts always establish HIP associations when they communicate with non-LHIP-aware hosts.

Like HIP, LHIP supports packet inspecting middleboxes. These middleboxes can learn the SPIs, HIs, and anchor values from the base exchange and UPDATE packets. Middleboxes can verify the LHIP UPDATE messages efficiently. However, the way of authentication differs from the way HIP authenticates UPDATE packets. Therefore, LHIP provides the same features for middleboxes but it requires modifying HIP-aware middleboxes. LHIP does not modify the way HIP works in general. The LHIP authentication layer can be adapted to support existing and future HIP extensions. LHIP supports the mobility and multihoming extensions. Other extensions, such as rendezvous servers and opportunistic HIP, are not supported yet.

LHIP can interact with unmodified HIP implementations and supports middleboxes. It reuses the HIP name resolution infrastructure and the host identity namespace. LHIP, therefore, achieves the third high-level goal *LH4 – compatibility*.

Part III

Infrastructure Support

9

Middlebox traversal

This chapter describes the problem and solutions of traversing HIP control and data packets through middleboxes, such as NATs and firewalls. Section 9.1 describes general issues with HIP traversal over legacy, i.e. HIP-unaware, middleboxes. Section 9.2 introduces HIP extensions for UDP-encapsulation of control and data packets to traverse legacy NATs. Section 9.3 describes design requirements for middleboxes that explicitly support HIP. As an example, Section 9.4 outlines an implementation of a HIP-aware firewall.

9.1 Requirements for traversing legacy middleboxes

Traditional IP architecture assumed that intelligence is located at end hosts, while the network is "dumb" and just attempts to deliver data packets to a destination host with a best effort. Such architecture implies that packets are not modified in transit, and routers do not inspect the packet content beyond the header. However, gradually the Internet became full of middleboxes to compensate for lack of IPv4 addresses, prevent attacks, perform data caching, and other reasons. By *middlebox* we refer to a device located on the data path between hosts that processes packets differently than standard IP routers, in particular by modifying packet content.

A strong advantage of HIP architecture, compared with other proposals, is its ability to function without changes to existing IP routers. However, some middleboxes can affect HIP packets or even prevent a successful association establishment. This is precisely the reason why middleboxes are seen as a harmful part of the Internet by the IETF. Designed to support only a very limited set of protocols, often only TCP and UDP over IPv4, middleboxes prevent deployment of new protocols in the Internet. Fortunately, middleboxes are mostly used only at the edges of the network, near the end hosts. The core of the Internet maintained global routing at the IP layer and remained free of middleboxes.

Middleboxes inspecting application payload traffic, such as TCP performance enhancing proxies, would not be able to operate with the ESP encapsulation for HIP, as it encrypts the entire packet content except for the IP header. Even if null encryption is used, a packet modification would fail the MAC and checksum verification at the destination. In the

Host Identity Protocol (HIP): Towards the Secure Mobile Internet   Andrei Gurtov
© 2008 John Wiley & Sons, Ltd

best case, the middleboxes would just forward HIP packets that are not understood without special handling. Then HIP connectivity would not be affected, although the performance may be worse than for plain TCP/IP packets. In the worst case, a middlebox drops HIP packets, which would prevent a direct HIP association establishment.

Traversing a middlebox implies using some well-known protocol understood by the middlebox, such as UDP, to contain HIP packets as payload. The problem statement (Stiemerling *et al.* 2008) of HIP interaction with middleboxes focuses of NATs and firewalls as the most commonly used middleboxes. It is a product of discussions in the HIP Research Group at the IETF.

Middleboxes can cause problems both to the HIP control packets, carried inside IP packets of protocol number 139, and data packets, encapsulated using IPsec ESP. The problem with ESP middlebox traversal is common to other protocols using IPsec, except that HIP utilizes a new IPsec mode BEET.

9.1.1 NAT traversal

HIP control packets

Here, we are most interested in the base exchange packets that are implemented differently for IPv4 and IPv6. For IPv6, a special extension header is used while for IPv4, the HIP header follows the IP header as payload. Only IPv4 NATs are widely deployed at the time of writing. IPv6 NAT implementations do exist, although in practice are not used.

Reaching a receiver located behind NATs is a common problem. Unless the receiver has first initiated communication and created some mapping state in the NAT, there is no possibility for the NAT to forward HIP control packets to the right destination. Some NATs can be configured to forward a particular protocol to one of hosts behind the NAT.

HIP base exchange packets for IPv4 do not include port numbers or SPI values that can be used for multiplexing for HIP-unaware NAT. Consequently, the base exchange cannot traverse NATs that perform port translation, which is a common NAT type (called NAPT). The basic NATs that translate only the IP addresses should not prevent the HIP base exchange. HIP base exchange over IPv6 uses packets with an extension header but without payload. While the behavior of IPv6 NATs is not well-known yet, it is likely that most NAT implementation would just support a few most common protocols thus preventing the HIP base exchange.

HIP data packets

The default mode of data encapsulation for HIP is IPsec ESP. The problems of NAT traversal of HIP data packets are similar to other IPsec applications, except that HIP utilizes a new BEET IPsec mode. In summary, NATs cannot observe the transport-layer headers in packets and can only modify fields in the IP header. Such changes can invalidate upper-layer checksums, even though HIP already uses HITs in place of IP addresses during checksum calculations.

Traversal of IPsec packets over NATs without encapsulation is possible but complicated. The SPI value contained in ESP packets has only one-way significance, i.e. it does not match the SPI in returning packets from a destination host. However, in one direction, the NAT can demultiplex flows using SPI instead of a port number. Some NATs offer a VPN pass-through feature where the NAT attempts to learn of SPI exchange in both directions and sets up a corresponding mapping. The NAT can even change the SPI in the packet in the case of a collision, when two hosts behind the same NAT happen to choose the same SPI value. However, ESP traversal over NATs tends to work robustly only for a small number of ESP hosts behind a NAT. Thus, it may not be a suitable solution e.g. if all hosts use HIP from behind a NAT.

9.1.2 Firewall traversal

A firewall is a middlebox typically installed between the internal network of an organization and the rest of the Internet. A firewall can either forward or drop a packet after inspecting its content against a set of rules. Firewalls can be configured to permit HIP control and data traffic, although the default rules appear to disallow incoming packets. It resembles the situation of Responder unreachability behind a NAT.

More conservative configurations often allow only a few known protocols such as TCP or UDP and only for certain well-known ports, such as port 80 for HTTP traffic. Such configurations would prevent HIP control and data packets even for the outgoing direction. Otherwise, ESP traffic containing HIP data packets can traverse firewalls well.

IPv4 firewalls typically block packets containing an IP option. This is likely behavior of IPv6 firewalls as well, which can block all IPv6 extension headers or, perhaps, uncommon extension headers. It would prevent HIP control and data traffic from crossing the firewall.

9.1.3 Strategies for legacy middlebox traversal

To provide HIP traversal over legacy middleboxes, a solution must provide some encapsulation format for HIP control and data packets to enable the middlebox to understand and demultiplex HIP traffic. UDP encapsulation suits well for this purpose, because it is supported by the majority of middleboxes and does not produce similar latency issues to TCP. The drawback of UDP encapsulation is header overhead and the possible exceeding of the path MTU value. Especially for small packets, e.g. containing VoIP traffic, UDP encapsulation can add more than 20% overhead. For large packets, adding a UDP header can result into fragmentation of the packet that has several undesirable consequences. In particular, one of multiple fragments is more likely to be lost than a single packet. A single lost fragment causes the whole packet to be discarded at the receiver. Furthermore, fragmentation is a known mechanism to attack the destination host. By sending plenty of small fragments of different IP packets, the attacker forces the host to allocate network buffers waiting for the missing fragments to arrive.

The second issue to be resolved for legacy middlebox traversal is reachability of a Responder located behind the middlebox. To enable reachability, the Responder should first register to a rendezvous server located outside of the middlebox in the public Internet. Such registration punches "holes" in the middlebox, enabling the rendezvous server to serve as a contact point for an Initiator that forwards packets through the middlebox to the Responder. Such holes typically expire after a short timeout, thus requiring the Responder to periodically send refresh messages to the rendezvous server to keep the holes open. Depending on the

NAT type, after the base exchange the Initiator might be able to send packets directly to the Responder through the same hole opened to the rendezvous server.

To support middlebox traversal of mobility and multihoming HIP extensions, the HIP host should first punch a hole through the new middlebox where it moves. Later, the host can inform its peer of the new locator and possibly of the new rendezvous server that has a hole open in the middlebox.

9.2 Legacy NAT traversal

This section describes encapsulation extensions to HIP to enable traversal of legacy HIP-unaware NATs according to IETF specification (Komu *et al.* 2007b; Schmitt *et al.* 2006) (Summer 2007). These extensions only traverse NATs supporting endpoint-independent mapping. Alternative approaches to NAT traversal (Nikander *et al.* 2006) do exist but are work in progress at the time of writing and are not covered here in detail. The latest trend within the NAT design team in IETF is the solution based on the Interactive Connectivity Establishment (ICE).

The NAT traversal extensions presented in this section utilize UDP tunneling of HIP control and data packets when the Initiator host, the Responder host, or both are located behind a NAT. The traversal extensions are needed when both hosts are located behind different NATs. Behind the same NAT, the HIP packets can flow directly between hosts. Only the ESP data encapsulation is currently supported by the NAT traversal extensions.

When the Responder is behind a NAT, but the Initiator has a globally routable (public) IP address, the Responder must use a rendezvous server supporting the NAT traversal extensions. The rendezvous server must have a public IP address.

When a HIP host moves from the public Internet to a private address space located behind a NAT, it can start using these extensions for executing the mobility update. Correspondingly, when moving to the public Internet, the host can stop using the NAT traversal extensions.

9.2.1 NAT detection

Before initiating the Base Exchange, a HIP host acting as Initiator needs to determine whether there is a NAT present between the itself and the Responder. Only if a NAT is present, the Initiator employs the traversal extensions, otherwise the standard base exchange is used. The case when both the Initiator and the Responder are located behind the same NAT requires special attention. If the NAT presence is not properly detected, both HIP hosts would use the traversal extensions unnecessarily sending packets via a rendezvous server located in the public Internet. Such a scenario is known as *hairpin translation*. To avoid such a scenario, the Initiator may first attempt sending an I1 message to the IP address of the Responder. If R1 arrives within a certain time interval, the normal HIP base exchange proceeds. Otherwise, after a timeout the Initiator starts using the NAT traversal extensions.

The Simple Traversal of UDP through NATs (STUN) specifications define a generic mechanism to detect the NAT presence (Rosenberg *et al.* 2003).

9.2.2 Header format

Figure 9.1 shows encapsulation of HIP control and data packets using a minimal UDP header. For control packets, the UDP header is followed by a HIP header with parameters. For data packets, the UDP header is followed by an ESP content in the BEET mode. The UDP length and checksum fields are computed as usual, following RFC768. The HIP checksum is not used and is set to zero. It is redundant because of UDP checksum and also would increase the complexity of the NAT translation.

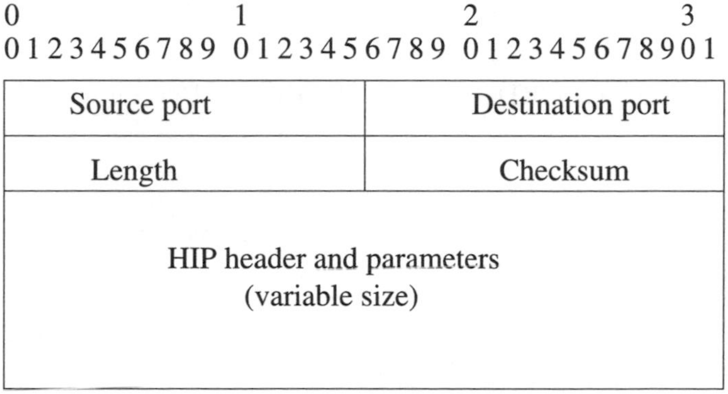

```
0                   1                   2                   3
0 1 2 3 4 5 6 7 8 9 0 1 2 3 4 5 6 7 8 9 0 1 2 3 4 5 6 7 8 9 0 1
+-------------------------------+-------------------------------+
|          Source port          |        Destination port       |
+-------------------------------+-------------------------------+
|            Length             |            Checksum           |
+-------------------------------+-------------------------------+
|                                                               |
|                   HIP header and parameters                   |
|                        (variable size)                        |
|                                                               |
+---------------------------------------------------------------+
```

Figure 9.1 UDP encapsulation of HIP control and data packets.

Keep-alive control packets maintain mapping in NATs when a HIP association is idle. HIP UPDATE messages encapsulated into UDP packets serve as keep-alive messages.

Figure 9.2 shows the format of the FROM_NAT parameter containing an IPv6 or IPv4-in-IPv6 address of the NAT. The VIA_RVS_NAT parameter has the same structure as FROM_NAT and serves a similar role as the VIA_RVS parameter, mainly debugging and diagnostic.

Traversal of ESP tunnel and transport mode packets over NAT using UDP encapsulation is described in RFC3948. HIP uses a new ESP mode BEET that requires extensions to the existing ESP traversal technique. With BEET transport-layer checksums in the packet are based on HITs and not IP addresses. The UDP header is inserted between the IP and ESP header as shown in Figure 9.3. The UDP checksum, in the opposite way to HIP checksum, is calculated using the IP addresses and not HITs. Destination IP options, if any, are placed after the UDP and ESP headers. The HIP checksum is set to zero when computing the UDP checksum.

The receiver checks the UDP checksum of an encapsulated BEET packet and drops the packet if the checksum fails. Afterward, the receiver performs BEET verification and decryption as described in Section 4.3.2 and Nikander and Melen (2006). The UDP checksum already protects the entire packet and, therefore, the HIP checksum is not verified by the receiver.

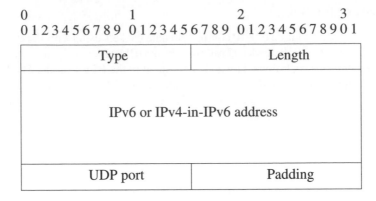

Figure 9.2 FROM_NAT and VIA_RVS_NAT parameters.

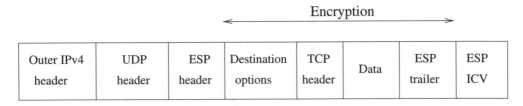

Figure 9.3 ESP BEET-mode packet encapsulated in a UDP datagram.

9.2.3   Initiator behind a NAT

This section describes a scenario when the Initiator is located behind the NAT but the Responder in a public Internet. Figure 9.4 shows the scenario and message headers in detail. The rest of this chapter uses the following notation for the content of IP and UDP headers. IP_i is an IP address, private or public, of the Initiator. IP_r is the IP address of the Responder. Public IP addresses of NAT and of the rendezvous server are denoted as IP_n and IP_s. The UDP port 50500 is used as a well-known port for HIP NAT traversal. Ephemeral UDP ports are shown as 11111 or 22222 and are assigned randomly in implementation. As an example, packet $I1(IP_n, IP_r, 11111, 50500)$ is a UDP-encapsulated I1 packet from the NAT to the Responder with an ephemeral source UDP port and destination UDP port of 50500.

When the Initiator starts the base exchange with an I1 message, it sets the destination UDP port to 50500. It is recommended to set the source port to 50500 as well, although a randomly chosen unused port can be set too. A randomly chosen port should be from the range of dynamic and private ports (49152–65535). Using a random source port instead of fixed 50500 has the advantage of supporting multiple HIP clients located behind the same NAT. Certain NAT types performing only address and not port translation cannot support multiple hosts using the same port number. The Initiator listens to the selected port and accepts incoming HIP control and ESP packets.

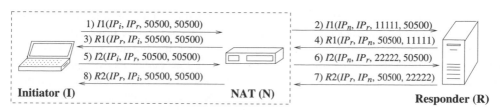

Figure 9.4 Initiator behind a NAT.

The HIP Responder processes a UDP-encapsulated I1 packet as for the normal base exchange after decapsulation. It sets its source port to 50500 and the source IP address to the address at which the message was received. The Responder is not allowed to reply with a plain HIP message to any UDP-encapsulated messages, but must use UDP encapsulation as well.

Often the NAT has sufficiently long state timeout that the same ports can be used for I1-R1 and I2-R2 messages. However, implementations must also support the scenario where the NAT state expires after the R1 message and a new port can be assigned to I2 and R2 messages. This scenario is illustrated in Figure 9.4.

Figure 9.4 shows an example of Initiator behind NAT operation after the Initiator has detected the presence of NAT. The Initiator sends an I1 packet encapsulated into a UDP datagram to the Responder's public IP address with source and destination UDP ports of 50500. The NAT translates the source UDP port to 11111 and replaces the source IP address with its own public IP address.

The Responder processes the I1 packet after decapsulating it from UDP and replies with an R1 message setting the destination IP address and UDP port to address and port taken from I1. The NAT replaces the destination address and port with the private address of the Initiator and default port 50500.

The Initiator processes the R1 packet and sends I2 with the same port and address as I1. The NAT replaces the source address and port as for I1. If the mapping state from I1 is still valid, the NAT would set the source UDP port to the same value as in I1, 11111. In this example, we assume that the mapping state has expired and a new port (22222) is assigned by the NAT. The Responder replies with an R2 packet to the new port taken from the I2 message. The old source port from the I1 message has expired and cannot be used anymore .

After the HIP base exchange over UDP completes, the data traffic is carried using the ESP BEET mode encapsulate to UDP. During the base exchange, ESP Security Associations with Security Parameter Indices (SPIs) are set up normally in each direction. The inner source and destination addresses are local and peer HITs, respectively. The outer source and destination addresses at the Initiator are the local private IP address and the public IP address of the peer. At the Responder, the outer source and destination addresses are the local public IP address and the public IP address of the NAT. In summary, the Initiator and Responder use the same IP addresses for ESP packets as for the base exchange.

The Initiator sends BEET packets encapsulated to UDP with a destination port of 50500. The source port can be either 50500 or randomly chosen; it is the same port as used in the I2 packet. The Responder sets the source port in ESP packets to 50500 and the destination port to the source port of the last ESP packet received.

Figure 9.5 shows the use of a rendezvous server when the Initiator is behind the NAT but the Responder is located in a public network. The rendezvous server listens on UDP port 50500 for incoming I1 packets from the Initiator. However, the Initiator cannot just relay the I1 packet to the Responder because the R1 from the Responder may not pass through all types of NATs. Only some NATs allow any host to send UDP packets through a pinhole open to the rendezvous server. Therefore, the rendezvous server replies to an I1 packet with a NOTIFY packet containing the Responder's address in a VIA_RVS parameter.

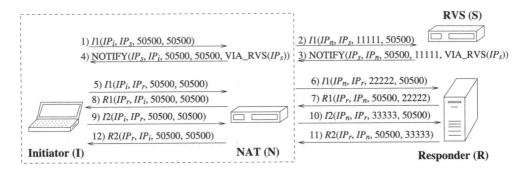

Figure 9.5 Initiator behind a NAT using a rendezvous server.

After receiving the Responder's address, the Initiator re-sends I1 to the Responder. However, retransmissions are made to the rendezvous server, because NOTIFY messages are unprotected by signatures and can be forged by an attacker. The Responder replies with R1 to the Initiator through a pinhole created in a NAT. Figure 9.5 shows a scenario where the NAT times out between I1 and I2 packets. Therefore, I2 arrives to the Responder from a different UDP port (33333) than I1 (22222).

9.2.4 Responder behind a NAT

In this section, we describe the use of a rendezvous server by an Initiator from a public network to access a Responder located behind a NAT in a private network.

A host behind a NAT that plans to be a Responder must first register to a rendezvous server. Figure 9.6 shows the process of registration. The process is similar to the usual registration described in Chapter 5, Section 5.2, except that messages are encapsulated to UDP datagrams. The host first acts as an Initiator and sends I1 to the UDP port 50500 at the rendezvous server. RVS replies with a UDP-encapsulated R1 that has the REG_INFO parameter. The host sends REG_REQ parameter in I2 and receives REG_RESP from the RVS server in R2. After a successful registration, the Responder sends keep-alive messages with the same source UDP port as in I2 to the rendezvous server to maintain a pinhole open in the NAT.

Figure 9.7 shows the base exchange towards the Responder behind a NAT. When some Internet host tries to initiate a base exchange to the Responder behind a NAT, it sends an I1 packet to the rendezvous server without encapsulation. The rendezvous server places I1 to a UDP datagram, zeros the HIP header checksum, and sends the datagram to the NAT.

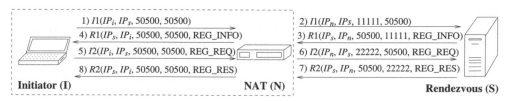

Figure 9.6 Registration to a rendezvous server from behind a NAT.

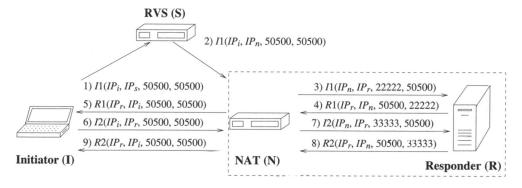

Figure 9.7 Responder behind a NAT.

The datagram has the destination address set to the public IP address of the NAT and destination UDP port to the value stored from I2 of the registration. The source IP address of the datagram is the public IP address of the rendezvous server and the source UDP port is 50500.

The rendezvous server adds a FROM parameter to I1 upon forwarding it to the Responder. The FROM parameter contains the public IP address of the Initiator protected by the RVS_HMAC. The IP header checksum is recalculated by the rendezvous server.

The NAT has the pinhole still open thanks to keep-alive messages and replays the datagram to the Responder. After processing I1, the Responder sends UDP-encapsulated R1 to the Initiator's public IP address. The Responder may add VIA_RVS_NAT parameter to R1 containing the IP address and UDP port that the rendezvous server used when forwarding I1. The R1 packet is sent to UDP port 50500 and a public IP address of the Initiator taken out of the FROM parameter. The recommended source port is also 50500 and the source IP address is the private address of the Responder.

The Initiator implementing NAT-traversal extensions listens to UDP port 50500. Upon receiving a UDP-encapsulated R1 message after sending a plain I1 message, the Initiator determines that the Initiator is located behind a NAT. After decapsulating and processing R1 according to base specifications, the Initiator replies with UDP-encapsulated I2 with destination address and ports set to the source address and port of R1. UDP-encapsulated I2 and R2 are forwarded by NAT as well.

When a base exchange completes, Initiator and Responder establish a Security Association in each direction using UDP-encapsulated BEET mode. The inner addresses in SAs are HITs. For outbound SA at the Initiator, the outer source address is set to its public IP address that was the source address of the base exchange. The outer destination address is set to the public IP address of the NAT taken from the R2 packet. The UDP source port is 50500 and the destination port is derived from the R2 packet.

For inbound SA at the Initiator, the outer source IP address is taken from the R2 packet. The outer destination address is the public IP address that the Initiator used during the base exchange. The UDP source port is taken from the R2 packet and the destination port is 50500.

For outbound SA at the Responder, the outer source address is its private (from the NAT address space) IP address used in the base exchange. The outer destination address is the public address of the Initiator used during the base exchange. The source UDP port is the same as the source port used in the R2 packet and the destination UDP port is 50500.

For inbound SA at the Responder, the outer source address is set to the public IP address of the Initiator used in the base exchange. The outer destination address is the private address as in the base exchange, as well. The UDP source port is 50500 and the destination port is same as the source port of R2.

In the RVS mechanism outlined above, the RVS server replies to the Initiator with the NOTIFY packet when receiving I1. NOTIFY contains VIA_RVS parameter with public address and port information of the Responder's NAT. The Initiator can use the NAT information to launch a DoS attack on the Responder early on. A more secure mechanism could relay I1 and R1 packets through the RVS server to avoid exposing NAT information to a possibly malicious Initiator until it has authenticated and provided a correct puzzle solution. However, this approach causes higher delay due to triangle routing through the RVS server.

9.2.5   Initiator and Responder behind a NAT

In this section, we describe the HIP base exchange and IPsec data traffic in a scenario when both Initiator and Responder are located behind two different NATs. The presented approach only works when both NATs support endpoint-independent mapping, that is the NAT translates a packet from a given internal address and port to any external address and port in the same way.

Figure 9.8 illustrates address and port translation by an endpoint-independent mapping NAT. The client initiates two parallel sessions to two separate servers simultaneously. The NAT uses the same public address and port 192.0.2.1:62000 for mappings for both servers. Therefore, a client can be identified by the address and port assigned by the NAT.

The scenario with both hosts located behind NATs combines the cases when only the Initiator or the Responder were behind a NAT. Figure 9.9 shows the packet flow when both hosts are behind NATs. After the Responder registers with its rendezvous server, the Initiator sends a UDP-encapsulated I1 packet to the rendezvous server. RVS listens on UDP port 50500 for incoming packets. Therefore, the destination port of I1 is set to 50500; it is also recommended to use the same source port.

The RVS forwards UDP-encapsulated I1 to the UDP port that the Responder stored during the registration from the UDP port 50500. RVS adds FROM_NAT parameter to I1 that contains the source address (the public address of the Initiator's NAT) and source UDP port (assigned by the NAT) from the original I1 packet. The RVS_HMAC field protects the integrity of the FROM_NAT parameter so that spoofing of address and port information by modifying the I1 packet is not possible. The source IP address of the forwarded I1 packet

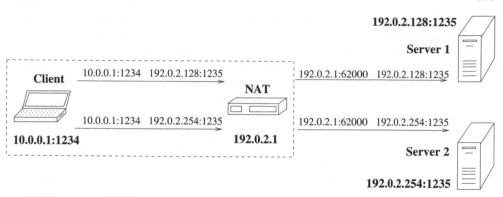

Figure 9.8   Address and port translation by a NAT performing endpoint-independent mapping.

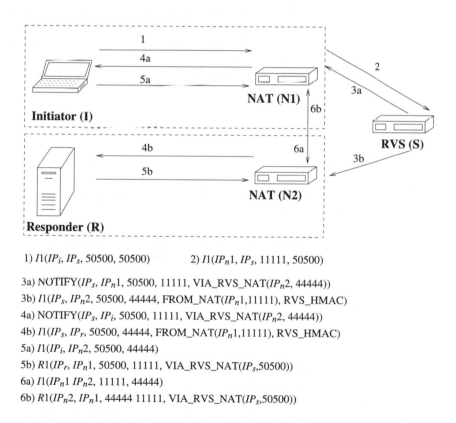

1) $I1(IP_i, IP_s, 50500, 50500)$ 2) $I1(IP_n1, IP_s, 11111, 50500)$

3a) NOTIFY(IP_s, IP_n1, 50500, 11111, VIA_RVS_NAT(IP_n2, 44444))

3b) $I1(IP_s, IP_n2$, 50500, 44444, FROM_NAT(IP_n1,11111), RVS_HMAC)

4a) NOTIFY(IP_s, IP_i, 50500, 11111, VIA_RVS_NAT(IP_n2, 44444))

4b) $I1(IP_s, IP_r$, 50500, 44444, FROM_NAT(IP_n1,11111), RVS_HMAC)

5a) $I1(IP_i, IP_n2$, 50500, 44444)

5b) $R1(IP_r, IP_n1$, 50500, 11111, VIA_RVS_NAT(IP_s,50500))

6a) $I1(IP_n1\ IP_n2$, 11111, 44444)

6b) $R1(IP_n2, IP_n1$, 44444 11111, VIA_RVS_NAT(IP_s,50500))

Figure 9.9 Initiator and Responder behind NATs.

is set to the public RVS address and the destination IP address to the public address of the Responder's NAT.

The RVS server also sends a NOTIFY packet to the Initiator containing the VIA_RVS_ NAT parameter. The Initiator obtains the public address of the Responder's NAT from the VIA_RVS_NAT and the UDP port number with which the Responder has registered to RVS. The Initiator sends own NOTIFY message to the Responder through its NAT. This messages creates a pinhole in the Initiator's NAT. The Responder sends a UDP-encapsulated R1 packet through the pinhole to the Initiator. R1 creates a pinhole in the Responder's NAT.

I2 and R2 packets flow directly between Initiator and Responder through pinholes created with previous packets. To prevent the translation state in NATs from expiring during puzzle calculations, the Initiator periodically sends a NOTIFY packet to the Responder, until R2 is received. Such keep-alive packets use the same addresses and ports as corresponding data packets. The recommended interval of NOTIFY transmission is 20 seconds; the receiver discards keep-alive packets. The base exchange exceeds a few seconds only in pathological cases (very large puzzle or very high RTT); sending keep-alives during a base exchange is seldom necessary.

After completing the base exchange, Initiator and Responder transmit data packets in UDP-encapsulated BEET mode. IP addresses and UDP ports for Security Associations are the same as in the base exchange. When temporarily no data is sent over a HIP association, keep-alives are necessary to prevent premature state expiration (by default, HIP associations remain valid for 30 minutes) in NATs. Since only outgoing packets refresh the NAT state, both hosts behind NATs should transmit keep-alive packets to each other.

9.2.6   Multihoming and mobility with NATs

Multihoming extensions for HIP allow the maintaining of several Security Associations between two hosts over multiple paths (Henderson 2007). Some paths can pass through NATs while others can have direct IP connectivity. HIP implementations are required to handle multihoming in such a way that some available paths use UDP encapsulation to traverse NATs and others use plain HIP packets.

Likewise, a host using NAT traversal extensions can change its IP address using the HIP mobility mechanism. At the moment, simultaneous host mobility also known as "double-jump" is not supported.

Possible host mobility scenarios include moving between a private (behind a NAT) address space and the public Internet, moving between private address spaces of two different NATs, or changing the IP address while staying behind the same NAT. The correspondent node can be either in a public or private network with or without a rendezvous server. Some of the above combinations may not work depending on the type of NAT, e.g. when the correspondent node is behind a NAT that is not endpoint-independent.

When changing its location, a mobile host attempts to detect the presence of a NAT on its new network. If the NAT is not detected, an UPDATE packet is sent as a plain HIP packet, otherwise it is sent UDP-encapsulated. The mobile host sends an UPDATE message to all its peers both through their rendezvous servers (if available) and directly to peers. The peers reply with a plain UPDATE or UDP-encapsulated UPDATEs depending on what the mobile host has used.

When relaying an UPDATE packet, the rendezvous server adds a FROM parameter for a plain HIP packet and a FROM_NAT parameter for UDP-encapsulated packets; both parameters are protected by HMAC. The peer host replies to the UPDATE to the mobile hosts adding VIA_RVS or VIA_RVS_NAT (for UDP-encapsulated packets) parameters.

Current specifications for NAT traversal disallow sending of private IP addresses (from a NAT address space) in the LOCATOR parameter. When the host has no public IP addresses, the LOCATOR parameter contains only the type and length fields. It is assumed that during encapsulation the UDP header contains the necessary information to access the private locator. Such restrictions may be harmful in the hairpin translation scenario, when both hosts are located behind the same NAT. In that case, both hosts are located within the same private address space and should communicate without involving the NAT. Then, sending private IP addresses in a LOCATOR parameter would be useful to support mobility within a single address space.

9.2.7 Traversing firewalls

Many existing firewalls support only TCP and UDP packets and drop any other packet types, such as HIP control or IPsec packets. Even when a user has control, e.g. over a host firewall in Windows XP, he or she cannot enable HIP traffic traversal selectively without disabling the firewall altogether. In other cases, the user cannot modify firewall rules and the firewall allows incoming UDP packets only if a packet from the same port and address has been recently transmitted in outbound direction from the firewall. If the firewall allows outgoing UDP packets to port 50500 from any internal port, and reply UDP packets in the reverse direction, then the described NAT traversal extensions can be directly used to traverse the firewall.

If the Initiator is located behind a stateless firewall, the firewall should be configured to allow all incoming UDP packets from UDP port 50500 to any internal IP address and UDP port. If the Responder is located behind a firewall, it should allow incoming UDP packets to port UDP 50500 from any IP address or port.

9.3 Requirements for HIP-aware middleboxes

This section discusses generic requirements and approaches for middleboxes (NATs and firewalls) that are HIP-aware, i.e., designed to understand the details of HIP messages (Tschofenig and Shanmugam 2006). Although middleboxes cannot inspect the protocol headers inside the IPsec ESP payload of HIP data packets, the HIP control protocol is specifically designed to reveal sufficient information in signaling packets. Using that information, HIP-aware middleboxes can separate and follow subsequent data flows, and potentially filter out unwanted traffic. That makes HIP more suitable to traverse middleboxes than other protocols that set up ESP Security Associations.

To traverse new "architectured" middleboxes, two approaches are possible. The first approach relies on HIP-specific modifications that monitor the HIP base exchange and learn SPI parameters in both directions. One example of such a middlebox is SPI-NAT (Ylitalo *et al.* 2005). SPINAT combines traversal of HIP control and data packets in a single solution.

The middlebox later identifies ESP data packets by the pair of their destination IP address and SPI value. The middlebox needs to verify that the HIP base exchange or update messages are authentic before creating internal state. Checking the signatures in HIP control packets would enable the middlebox to avoid certain attacks. The middlebox learns the necessary flow identifiers from ESP_INFO parameters in I2 and R2 messages of base exchange or from the LOCATOR parameter in an UPDATE message.

HIP hosts located behind middleboxes and willing to be reachable from outside of their private domain would need to perform an explicit registration procedure to the middlebox using the HIP registration protocol. Afterward, the HIP host should maintain the registration state in the middlebox by sending keep-alive messages.

A second approach is based on a generic signaling mechanism to middleboxes, such as one developed by IETF NSIS or MIDCOM working groups. Extensions to these mechanisms have already been proposed that support signaling of SPI values from host to middleboxes. This is sufficient to enable traversal of HIP data traffic through middleboxes. To enable traversal of HIP control traffic, extensions to the signaling mechanism are needed to include HIP control traffic. The NSIS NAT/FW traversal protocol can set up the traversal state in all middleboxes along the path from an Initiator to the Responder.

Implementations of a HIP-aware middlebox must not introduce new possibilities of traffic attacks, such as redirection Denial-of-Service attacks. The middlebox must be able to intercept and authenticate HIP signaling messages before creating internal state. The implementation should drop any HIP data and control traffic that does not have a properly set up state in the middlebox or fails authentication (such as incorrect signatures).

Ideally, a HIP-aware middlebox should function in a scenario where due to routing asymmetry or multihoming packets can go through different routes to the HIP host. For instance, packets can enter the local network of the HIP host through one firewall and leave through another firewall. The middleboxes may employ a state synchronization protocol to agree on common flow identification parameters, such as the SPI values. The HIP hosts should be able to decide which of the several present middleboxes they should register to.

9.4 HIP-aware firewall

A firewall typically separates an organization's network from the rest of the Internet. An Access Control List (ACL) specifies packet forwarding policies in the firewall. Current firewalls can filter out packets based on IP addresses, transport protocol, and port values. These values are often unprotected in data packets and can be spoofed by an attacker. By trying out common well-known ports and a range of IP addresses, an attacker can often penetrate the firewall defenses. Furthermore, legacy firewalls often disallow IPsec traffic and drop HIP control packets.

9.4.1 Flow identification

Designing a firewall that understands HIP messages and authenticates hosts based on their host identities would solve these problems and promote deployment of HIP. In this section we describe a HIP-aware firewall that was designed and implemented for HIPL (Vehmersalo 2005).

Firewalls can be stateless, filtering packets based only on the ACL, and stateful, which can follow and remember packet flows. Stateless firewalls are simple to implement but provide only coarse-grain protection. However, their performance can be high since packet processing requires little memory or CPU resources. A stateful firewall determines if a packet belongs to an existing flow or starts a new flow. A flow identifier combines information from several protocol headers to classify packets. A firewall removes the state when the flow terminates (e.g., a TCP connection is closed) or after a timeout. A firewall can drop suspicious packets that fail a checksum or contain sequence numbers outside of the current sliding window. A transparent firewall does not require that hosts within the protected network register or even know of the existence of the firewall. An explicit firewall requires registration and authentication from the hosts.

A HIP-aware firewall identifies flows using HITs of communicating hosts, as well as SPI values and IP addresses. The firewall must link together the HIP base exchange and consequent IPsec ESP data packets. The firewall, therefore, must be stateful. During the base exchange, the firewall learns the SPI values from I2 and R2 packets. Then, the firewall only allows ESP packets with a known SPI value and arriving from the same IP address as during the base exchange. If the correspondent host changes its location and the IP address, the firewall learns about the changes by following the mobility update packets.

A HIP host can register to the firewall using the usual procedure (see Section 5.4). The registration enables the host and the firewall to authenticate each other. In a common case where the Initiator and Responder hosts are located behind different firewalls, the Initiator may need to register with its own firewall and afterward with the Responder's firewall.

9.4.2 Advanced extensions

Opportunistic mode

The HIP opportunistic mode presents a challenge to the firewall authentication since the I1 packet does not contain the destination HIT. The firewall can still allow such packets based on the source and destination IP address rules. The firewall creates an initial state based on I1 addresses and updates the state to include HITs after receiving the R1 packet. If opportunistic I1 packets are disallowed by the firewall, it can generate a NOTIFY packet to the registered HIP host that sent I1.

The HIP architecture has an advantage by enabling the firewall to authenticate packets using the signatures in addition to HIT-based ACL. The R2 packet includes the Responder's identity in plain text. This enables the firewall to verify the Responder's signatures in HIP control packets. To protect the Initiator's privacy, its identity is encrypted in the I2 packet, although some protocol designers suggested sending the identity in plain text. This is a design issue involving a trade-off between packet authentication and identity protection.

Rendezvous server

The presence of a firewall complicates the use of a rendezvous server. The rendezvous server changes the IP addresses in the I1 packet before forwarding it to the Responder. Therefore, I1 and R1 packets can have different addresses arriving to the firewall. The firewall can look inside the I1 packet and take the Responder's address placed to the FROM parameter. For final state configuration, the firewall takes IP addresses and SPI values from I2 and R2 packets that do not travel through the rendezvous server.

Mobility and multihoming

HIP mobility and multihoming extensions enable the host to add and remove IP addresses to the active address set. The firewall needs to follow control messages related to address changes to prevent address hijacking and redirection attacks. By default, HIP hosts perform a mobility with rekeying and address verification always when moving to a different network. In this case, the mobility signaling consists of three UPDATE messages that are sufficient for a firewall to learn new SPI values and IP addresses. However, if the host performs an update without rekeying, the firewall is not able to verify the update and likely to drop subsequent packets coming from a different address.

A HIP host can move to a network protected by a firewall that previously has not seen any traffic from that host, as shown in Figure 9.10. In this scenario, it is important that the host sends the first UPDATE packet after moving to the new network. Otherwise, the firewall could not observe the complete mobility signaling and correctly establish the forwarding state for the host. The firewall extracts the information shown in italics from the first and second UPDATE packets.

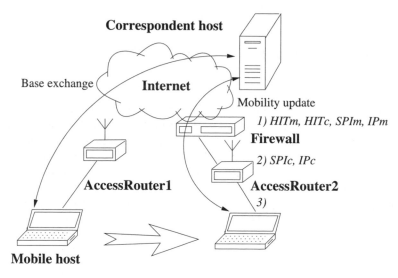

1)UPDATE(ESP_INFO, LOCATOR, SEQ)

2)UPDATE(ESP_INFO, SEQ, ACK, ECHO_REQUEST)

3)UPDATE(ACK, ECHO_RESPONSE)

Figure 9.10 A mobile host moves behind a firewall.

The details of mobility update processing at the firewall depend on whether IP addresses are added to the existing Security Association or a new SA is created. The recommended policy is that hosts create a new pair of SAs for each pair of their IP addresses.

Certificates

The HIP specifications include a possibility for sending certificates in a special parameter in base exchange messages. Certificates follow the Simple Public Key Infrastructure (SPKI) format and can be used during the registration process for authentication. A HIP-aware firewall can use certificates to simplify the authentication process. Without a certificate, the firewall ACL includes public keys of each host that is allowed to pass the firewall. When hosts supply a certificate, storing only the public key that issued the certificate is sufficient. During the base exchange, the firewall extracts the host public key from the certificate and verifies it. After closing the connection, the firewall can purge the host public key from the memory.

The limited lifetime of a certificate would control when the firewall approves the authorization. A short lifetime forces the host to apply for new certificates more frequently, but reduces the need for the firewall to check Certificate Revocation Lists (CRL) for compromised certificates.

9.4.3 Asymmetric routing

Internet routes can be asymmetric in such a way that packets flowing between two hosts do not pass the same routers and middleboxes in two directions. This scenario creates a problem with creating a proper state in middleboxes such as NATs and firewalls. Consider a case when outgoing and incoming HIP packets would go through two separate firewalls (Tschofenig *et al.* 2007b). As the firewall 1 sees only packets from Initiator to Responder, it does not know the SPI value sent by the Responder. Therefore, it cannot pass through subsequent ESP packets with that SPI.

One proposed solution for asymmetric firewalls would be to extend the HIP base exchange with a fifth packet, I3, sent from Initiator to Responder. However, the I3 packet would incur the additional signaling cost for all HIP associations even without firewall presence. An alternative solution would be to use an SPISIG packet sent from a HIP host to the firewall to signal the SPI value from the opposite direction. The SPISIG packet assumes that both HIP hosts are aware of the firewall presence and requires that hosts register to the firewall. Figure 9.11 illustrates the asymmetric routing scenario.

9.4.4 Security risks

The Initiators of HIP associations create state at the firewall and could mount a DoS attack. In the typical deployment scenario, a firewall allows connections from the local network to the Internet, but drops all connection attempts from the Internet to the local hosts. Only incoming traffic that is related to existing state initiated by a local host is permitted. In this scenario, hosts in the local network are trusted and their authentication may be unnecessary. This reduces the load and the risk of Denial-of-Service for the firewall.

The Responders in this scenario are untrusted, and their packets need authentication in the firewall. Fortunately, the HIP base exchange contains public key of the Responder in plain text that enables the firewall to check the signature in the Responder's packets. The firewall does not need to store the Responder's host identity locally, which reduces space requirements.

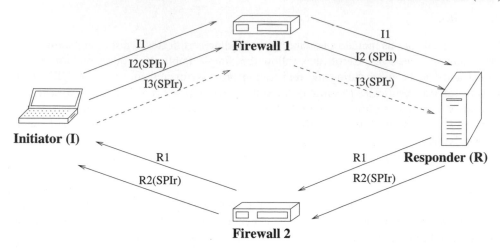

Figure 9.11 Asymmetric routing with firewalls. The I3 packet is introduced in HIP base exchange to deliver Responder SPI to firewall 1.

A prototype of HIP-aware transparent firewall is implemented for HIPL on Linux. The details of firewall design and implementation are available in the Master's thesis (Vehmersalo 2005).

10

Name resolution

Section 5.3 and Section 5.2 have already introduced the basic HIP extensions to DNS servers and a new redirection element called a rendezvous server. This chapter focuses on advanced aspects of naming resolution, such as reverse mappings and privacy of lookups. Section 10.1 begins the chapter by stating generic requirements for HIP name resolution. Section 10.2 presents an overview of Distributed Hash Table (DHT) technology. Section 10.3 describes an interface between a public DHT service OpenDHT and a HIP host. Section 10.4 contains an overview of overlay networks, focusing on Internet Indirection Infrastructure (i3). Section 10.5 presents a hybrid architecture of HIP and i3 called Hi3.

10.1 Problem statement of naming

The problem description for HIP resolution and rendezvous mechanisms is given in Eggert and Laganier (2004). The description lists generic issues that are common for any proposed naming mechanism.

Figure 10.1(a) shows name resolution in the current Internet, a logical resolution process for HIP, and the model proposed in current HIP specifications. In the current Internet, a host resolves a FQDN to a set of IP addresses using a DNS lookup. An IP address of a local DNS server is pre-configured in the host or obtained from a DHCP server. The reverse DNS lookup maps the IP address to one of the host FQDN names.

The introduction of HIP splits the resolution process in two logical parts, as shown in Figure 10.1(b). First, the host FQDN name is resolved to a set of host identities. Second, each host identity can be resolved to a set of IP addresses. The DNS system is well-suited for resolving hierarchical FQDN names and is perhaps the best choice for the first resolution step. The second step can be implemented, for example, using a DHT lookup as discussed further in this chapter. A reverse DNS lookup maps a host identity to one of the host FQDNs and reverse HIP lookup maps a host IP address to one of its host identities. Reverse lookup is a useful tool for diagnostic and debugging purposes.

Figure 10.1(c) shows a resolution model suggested by HIP DNS extensions (Nikander and Laganier 2008). In this model, two logical steps are combined into one DNS lookup that returns both the host identity and a set of IP addresses of a host. The host identity can be stored in a new DNS Resource Record in a HIP-aware DNS server, or in a form of HIT in

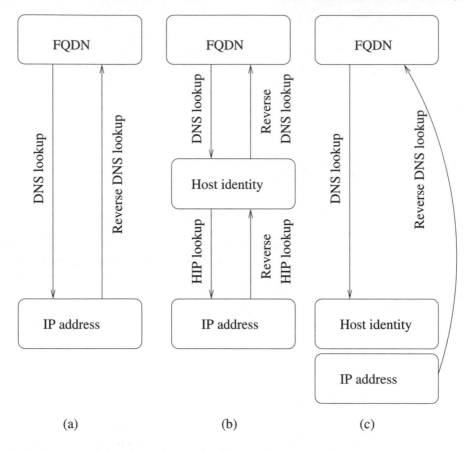

Figure 10.1   Name resolution mechanisms. (a) DNS resolution in the current Internet; (b) logical resolution for HIP; (c) resolution proposed in HIP DNS extension.

an AAAA field (meant for storing IPv6 addresses) in legacy DNS servers. This approach has benefits of simplicity and reduced latency compared with the logical two-step lookup, but raises several issues, listed below.

- *Possibility of direct communication.* In the current Internet, the DNS system can be considered an optional component because direct communication based on IP is always possible. However, in practice, DNS is necessary for most users. Who, for example, remembers by heart the IP address of www.linux.org? HIP hosts experience a similar problem if only the host identity of the destination host is known. A resolution mechanism to obtain the current IP address of the destination host becomes a critical component to enable direct communication.

- *Dependency on DNS upgrade.* HIP hosts can always revert to communication based on plain IP or attempt association establishment in an opportunistic mode. However, gaining full benefit of HIP requires uploading of host identities to DNS servers and preferably the use of DNSSEC. Not all hosts have sufficient authority even to change their DNS information, let alone upgrade the DNS server to support HIP Resource Records.

- *Reverse host identity lookup.* The simple HIP resolution model in Figure 10.1(c) does not provide reverse HIP lookup from a host identity to a FQDN or from an IP address to a host identity. Such lookup can be emulated by first performing reverse lookup from an IP address to a FQDN and then from a FQDN to the host identity. However, there is no guarantee that the FQDN used to store the host identity is the same as a FQDN returned by the reverse IP lookup. IP addresses do have structure that enables reverse IP lookups from a DNS root in.arpa. The host identity represented by HIT is random-looking flat data making it difficult to reproduce the reverse IP mapping mechanism for HITs.

- *DNS lookups over HIP.* DNS lookups are typically short operations carried out using connectionless UDP rather than connection-oriented TCP protocol. Connection establishment would introduce additional latency to DNS resolution that can double the duration of a lookup. HIP association establishment is even more heavyweight than TCP, and therefore access to DNS servers over HIP is unlikely to be popular. On the other hand, when using a local DNS server that performs recursive lookups on behalf of a client, performing DNS lookups over HIP may be an attractive option. It requires that the host identity of the DNS server is provided with its IP address by manual configuration or a DHCP extension.

- *Middlebox traversal for lookups.* Middleboxes, such as NATs and firewalls, can often prevent lookups from a local network. DNS requests typically traverse middleboxes well because DNS resolution is widely used in the current Internet. However, other resolution mechanisms, such as DHT lookups, can be blocked by a middlebox. If a HIP host is unable to resolve the host identity to an IP address, direct communication between hosts is not possible. Therefore, proposals for name resolutions should traverse through middleboxes.

- *Privacy of lookups.* By performing a lookup, a host reveals its intention to communicate with a given host. This privacy-violating issue is present in the current Internet with DNS lookups. HIP lookups can aggravate the problem, as host identities are long-lived names that can be cryptographically proved to belong to a host storing a private key. Proposals for HIP name resolution should take this issue into account, by possibly encrypting lookups to prevent eavesdropping. An attractive option would be to use HIP between a host and the resolution server. A host can also blur its lookup history by executing frequent random lookups or periodically changing its host identity.

- *Efficient mobility.* While HIP enables mobility of one association end point for established associations, initial rendezvous with a mobile host or simultaneous host mobility requires infrastructure support. The simple HIP resolution model in Figure 10.1(c) cannot provide fast updates to the IP address of the mobile host. The two-step resolution process in Figure 10.1(b) can be implemented with a method that permits faster updates than DNS, such as using a DHT. A HIP host can substitute its IP address with an address of a rendezvous server that relays packets to the mobile host. The rendezvous server can relay a single packet to a mobile host that afterward sends packets directly to a peer host, or relay all data packets. Relaying all packets can provide a higher delivery rate when both hosts frequently move at the cost of increased latency of triangular routing.

- *Interoperation with legacy hosts.* Designers of name resolution mechanisms should consider whether communication between legacy HIP-unaware hosts and HIP hosts is possible. As an example, the HIP rendezvous extension (Laganier *et al.* 2008) stores an IP address of the rendezvous server in a DNS server using a RVS Resource Record. It can be the only record stored for a HIP host. In that case, a legacy host cannot obtain an IP address of a HIP host for sending packets. Furthermore, if HITs are stored in AAAA fields of a legacy DNS server, a legacy host can misinterpret a HIT for an IP address and try to send packets to a HIT. Since currently HITs are not routable, the legacy host would not be able to contact the HIP host, unless the application can try other available AAAA fields.

10.2   Distributed Hash Tables

In this section, we review the basic principles behind Distributed Hash Tables (DHT) and mention the most popular implementations. Afterward, we describe the interface between HIP daemon and the publicly available DHT deployment, OpenDHT.

10.2.1   Overview of Distributed Hash Tables

A traditional *hash table* is a data structure associating keys with values with a hash function. A hash function maps a key to a bucket with a simple operation. Collisions appear when multiple values are hashed to the same bucket and can be resolved by creating a list of values in a bucket. A good hash function is computationally inexpensive and maps keys to values uniformly, without bias to some portion of key space. Typically the number of potential collisions is large, which creates a useful property of reverse mapping of a hash function. Namely, a bucket can only be associated with a set of possible keys, not with the exact key that produced the mapping.

A Distributed Hash Table is an extension of traditional hash tables for distributed applications running on multiple Internet hosts. A host serves as a node in the DHT by being responsible for storing values of a certain range of keys. To lookup a key in a DHT, an Internet host can contact any DHT node. The DHT nodes route the query towards a node responsible for storing the value corresponding to the key.

CAN, Chord, Pastry, and Tapestry were the first DHTs introduced in 2001. Later, researchers developed DHT variants such as Kademlia and Bamboo that optimize certain characteristics of the first DHTs. Many distributed applications including file sharing, cooperative web caching, and instant messaging utilize DHTs.

DHTs are typically designed to be scalable, supporting thousands and more nodes. Nodes can dynamically join and leave the DHT. Join and leave operations disrupt the current DHT structure and trigger recovery. A frequent change of node membership in a DHT is called *churn*. Successfully completing lookups under churn is a challenging design requirement for DHTs.

10.2.2   OpenDHT interface

OpenDHT (Rhea *et al.* 2005) is a publicly available DHT service running on PlanetLab, a world-wide testbed of several hundred servers. Unlike other DHT systems, the user does not have to run a local DHT node to be able to access OpenDHT. There is no registration or accounts required to store and lookup data in OpenDHT; available storage is shared fairly among all users. Applications can access OpenDHT using Sun RPC and XML RPC interfaces. The TCP protocol is used to contact DHT nodes from a client that allows access from behind most NATs and firewalls.

OpenDHT is based on Bamboo DHT servers. While its code is publicly available to set up its own set of DHT servers, it is more convenient for most users to rely on a publicly maintained set of servers instead of its own set. Some advanced HIP extensions, such as proving HIT ownership to DHT, require modification to OpenDHT code to operate. Then, running a private set of modified OpenDHT servers is the only alternative until the proposed changes have been incorporated to the official OpenDHT release.

OpenDHT stores values using a key that can be up to 20 bytes long. A key is typically a fixed-length hash of an actual identifier, such as a DNS host name. Each value has a limited Time-to-Live of a maximum of 604 800 seconds, which is one week.

The basic interface provided by OpenDHT implements *put* and *get* operations. In our experiments we are using an XML-RPC interface. XML tags corresponding to field names are used to encapsulate data values in the interface messages.

Table 10.1 shows the content of all OpenDHT put operations. OpenDHT uses application and client_library fields for the logging of requests. The application is given as a string name, and the client_library refers to the particular name and version of the XML-RPC library used in the query. The key supplied by an application selects an OpenDHT server to store the value.

Table 10.1   General content of a put operation to OpenDHT

Field	Type
application	string
client_library	string
key	max 20-byte array
value	max 1024-byte array
ttl_sec	four-byte integer (max 604800)

The OpenDHT server can reply to a put operation with three values. "Zero" refers to a successful put, "one" indicates failure because the capacity of the server is exceeded, and "two" indicates a temporal failure and a suggestion to re-try the put operation.

Table 10.2 shows the content of a get operation to OpenDHT server replies with values in an array and a placemark. The placemark can be used in a subsequent query to obtain additional values stored for the key.

To select the closest available server to the client, OpenDHT employs Overlay Anycast Service InfraStructure (OASIS) (Freedman *et al.* 2006). By performing a DNS lookup on the

Table 10.2   Content of a get operation to OpenDHT

Field	Type
application	string
client_library	string
key	max 20-byte array
maxvals	four-byte signed integer (max $2^{31}-1$)
placemark	byte array (100 bytes max)

host name "opendht.nyuld.net", the OpenDHT client obtains an IP address of the server that has the lowest RTT to the client's host. Alternatively, the client can retrieve a list of available OpenDHT servers from a well-known location and itself select the server to use.

Since OpenDHT is a public open data repository, anyone can place data under any key using the simple put/get interface. This opens a way to drowning attacks, when an adversary places a large number of garbage values under the same key as the client being attacked. To prevent such attacks, OpenDHT supports two mechanisms: immutable puts and signed puts. *Immutable puts* are of the form $k = H(v)$ (the key is a hash of the stored value) and cannot be removed until their lifetime expires. Clearly, immutable puts are not a suitable mechanism to secure HIP name resolution as the key value is fixed for such operations.

Signed puts include a value and nonce signed by the private key of the client storing the value. When another client retrieves a value stored under the given key, it also provides a hash of the public key of the client that stored the value. Therefore, that client would obtain only values stored with the private key matching the supplied public key. This prevents the client from drowning attacks. Signed puts are a suitable mechanism to secure HIT to IP address mappings for HIP hosts, as the HIT effectively is a hash of the public key of the client that stored the mapping.

10.3   HIP interface to OpenDHT

This section presents an instantiation of the OpenDHT interface to store HIP data (Ahrenholz 2007). The main operations defined by the specification draft are publishing and lookup of an IP address using HIT as a key. A mobile host can re-publish its current IP address to the DHT after moving. Compared with updating DNS, a DHT update is a fast operation. An Initiator can use the lookup operation to obtain the Responder's current IP address when a first connection attempt is made or if the peer location information is lost, e.g. after both hosts move simultaneously.

Table 10.3 shows the content of a publish operation using OpenDHT put. Base64 encoding allows presenting binary data using printable ASCII characters, in particular, letters A–Z, a–z, digits 0–9, and + and / symbols. To convert binary data to base64 encoding, the data is concatenated together and six bits at a time are used as an index to a string "ABCDEFGHIJKLMNOPQRSTUVWXYZabcdefghijklmnopqrstuvwxyz0123456789+/" to determine the substitution character. Correspondingly, at the receiver end the string is converted back to the binary form.

Table 10.3   General content of a put operation to OpenDHT
for HIP

Field	Value	Data type
application	hip-addr	string
client_library	implementation-dependent	string
key	128-bit HIT	base64 encoding
value	struct sockaddr	base64 encoding
ttl_sec	lifetime of address	numeric string

Figure 10.2 presents an example publish operation using a test stub. It submits a mapping from a HIT (2001:0014:766e:fbee:f74d:ec73:d6c5:28c0) to an IP address (193.167.187.132). The test stub used a publicly available OpenDHT service on PlanetLab.

```
POST /RPC2 HTTP/1.0
User-Agent: hipl
Host: 0000:0000:0000:0000:0000:ffff:c1a7:bb84:5851
Content-Type: text/xml
Content-length: 358

<?xml version="1.0"?>
<methodCall><methodName>put</methodName>
<params><param><value>
<base64>IAEAFHZu++73Texz1sUowA--</base64>
</value></param><param>
<value>
<base64>MDAwMDowMDAwOjAwMDA6MDAwMDowMDAwOmZmZmY6YzFhNzpiYjg0</base64>
</value></param>
<param><value><int>120</int></value></param>
<param><value><string>HIPL</string></value></param>
</params></methodCall>

HTTP/1.0 200 OK
Date: Wed, 19 Nov 2003 01:49:00 GMT
Server: bamboo
Connection: close
Content-Type: text/xml
Content-Length: 135

<?xml version="1.0" encoding="ISO-8859-1"?>
<methodResponse><params>
<param><value><int>0</int></value></param>
</params></methodResponse>
```

Figure 10.2   Example of XML-RPC put operation for HIT and the server's response. HIT (2001:0014:766e:fbee:f74d:ec73:d6c5:28c0) (HIT) serves as a key for an IP address (193.167.187.132).

Table 10.4 illustrates the content of a lookup operation for HIP using OpenDHT. An example of the lookup operation with a test stub is shown in Figure 10.3.

Table 10.4   Content of a get operation to OpenDHT for HIP

Field	Value	Data type
application	hip-addr	string
client_library	implementation-dependent	string
key	128-bit HIT	base64 encoding
maxvals	implementation-dependent	numeric string
placemark	NULL or copied from server reply	base64 encoding

The following HIP-related additional operations are under discussion for OpenDHT, although not yet a part of a specification.

Lookup of a FQDN to HIT (and possibly also to an IP address) supplements DNS because many users do not have access to the organization's DNS server or do not have the expertise necessary to change the DNS configuration (e.g., using the Dynamic DNS). Furthermore, mapping symbolic names instead of FQDN would allow users to give human-friendly names to their devices (as in Unmanaged Internet Architecture (Ford *et al.* 2006)).

Reverse lookup from IP address to HIT (and possibly to a FQDN) supports security and debugging in a similar way as reverse DNS does today. If an Initiator only has the IP address but not the HIT of the Responder, the Initiator can attempt to use HIP opportunistic mode to create a HIP association. However, that opens the possibility of Man-In-The-Middle attacks. An Initiator can lookup the HIT in the DHT using the IP address as a key and check that also the HIT maps correctly to the given IP address.

To reduce the amount of configuration state needed at a host, it may be useful to store LSI to HIT mapping in a DHT. However, LSIs have local meaning only. Managing the LSI to HIT mapping and additional delay for such lookups presents significant arguments against such a design. Therefore, LSI to HIT mapping was removed from the OpenDHT HIP specification.

When a host publishes a HIT, it is better to verify that the host owns the HIT, i.e., has a private key matching the public key that produces the HIT. The DHT server performs verification by requiring the host to execute the HIP base exchange before its HIT is published. This requires updates of the operating system software on DHT servers to enable HIP support, and possibly insertion of a few checks to the DHT implementation.

Specifications for the HIP-DHT interface suggest that a HIP daemon can query IP addresses of all known HIP upon startup. Querying each address from OpenDHT when an association setup is requested would introduce noticeable delay to the user. Since the IP addresses can change often for a mobile host, the Initiator can send I1 to the latest known IP address and launch a DHT lookup in parallel. Then if the I1 timer expires, I1 can be retransmitted to a current address returned from the DHT lookup.

Initially, OpenDHT did not support the remove operation. The inserted value could not be removed from the DHT until its Time to Live expired. When an IP address changed, the old value remained in the DHT and the new value was added to the end of the list of values. For backward compatibility, the last value should be considered as the currently reachable IP address.

10.4   Overview of overlay networks

The term *overlay network* refers to a set of virtual links established between Internet hosts on top of existing IP connectivity. Each virtual link can span several IP hops or physical

```
[gurtov@hipserver ~]$
hipconf dht get 2001:0014:766e:fbee:f74d:ec73:d6c5:28c0
name='opendht.nyuld.net' service='5851'
Connected to OpenDHT gateway 132.68.237.34.
Value received from the DHT 193.167.187.132

POST /RPC2 HTTP/1.0
User-Agent: hipl
Host: 127.0.0.1:5851
Content-Type: text/xml
Content-length: 305

<?xml version="1.0"?>
<methodCall><methodName>get</methodName>
<params><param><value>
<base64>IAEAFHZu++73Texz1sUowA==</base64></value></param>
<param><value><int>10</int></value></param>
<param><value><base64></base64></value></param>
<param><value><string>HIPL</string></value></param>
</params></methodCall>

HTTP/1.0 200 OK
Date: Wed, 19 Nov 2003 01:49:00 GMT
Server: bamboo
Connection: close
Content-Type: text/xml
Content-Length: 566

<?xml version="1.0" encoding="ISO-8859-1"?><methodResponse>
<params><param><value><array><data><value><array><data>
<value><base64>MDAwMDowMDAwOjAwMDA6MDAwMDowMDAwOmZmZmY6YzFhNzpiYjg0
</base64></value>
<value><base64>MDAwMDowMDAwOjAwMDA6MDAwMDowMDAwOmZmZmY6YzBhODo0ZDAx
</base64></value>
<value><base64>MjAwMTowNzA4OjAxNDA6MDIyMDowMjExOjExZmY6ZmU4NDpiNzkkx
</base64></value>
<value><base64>MDAwMDowMDAwOjAwMDA6MDAwMDowMDAwOmZmZmY6YWMxMDo2YzAx
</base64></value></data></array></value>
<value><base64></base64></value></data></array></value></param>
</params></methodResponse>
```

Figure 10.3 Example of XML-RPC get operation for HIT and the server's response. Note that HIT in request and the first IP address in the response match those in Figure 10.2.

links. Overlay networks provide a popular solution for various peer-to-peer systems and advanced network services such as multicast and QoS. The advantage of overlays is easier deployment, as they do require changing of existing routers or other collaboration from ISPs. The disadvantages include worse performance due to indirect routing and higher protocol overhead. Often the *stretch* is defined as the ratio of the number of IP hops or the latency in the overlay route versus in the direct IP route. The use of overlays often presents an "unclean" solution from the Internet architecture viewpoint, when the protocol stack is not layered in a classical way; in that case some functions can be duplicated on even in a conflict at different protocol layers.

To ease the deployment of services, Stoica et al. proposed an Internet Indirection Infrastructure (i3) overlay network that offers a rendezvous-based communication abstraction (Stoica *et al.* 2002). Instead of explicitly sending a packet to a destination, each packet is associated with a destination identifier; this identifier is then used by the infrastructure to deliver the packet. As an example, a host R can insert a trigger (id, R) in the i3 infrastructure to receive all packets that have the destination identifier id.

i3 provides natural support for mobility. When a host changes its address, the host needs only to update its trigger. When the host changes its address from R_1 to R_2, it updates its trigger from (id, R_1) to (id, R_2). As a result, all packets with the identifier id are correctly forwarded to the new address. Note that this change is completely transparent to the sender.

The primary aim of the Secure-i3 proposal (Adkins *et al.* 2003) was to provide a network architecture that is more robust against DoS attacks than today's networks. The basic idea is to protect against DoS attacks by hiding the IP addresses of the end-hosts from other users of the network. The indirection approach provides straightforward implementation for multicast, mobility, and multi-address multihoming. In Secure-i3, there are two types of triggers, public and private. Public triggers are used to announce the existence of a service and are well known (announced on web pages, in the DNS, or on other public media). Private triggers are used for the actual communication between sender and the receiver(s), which are the only ones that know the private triggers.

Finally, we describe three advanced capabilities of Secure-i3. In Secure-i3, a public trigger cannot point to the end-host, but only to a private trigger to prevent cycles in the infrastructure and malicious misuse of triggers. Therefore, a *trigger chain* of two right-constrained triggers is used to insert a given identifier into the infrastructure. To run legacy applications over i3, a *proxy* located on the client and the server must be used. The proxy transparently intercepts DNS requests and forwards data packets to the i3 infrastructure. Recently, a capability to send data directly between the client and the server has been added to i3. Known as *shortcuts*, it allows efficient data transfer between hosts, but does not offer currently any cryptographic data protection.

Since Hi3 relies on features from the basic i3 architecture and the Secure-i3 extension, from here on we do not differentiate between them. Hence, whenever we write i3, we refer to Secure-i3.

10.5 Host Identity Indirection Infrastructure[1]

In this section we describe the Hi3 architecture in detail. More specifically, we consider the particulars of separating session control, actual data delivery, and service naming. We analyze problems induced by this separation and present the solutions. Additionally, we qualitatively analyze the key advantages of Hi3 and discuss perspective of the design (Korzun and Gurtov 2006).

---

[1]Reproduced by permission of © 2006 Elsevier: Korzun D and Gurtov A. On scalability properties of the Hi3 Control Plane, *Elsevier Computer Communications*, **29**(17): 3591–3601, November 2006. doi:10.1016/j.comcom.2006.05.014

The original concept of Hi3 (Nikander *et al.* 2004) observed that a HIP rendezvous server and a single i3 server are functionally close. Therefore, the basic idea is to allow direct, IP-based end-to-end traffic while using an indirection infrastructure to route the HIP control packets.

We enhance this proposal in the following way. First, we generalize the i3 representation of the host identity namespace such that peers can use separate identifier layers for service and for hosts. Second, we concentrate on the Hi3 design for the control traffic and introduce all available control messages. Third, our discussion of Hi3 advantages and perspective gives the most comprehensive overview available at present.

10.5.1 Separating control, data, and naming

In HIP, a rendezvous server is used to fully support association setup between two end-hosts, simultaneous movement, location privacy, and third party referrals. The concept of the HIP rendezvous can be enhanced to an overlay rendezvous infrastructure, a distributed and decentralized instantiation of the HIP rendezvous server. In the Hi3 solution, an i3 network implements this infrastructure. Figure 10.4 illustrates this idea.

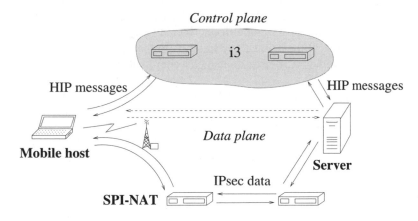

Figure 10.4 i3 as an instantiation of the HIP rendezvous server.

The infrastructure forms *the Hi3 control plane*, relaying HIP signaling messages. Data traffic flows directly between end-hosts, using plain IP routing and forming *the Hi3 data plane* (shown with dashed lines in Figure 10.4). HIP and i3 provide secure communications for the control plane; IPsec and IPsec-aware NAT (SPINAT) give basic protection to the data plane.

Inherited from HIP, Hi3 uses separate identifiers for location (IP addresses) and endpoint (HIT). HIT acts as an identifier for a public i3 trigger, reflecting in the infrastructure the HIP-based separation of naming. Note that if i3 identifiers are longer than 128-bits, then prefixes to HITs are applied.

For a host A, the pair of triggers [HIT_A | ID_A] and [ID_A | IP_A] can be stored in i3, where HIT_A is a public i3 identifier of A and ID_A is its private i3 identifier. The latter is constructed by *A* according to the constrained IDs technique (Adkins *et al.* 2003; Lakshminarayanan *et al.* 2005).

We enhance this mechanism to support a higher naming layer. Public trigger identifiers are used for association setup between a client and a server. When the association has been completed, the server creates a private identifier for the client (or uses an existing identifier for a group of clients). After that, only the private trigger is used to relay control messages from the client to the server.

Therefore, the i3 public identifiers resemble the service identifiers in the layered naming architecture (Balakrishnan *et al.* 2004), while the private trigger identifiers clearly form some "lower" naming layer. Utilizing HIP, we can use fresh, newly generated host identifiers (or identifiers for a group of hosts) as private trigger identifiers. To secure the binding between the public and private triggers, i.e. between the service and host identifiers, the cryptographic delegation is used (Nikander and Arkko 2002).

10.5.2   The data plane

Protecting end-to-end data traffic

For basic end-to-end data protection we use HIP. In its simplest form, HIP encapsulates all data traffic in ESP, protecting integrity, authenticity, and (optionally) confidentiality. However, HIP alone does not protect against distributed Denial-of-Service attacks. In plain HIP, the hosts always reveal their real IP address(es) to their potential peers. Therefore, a host could tell a large number of zombies to launch a coordinated bombing attack against the target host.

To protect against distributed Denial-of-Service, we extend the notion of using IPsec-aware middle boxes (Nikander *et al.* 2004). A number of SPINATs[2] (IPsec-aware middle-boxes) are placed on or close to the possible data paths. These provide a fast-path barrier against bombing Denial-of-Service, simultaneously hiding the actual IP address of the servers. The method structurally resembles i3 shortcuts (Stoica *et al.* 2002) but is more secure than using shortcuts and works independently of the rendezvous infrastructure.

To employ SPINATs at the time the client and server inform each other about the IP addresses to be used for data traffic, they tell the addresses of SPINATs serving them instead of telling their real IP addresses. In other words, the use of SPINAT is completely controlled by the involved host, independently of the rendezvous infrastructure. In practical terms, in most cases the SPINAT can act by inspecting HIP base and mobility exchange packets flowing through it; see Section 10.5.2. Mobility performance and DoS resistance of SPINAT has been measured by Ylitalo *et al.* (Ylitalo *et al.* 2004). The results suggest that the efficiency of the data plane is not significantly reduced by the presence of SPINATs.

Figure 10.4 illustrates the use of ESP envelopes and SPIs to implement the Denial-of-Service protection for the data traffic. As described in Tschofenig and Shanmugam (2006), it is easy to design such a middle box that forwards and filters traffic based on <dst,SPI> pairs. The filtering can be extended to include source addresses. In the typical case of the control

---

[2]Note that even though the SPINATs in their basic form translate network addresses in order to hide the real IP address(es) of the server, that translation may still happen between IP addresses belonging to the same IP realm instead of distinct IP realms. Alternatively, if placed always on the path (instead of close to the path), they can function as plain filters that do not perform address translation at all. The following discussion mostly applies to all cases, with just minor differences.

packets passing through the middlebox, the middleboxes can securely learn the appropriate mappings by listening to the signed control packets. If the control and data packets take completely different paths, there must be explicit signaling between policy points at the control and data path. For example, the hosts can use the HIP registration protocol (Laganier *et al.* 2008) to create suitable initial state at some SPINAT.

As the SPINAT knows the allocated SPI mappings, including the source and destination IP addresses, for its basic functionality, it can easily filter out most unwanted traffic. A random attacker can't learn the real IP address of the server; it can only learn the IP address of the SPINAT. Getting packets through the SPINAT requires that the attacker knows a valid SPI, causing random packets to be effectively filtered. However, an attacker that establishes an (opportunistic) HIP association with the server learns a valid SPI, which it can communicate to a large number of zombies. Hence, source address spoofed traffic from zombies that have learned a valid SPI still forms a potential problem. Applying heuristics based on ESP sequence numbers makes such coordinated attacks harder but not impossible; the zombies can increase the sequence number in rough synchrony, resulting in unwanted high-volume traffic where the sequence numbers mostly fall within the replay window.

An obvious means to protect against zombie-based synchronized bombing attacks is to deploy source address filtering everywhere in the network. That would prevent zombies from sending valid-looking packets; the packet's source address would necessarily be different, resulting in the packets being dropped at the first SPINAT on the path.

In Hi3 the IP source address field is no longer needed[3]. The control packets are explicitly routed by the identifiers; there the source HIT takes the function of the source IP address. The data packet destination is always based on the local by-HIP-created IP-layer state, and the source address is always ignored (Moskowitz *et al.* 2008). Hence, we surmise that the source address field could be used to record the actual path taken by the packet (Candolin and Nikander 2001).

Utilizing the possibility of using HIP-based mobility, a server under an attack can move the legitimate traffic to other available SPINATs. Hence, a multihomed site with multiple entry SPINATs or a host with suitably selected independent SPINATs can move legitimate traffic from the SPINAT under attack to another one. The server can also use the HIP control packets to tell the attacked SPINAT to drop forwarding all traffic on the attacked SPI. This is structurally similar to a host dynamically changing its private trigger in i3.

All these mechanisms still leave the SPINATs themselves vulnerable to distributed Denial-of-Service attacks. However, a well designed SPINAT can handle the packets at the fast path, being able to process orders of magnitude more packets than a typical host can. Secondly, SPINATs can be placed in the network at high capacity links that are harder to flood completely. Hence, even though the outlined mechanisms do not necessarily prevent zombie-based coordinated flooding Denial-of-Service attacks from taking place, they greatly mitigate their destructiveness by redirecting the attacks from their current targets to nodes with higher-capacity links and ability to handle a much larger number packets.

Compared with i3 shortcuts and private triggers, the main difference is that the protection works independently of i3. Hence, different parts of the network can use different protection mechanisms. The hosts do not need to learn the location of i3 servers (which is better kept hidden anyway), making the selection of a suitable forwarder easier. Furthermore, we surmise

---

[3]The source addresses are often still useful, if for no other purpose than learning the identities of certain i3 servers; see Section 10.5.3.

that the resulting i3 infrastructure is easier to protect than the original one as it need not be designed or provisioned to carry all traffic.

Supporting multiple IP realms

In the discussion above we have glossed over problems caused by multiple IP realms and the resulting partial connectivity. For the system to work properly in the current multi-realm IP reality, two requirements must be fulfilled. First, all hosts must be reachable through the i3 infrastructure. Second, the hosts must know at least one public IP address of a SPINAT serving them so that they can tell that address to their peers at or behind the public Internet. There are multiple ways to fulfill the requirements.

We first consider the requirement of knowing a public IP address of a serving SPINAT. As the SPINATs are assumed to form a new piece of infrastructure, an anycast-based mechanism can be used to learn suitable nearby SPINATs. Alternatively, in a corporate environment SPINAT-related information could be naturally distributed along with other managed configuration data. Additionally, on-path, passive plain NATs could be detected directly, and the necessary state in them can be created with methods similar to STUN (Rosenberg *et al.* 2003) or ICE (Nikander *et al.* 2006).

To make hosts reachable by the i3 infrastructure, the simplest way seems to be to locate the infrastructure in the public Internet, requiring the hosts in other IP realms to maintain active connectivity with that/those i3 server(s) that hold their private trigger(s). In that way the packets sent to the private trigger can be always passed to the hosts over active connections. Alternatively, if a host is able to create semi-permanent state at some SPINAT with a public IP address, it can list the SPINAT's IP address at the private trigger, again resulting in the packets coming to the right host. However, in this case the i3 server does not use an existing connection for sending the packet but sends it to the SPINAT, which in turn forwards it according to the state associated with the HIT.

In any case, multiple realm support requires reachability state to be created at the SPINATs between the realms. This state can be created explicitly, by hosts registering their identifiers at the cross-realm SPINATs. The resulting infrastructure resembles proactive hop-by-hop host routing, but takes place on a layer above the current IP routing layer. Alternatively, assuming the existence of a single most preferred realm (i.e., the public Internet), SPINATs at the realm boundaries can learn the identifiers of the hosts behind them. In order to remain reachable, the hosts must keep sending packets towards the preferred realm. In this case, the resulting infrastructure resembles link layer bridging.

Handling alternative encapsulation formats

The HIP architecture is planned to be extended to support other encapsulation mechanisms in addition to ESP (Moskowitz *et al.* 2008). Independently of the encapsulation method, HIP requires that there is enough of information in the data packet that it can be successfully de-multiplexed and tagged with source and destination HITs. Furthermore, de-multiplexing must work independently of the source address in the packet. We surmise that as long as the information used for de-multiplexing is sufficiently hard to predict but easily verifiable with a state that can be formed from the public information in the HIP control packets, the above-outlined SPINAT-based protection can be easily generalized to future HIP encapsulation methods.

Delegating part of the processing to the infrastructure

When a service is registered to the i3 infrastructure by creating a trigger for the service identifier, instead of pointing to a server host that directly implements the service, the trigger can point to an infrastructure node that handles the I1 packets directly. In HIP, the hosts handle I1s by replying with precomputed R1 messages. To implement service/host separation, we assumed above that the R1 messages use a host identity different from the service identity, explicitly denoting delegation from the service identifier to the host identifier.

We now propose a separate *service distribution function, SDF,* which can be integrated with the actual service nodes, if desired. As in HIP, the function handles I1 messages sent to the service by replying with suitable pre-computed R1 messages. In contrast to HIP, the R1 packets are not signed by a host key but by the service key. Instead of carrying just one public key, they carry two: the service key corresponding the HIT and used for signing the packet, and a host key. The host key is stored in a HIP parameter with a new parameter type, denoting delegation. The host key identifies a host that will provide the service to the client[4]. The signature in the packet allows the client to verify that the host is indeed authorized to provide the service[5].

The service distribution function allows I1 packets to be processed completely by the infrastructure. The servers need not see the I1 packets in the first place. Furthermore, the function of forming the R1 packets and replying to I1 packets with R1 packets can be separated. Hence, the R1 packets can be created and signed by a well protected node that has authority over the service while the R1 distribution can take place directly by the i3 server that stores the public trigger. All that is required is that the R1 generating node supplies the i3 servers with pre-computed R1 packets.

Compared with the previous proposals (Nikander *et al.* 2004), the service distribution functions delegates the R1 *formation* from the server host to another node. This modifies the expected semantics of the HIP puzzle, requiring slight modifications. In HIP, the server (Responder) creates the puzzles in such a way that it can easily verify that a given puzzle has indeed been recently created by it. That allows the Responder to reject correctly solved puzzles that have not been created by it, thereby protecting the host from pre-computation attacks. By delegating puzzle creation to another function, possibly located in another node, we lose the ability to verify the puzzle origin. On the flip side, while the previous proposals suggested that the puzzle solution is verified by the infrastructure instead of the Responder, the infrastructure could only verify that the puzzle has been correctly solved, not that it had indeed been posed by the Responder.

In our enhanced processing model the situation changes. Note that the host can no longer verify the puzzle alone, as it would become vulnerable to pre-computation attacks. Hence, the infrastructure *must* verify the puzzle origin; at the same time, it can easily verify that the puzzle solution is correct. For this, the node that generates the R1 messages can provide the i3 server holding the private trigger with the necessary information for verifying the puzzle's origin and freshness.

---

[4]We surmise that the host would typically be a *virtual* host and that the host key would be used only for serving one client. However, the architecture does not impose these properties; there are clearly cases where a different pattern offers better utility.

[5]Compare this with the SPKI (Ellison *et al.* 1999) use of signatures.

The remaining problem of providing the server assurance of puzzle verification can be solved in several different ways. The simplest way is to let the server assume that any I2 packets coming from the i3 server holding the private trigger have passed puzzle verification. If the server fully trusts the infrastructure and if there is a secure channel between the nodes, this can be deemed secure. Another solution is to let the service distribution function and the actual servers share data about puzzle creation, thereby allowing the servers to (re)verify the puzzle[6].

Figure 10.5 demonstrates the relaying of HIP handshake over i3 with delegation of I1 processing to the infrastructure (see also Section 10.5.3).

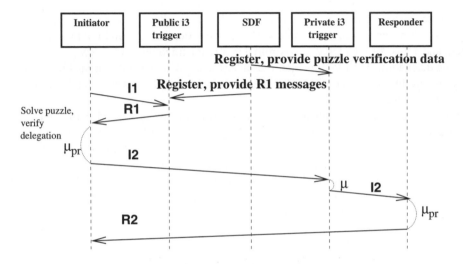

Figure 10.5 Hi3 base exchange with delegation.

Once the Initiator has processed the R1 packet and produced the puzzle solution (in time μ_{pr}), it sends an I2. The I2 is now sent to the private trigger. The i3 server that keeps the private trigger verifies that the puzzle solution is correct and created by the SDF before passing the I2 packet to the server. Effectively, this distributes the proposed i3 DoS-filter function over all i3 server nodes, allowing the puzzle to be formed and verified by different nodes. In sum the server and the Responder spend time μ_{pr} in verifying the puzzle and authenticating the client.

Mobility

In Hi3, basic mobility between already communicating hosts can be provided directly at the HIP and SPINAT level, without involving the i3 infrastructure. Only if the hosts lose direct reachability, do they need to revert back to the infrastructure (see also Section 10.5.3). Even in that case hosts will use private triggers, being safe from attacks launched by third parties. For this to work, the hosts must keep the infrastructure updated with their current location information.

---

[6]For example, the service distribution function and the servers can use the same keyed pseudo-random number generator, run in synchrony, to drive puzzle generation.

By combining the end-to-end mobility provided by HIP and the indirect mobility support provided by the infrastructure, the resulting mechanism is highly efficient (no triangle routing for regular data) and robust (a property inherited from i3).

In addition to plain end-host mobility, Ylitalo has suggested how to apply HIP for network mobility (Ylitalo 2005). Furthermore, the signaling delegation ideas by Nikander and Arkko (2002) can be applied more generally to HIP, resulting in savings in the air interface. The support of these ideas in Hi3 is left for future study.

Signaling multicast and anycast

Inheriting from i3, Hi3 allows the HIP control messages to be replicated and distributed to multiple receivers. While we can't imagine any immediate benefit of this for the current HIP control packets, we could imagine a HIP-based group key distribution protocol. Such a protocol would most probably benefit from a control plane multicast facility.

The proposed service distribution function forms a rudimentary control-layer anycast service. It allows a service to be identified by a single identifier while allowing the actual service to be provided by multiple nodes. By providing the i3 node serving the public trigger with IP-layer topology information, it becomes possible to create a service that is always provided by a node topologically close to the client.

10.5.3   The control plane

The control plane is used to relay HIP messages in two cases. The first is when, before direct end-to-end communication, two end-hosts establish a HIP association. In this case the main benefit of using the control plane is DoS protection; the IP addresses of both hosts are not revealed until mutual authentication is completed. The second case is when, having an established connection, the hosts lose the direct end-to-end connectivity. Such a case is important for end-host mobility, e.g., the connectivity is lost after a simultaneous movement of both hosts. Therefore, the control plane is a trusted third-party that aims in establishing and keeping the data plane connectivity between communicating peers.

Let C and S be two communicating end-hosts. We assume that C is a HIP Initiator (e.g., a mobile client) and S is a HIP Responder (e.g., a stationary Internet server). Figure 10.6 illustrates a distribution of their HIT-based public and private triggers in i3; for first contacts the peers use neighbor i3 nodes that they happen to know ($S2$ and $S5$).

In the rest of this section, available request types to the control plane are described. Diagrams of packet flows of the requests are shown in Figure 10.7. Arrows represent paths, by which packets of requests follow. An arrow label is a packet name (e.g., I1 or R2) and a sequence mark (i.e., "a" is for the first part of the flow, "b" is for the next part, etc.). Thick arrows denote possible multi-hop Chord lookups (Stoica *et al.* 2001), when the destination i3 node is not cached in the source i3 node.

Pure HIP association setup

Figures 10.7(a) and 10.7(b) show the Hi3 establishment of a HIP association between C and S.

To establish a connection with a server, a client C sends an I1 packet to the IP address of a random i3 node it happens to have; in our case this node is $S2$ (path I1.a). The public trigger

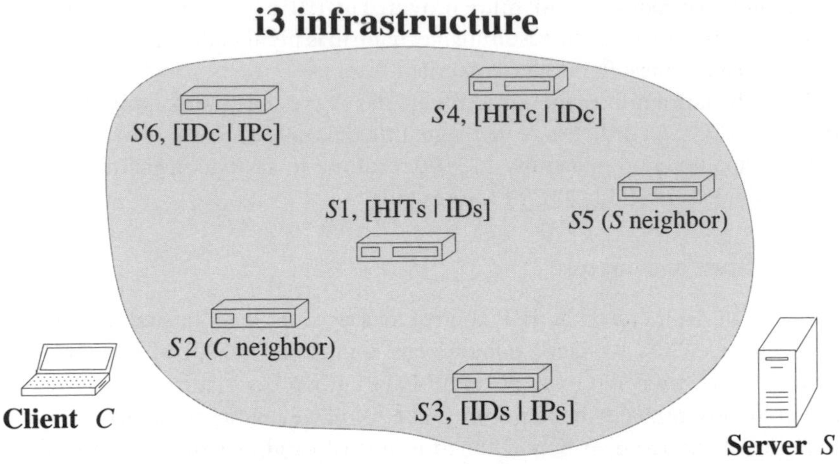

i3 infrastructure

*S*6, [IDc | IPc]

*S*4, [HITc | IDc]

*S*1, [HITs | IDs]

*S*5 (*S* neighbor)

*S*2 (*C* neighbor)

Client *C*

*S*3, [IDs | IPs]

Server *S*

Figure 10.6 Schematic distribution of HIT-based triggers in i3 for *C* ↔ *S* communication.

for *S*, HITs, is stored in *S*1, and *S*2 forwards the packet to *S*1 via i3 (path I1.b). The client obtains the correct i3 node for future contacts to *S* (path I1.c', in parallel with the primary branch I1.c–I1.d). The private trigger of *S* resides on *S*3, to which *S*1 forwards the packet (path I1.c), and finally *S*3 delivers it to *S* (path I1.d).

 A similar procedure is followed by *S* to send an R1 reply packet to *C*. The neighbor i3 node *S*5 is contacted first (path R1.a). The public trigger for the client *C*, HITc, is stored in the node *S*4, to which *S*5 forwards the packet via i3. Then *S*4 notifies *S* about the correct i3 node for communicating with *C* (path R1.c', secondary branch) and forwards R1 to *S*6, which keeps the private trigger for the client (path R1.c). Finally, *S*6 delivers the packet to the client (path R1.d).

 The consequent I2–R2 exchange occurs in a similar manner, see Figure 10.7(b). The only difference is that the packets are sent straight to the i3 nodes keeping the public triggers, *S*1 and *S*4.

Optimized HIP association setup

In the pure form of association setup, the public and private triggers of both hosts must have been inserted to i3. It is, however, unreasonable to require a client to keep its triggers in i3 even temporarily. The solution is to delegate an initial part of the setup from *S* to i3. This way, the node *S*1, which keeps a public trigger of *S*, caches pre-computed R1 packets. Figure 10.7(c) shows the optimized form of setup.

 As in the pure association setup, *C* sends an I1 packet to the *S*2, its neighbor (path I1.a), and the packet is forwarded to *S*1 (path I1.b). Then, unlike the pure setup, *S*1 replies directly to the client with an R1 packet since it has been cached (path R1). In this reply, *S*1 also notifies *C* about the correct node to contact *S* via i3.

 The I2 packet is sent to *S*1 (path I2.a), then forwarded to *S*3 via i3 (path I2.b), and finally delivered to *S* (path I2.c). The packet is expected to contain the HIP locator parameter,

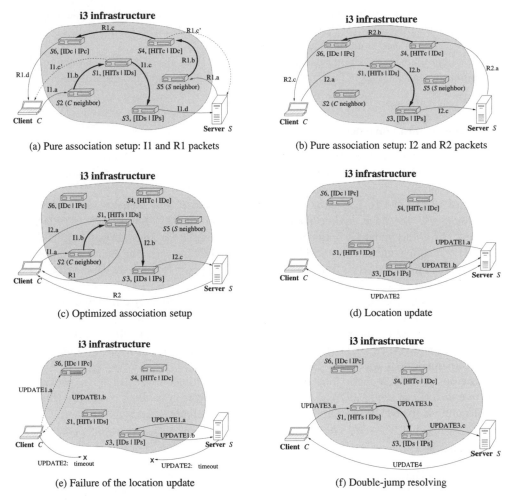

Figure 10.7 Request types in Hi3.

listing the client's real IP address. The server replies with an R2 packet directly to the client (path R2). That is, the control plane is not involved in R2 delivery.

The optimization reduces the load of S, since it receives fewer packets and does not check the puzzle solution of C.

Location update

Both C and S can change their locations and, consequently, their IP addresses during the communication.

Typically, only one host at a time changes its IP address and performs a *location update*. If the change is due to the server, then S updates its private trigger in i3 (Figure 10.7(d)). The location update also causes the *HIP update exchange* (Nikander *et al.* 2008) running over the data plane between C and S.

One update to i3 is sufficient independently of the number of hosts communicating with S via the private trigger. UPDATE1 and UPDATE2 packets can be sent in parallel. Therefore, no significant overhead is introduced, compared with HIP.

For client C, having a trigger pair in i3 is optional. Thus, if C changes its location, then the signaling packets, UPDATE2, run directly between hosts, and the control plane is not used.

Simultaneous host movement

It may happen that both hosts change their locations at once, an event known as the *double-jump problem*. Note that simultaneous mobility of C and S is rare compared with the usual location update.

Figure 10.7(e) and 10.7(f) show what happens if both C and S change their addresses simultaneously. The hosts update their private triggers in i3 (Figure 10.7(e), UPDATE1 packets). For C this update is optional (dashed arrows in the diagram). In parallel, the hosts start a HIP location update over the data plane, UPDATE2 packets. The exchange fails since each host uses the out-of-date IP address to contact the peer. This failure can be discovered by a timeout.

At this point the hosts need the control plane to recover from the double-jump (Figure 10.7(f)). The double-jump can be discovered by both hosts, but the client is responsible for starting the recovery. It sends the first packet of the HIP update exchange (UPDATE3, addressed to HIT_S) to S via i3. After receiving this packet, S continues the update talking directly to C.

Obviously, in this scenario most of the delay is due to timeout for UPDATE2 packets. To improve this, C can send UPDATE2 and UPDATE3 packets in parallel, especially when the double-jump is likely. Unnecessary UPDATE packets are ignored by S.

HIT insertion

Let HIT_A be a HIT of a host A and IP_A be a recent IP address of A. There are two reasons for A to insert HIT_A into i3. First, A is a server, thus HIT_A has to be in i3 permanently; second, A is an Initiator of a pure association setup, thus HIT_A is used by the Responder.

To insert HIT_A, A constructs a private i3 identifier ID_A. Then A sends two requests to i3, namely inserting the public trigger $[HIT_A \mid ID_A]$ and private trigger $[ID_A \mid IP_A]$. Each request requires a Chord lookup to assign an i3 node, say $S0$, for storing the trigger. In a successful case, $S0$ sends directly to A the acknowledgment, which contains a recent node's IP address.

HIT refreshment and HIT re-insertion

At trigger's Time to Live is limited. For instance, the recent implementation[7] applies the value of 30 sec. Keeping HIT_A alive in i3 requires the host A to refresh regularly both public and private triggers, sending the refreshment requests directly to the corresponding i3 nodes.

If the i3 node crashes, the trigger is lost. The end-host can re-insert it by sending the insertion or refreshment request to any i3 node. Similarly, if the trigger has been removed

---

[7]http://i3.cs.berkeley.edu/impl/

before the refreshment was received, then the trigger is re-inserted. However, the trigger might be located to another node, since i3 undergoes node joining and leaving.

In a successful case, the acknowledgment is received. It always contains a recent IP address of the i3 node that keeps the trigger.

HIT removal

HIT removal happens either automatically after the public and private triggers expire, or explicitly by sending two requests. Although the latter case seems redundant, it can be important for a host to remove HIT as soon as possible when an attack via the triggers has been discovered.

A case with separate naming for hosts and service

Although the requests were designed without taking separating naming into account, the corresponding modification is clear. A public identifier (HIT) is used only for a first contact with the server. Whenever a client is trusted, it uses a given server's private trigger, e.g., for resolving from a double-jump or for establishing new connections.

10.5.4   Discussion of the Hi3 design

Let us summarize the key design ideas of the Hi3 architecture and their consequences. We also compare the pros and cons of using HIP and i3 versus their combination, Hi3.

The integration of the i3 triggers mechanism allows using two name layers; private triggers identifiers form a host namespace like in HIP, and public triggers are considered as a service namespace. The latter is close to endpoint descriptors for applications in the namespace model of Komu *et al.* (2005). Hence, this Hi3 feature can be used to fill the gap between HIP and application layers.

When this separating naming is used, the control plane performance increases since a client talks not via the public trigger's node but directly via the node with the private trigger. In this case, the security can suffer since the rule of compulsory usage of a public/private trigger pair is violated. This rule is due to Secure-i3, and its violation substantially reduces the infrastructure to that of basic i3. Hence, extra security mechanisms should be applied, like the cryptographic delegation (Nikander and Arkko 2002) or the HIP registration extension (Laganier *et al.* 2008).

Any successful request ends with the acknowledgment to the end-host. Some requests can fail owing to packet losses in a network or i3 inconsistency and unavailability. This problem is not very significant since i3 is based on top of Chord DHT, which is known as resilient to massive failure of nodes (Stoica *et al.* 2001), to dynamic node joining and leaving (Li *et al.* 2004; Stoica *et al.* 2001), and to pathological states (Liben-Nowell *et al.* 2002). Therefore, the problem happens rarely, and a simple resubmission of a request after a timeout is adequate.

The basic HIP provides efficient and secure end-to-end connectivity. If the HITs and IP addresses of end points are known, it can work without additional infrastructure, thus having no issues with infrastructure cost, accountability, trust, or fault-tolerance. Basic HIP provides limited DoS protection by enabling the Responder to make the Initiator solve a computationally substantial puzzle before creating state in the Responder. Mobility of one end point at a time is supported, but there is no way to perform the reverse mapping support[8]. HIP with a rendezvous server enables mobility of both end points, while preserving accountability and the trust model, since the rendezvous server is chosen by the Responder.

The advantages of i3 include better DoS protection, support for simultaneous mobility, and higher fault-tolerance when using a DHT with data replication. Disadvantages of i3 include reliance on an extensive infrastructure, server scalability, use of UDP, lack of traffic encryption, and considerable complexity of i3 as an overlay network. There is limited experience with widespread i3 deployment, thus it is difficult to assess how scalable the servers are. The latency of relayed control traffic will mostly be affected by forwarding and network delays. However, relaying all control and data traffic through the i3 infrastructure would likely prove burdensome, and by mutual agreement, the client and the server could use i3 only for initial contact and afterward exchange the data directly using shortcuts.

The basic i3 system does not provide data encryption, although it could be implemented as an add-on feature. There is no encryption and privacy for control packets. When a public infrastructure is used, i3's extensive infrastructure requirements bring other serious security issues including the possibility of malicious or misbehaving i3 nodes that do not forward correctly and a lack of trust of arbitrary i3 nodes from end points. Note that Secure-i3 introduced several constraints on the structure of triggers to prevent misuse of triggers by third parties and formation of loops in the topology. Finally, diagnosing problems in a distributed Internet system is always challenging, and the indirection introduced by i3 further complicates the situation.

The combination approach of Hi3 helps to address some of the separate shortcomings of HIP and i3. The advantages of using i3 as a control plane for HIP include protection from DoS attacks, solving the double-jump problem, and providing an initial rendezvous service. By hiding parties' IP addresses until the HIP handshake partially authenticates them, Hi3 provides additional protection against DoS attacks. Although some DoS protection could be provided by a HIP rendezvous server, the client's IP address is revealed to a server in the first control packet. Simultaneous mobility of both hosts in i3 is supported by sending update control packets via i3 when end-to-end connectivity is lost. Hi3 inherits the challenges of the extensive i3 infrastructure, including trust, accountability, and cost issues.

---

[8]A reverse lookup from an IP address to HIT (similar to reverse DNS) provides additional functionality, for example, for security purposes.

11

Micromobility

The HIP mobility extensions presented in Section 5.1 enable end-to-end host mobility. Upon each change of the IP address, the mobile host needs to inform all its peers. When the communicating hosts are located far apart (e.g. on different continents) the end-to-end latency is high, which increases the probability of packet loss during a handover. Furthermore, with frequent handovers and a large number of established HIP associations the amount of mobility signaling can be significant. It is especially a concern for mobile hosts located behind a slow or expensive wireless link, e.g., in a cellular network.

The use of the rendezvous server (Section 5.2) helps to establish initial connectivity to a mobile host and locate it when two hosts move simultaneously. However, the rendezvous server only relays initial I1 or UPDATE packets. It does not eliminate the need for end-to-end updates when the IP address changes. The use of a rendezvous server can even increase the signaling load as the updates are sent to RVS and correspondent hosts.

Micromobility is a term that refers to a mobility architecture that attempts to handle host mobility with a certain network area locally. It is typically implemented by supplying the correspondent hosts with a stable anchor point representing the network area. When a host moves within the area, only the local access point needs to be updated with a new address. However, when the host moves out of the area, the new anchor point is selected and all correspondent nodes are updated with its address. Existing micromobility proposals include Hierarchical Mobile IP.

11.1 Local rendezvous servers

This section presents a micromobility architecture for HIP based on local rendezvous servers (LRVS) that extends the concept of the normal HIP rendezvous server (Novaczki *et al.* 2006). Each LRVS is responsible for host mobility in its own micromobility domain. Hosts in a domain use private (not globally routable) IP addresses. After registering to an LRVS using the normal rendezvous registration protocol, a host can access other hosts outside of its domain. The LRVS acts as a NAT by translating local private addresses to its external public address. Figure 11.1 illustrates the micromobility architecture.

Host Identity Protocol (HIP): Towards the Secure Mobile Internet   Andrei Gurtov
© 2008 John Wiley & Sons, Ltd

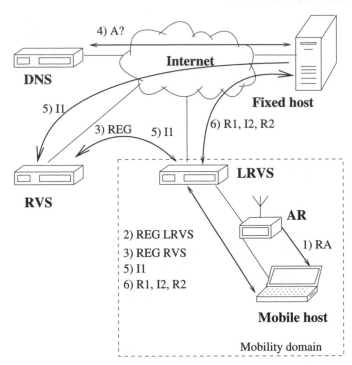

Figure 11.1 HIP micromobility architecture.

11.1.1   Intra-domain mobility

After joining the domain, a host obtains a private IP address through DHCP configuration (IPv4) or address autoconfiguration (IPv6). The IP address and HIT of the domain LRVS are obtained from an Access Router (AR). This information is periodically broadcast by the AR as extended ICMPv6 Routing Advertising messages (1). The host can accelerate the configuration process by broadcasting the ICMPv6 Router Solicitation message that triggers a Router Advertisement explicitly.

After configuring, the host registers to the LRVS using the normal HIP registration protocol (2). During the registration, LRVS creates an entry in its mapping database from the host local IP address to the LRVS external public IP address. Afterward, the host registers to the RVS server to ensure that it is reachable by other hosts (3). The host sends its HIT and the public address of the LRVS to the RVS server through LRVS. The source address is changed from the private host address to the public address by LRVS. At this point, the mobile node is registered to LRVS and RVS and can be contacted by another HIP host.

Suppose a fixed host in the Internet would like to communicate to the mobile node. It starts by performing a DNS query to find out the IP address of the RVS of the mobile node (4). When a DNS server has replied, the fixed host sends an I1 packet to the RVS of the mobile host (5). The RVS forwards I1 to LRVS, which further forwards it to the mobile node after performing the address translation (6). The rest of the base exchange between the fixed host and the mobile host goes through LRVS without the involvement of RVS (7).

Figure 11.2 shows the process of host mobility within a mobility domain. In this case the host performs an intra-domain handover to a new Access Router. However, the LRVS address does not change. After moving to a new location, the host receives advertisement messages from a new Access Router that contain the same HIT and IP address of LRVS as before. Therefore, the host learns that the handover is intra-domain and sends its new IP address to the LRVS to update its registration record. The mobile host does not need to inform the correspondent hosts nor own RVS about the address change. Mobility within the domain is completely handled by LRVS, which reduces signaling load and handover delay for mobile hosts. For hosts outside the domain, the host still appears to be located at the public address of LRVS.

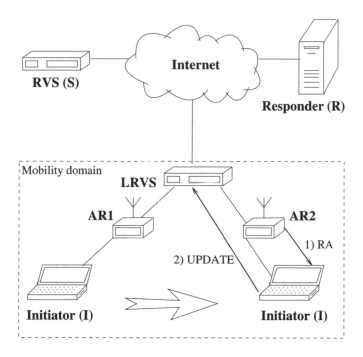

Figure 11.2 Intra-domain mobility update.

11.1.2 Inter-domain mobility

The inter-domain handover is more complex than the intra-domain handover as several bindings need to be updated. Figure 11.3 shows the handover when a host moves from domain 1 to domain 2. The domains can be, for example, networks of different organizations. The host knows that it has changed domain by receiving a Router Advertisement message (1) with a different HIT and IP address of LRVS than before. The host registers to the new LRVS2 (2). Afterward, the host informs its old LRVS1 of its new location (3). The old LRVS1 forwards packets to the mobile host through the new LRVS2 to the host. The host also updates its registration at RVS to point to the new LRVS2 (4). All correspondent hosts need to be updated as well (5). According to HIP specifications, the host should perform a rekeying update when moving to a new network. During rekeying, LRVS2 can record the SPI values of Security Associations of the host. When all updates are completed, the mobile host can close its registration with the old LRVS1.

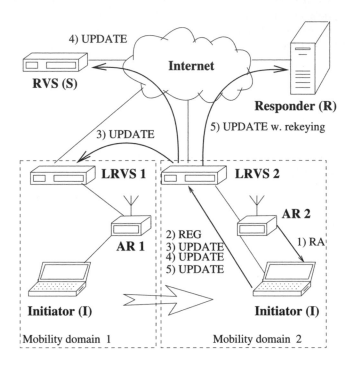

Figure 11.3 Inter-domain mobility update.

In this section, we describe the micromobility architecture that enables HIP hosts to perform local mobility in the private address space of an LRVS. The architecture introduces one level of hierarchy where LRVS is located below the RVS server of the host. A single level may not be sufficient in networks with a large number of mobile hosts (for example, a national cellular wireless network). The architecture can be extended to enable several levels of LRVS with nested subdomains to reduce the load on upper-layer LRVS. When a host moves within a subdomain, only the nearby LRVS needs to be involved in mobility updates.

11.2 Secure micromobility

In Section 11.1, we described micromobility architecture for HIP that uses local rendezvous servers as mobility anchor points. A mobile host registers to an LRVS to authenticate itself. The registration procedure is heavyweight and can present a challenge for mobile hosts changing domains often. In this section, we present an approach for securing micromobility that is more lightweight and better suited for hierarchical mobility domains

(Ylitalo *et al.* 2004). The approach does not require registration to the mobility anchor points, but the micro and macro mobility updates use the same messages.

11.2.1 Hash chain authentication

The micromobility solution that we describe in this section does not use public keys nor require the mobile anchor points to perform demanding computations. The trust between the anchor point and the mobile host is derived from the trust between the mobile host and its correspondent host during macromobility exchange. The anchor point verifies that the IP address indeed belongs to the mobile host to prevent hijacking of addresses and DoS attacks to other hosts.

The solution is based on the use of Lamport one-way hash chains (see Section 2.4.3) and secret splitting techniques. When a mobile host changes location, it sends a next value of the hash chain encrypted to the peer. Messages are protected by HMAC codes using values from the same hash chain. One or more of mobile anchor points are located on the path and observe the mobility exchange between two hosts. The anchor point can authenticate the message sent by the mobile host after it has received the reply message from the correspondent host.

The anchor point does not care about the ownership of the key chain as long as hash chain values remain correct within one communication context. However, simply using host identifiers as authenticator with hash chains would allow an attacker to hijack the identifier by creating a state at the middlebox that associates a hash chain with the identifier. To avoid such spoofing of an identifier, it is concatenated with a random number that can be changed later if that combination is already reserved in the middlebox.

The protocol implementing delayed authentication consists of three messages. The first message is sent from the mobile host to the peer host through the anchor points. It includes the mobile host identifier, encrypted value of a hash value two steps ahead, the current hash value in plain text, and the HMAC of concatenated identifier and hash value using the next hash value as the key:

$$ID_m, Enc(H_{i+2}), H_i, HMAC(H_{i+1}, ID_m||Enc(H_{i+2}))$$

The correspondent host replies with a message containing own identifier, the next hash value in plain text and HMAC taken from the identifier using the hash value from two steps ahead as the key:

$$ID_c, H_{i+1}, HMAC(H_{i+2}, ID_c)$$

The mobile host should always wait for this reply message to arrive before transmitting the next hash chain value to the correspondent host. Ignoring this rule allows an attacker to learn several hash values by delaying the messages.

The final message includes the mobile host identifier and the hash value of two steps ahead in plain text:

$$ID_m, H_{i+2}$$

To bootstrap the secure communication between two hosts, a HIP host creates a hash chain containing n hashes. Each micromobility exchange consumes three hash values. When almost all values are used, the host must create a new hash chain and link it to the old hash chain. The old starting value of a hash chain is replaced with the new starting value after the message exchange listed below. The HMAC securely binds the new hash starting value to the old starting value. During the next mobility update that is run as the macromobility update all middleboxes on the path learn the new anchor hash value. The first message is sent to the corresponding host, which replies with the second message:

$$H_0^{new}, H_i^{old}, HMAC(H_{i+1}^{old}, H_0^{new})H_{i+1}^{old}$$

The middleboxes verify that the new hash value is correctly linked to the existing hash chain. If a spoofed value is detected, the middlebox drops the UPDATE packets for the host.

11.2.2 Secure network attachment

The micromobility management protocol utilizing the above ideas is an extension of HIP macromobility messages. The UPDATE packet is augmented to include the hash values and parts of the session key for the middlebox. When attaching at a new network, the HIP host executes a macromobility exchange with its peers. Middleboxes on the path learn the hash values that can later authenticate messages belonging to the same HIP host. Afterward the host can move within the local mobility area transparently to the peer host. The middleboxes handle the necessary signaling locally. The HIP host uses the same sequence of three UPDATE packets for micromobility as for macromobility to prevent address hijacking and flooding attacks.

The hash chain is used by middleboxes on the path to determine that update messages belong to the same host. Additionally, a technique called key splitting is used to share a secret between a HIP host and the middlebox nearest to it. The secret key is split into two parts, K1 and K2, which are delivered separately to the middlebox. A mobile host sends the K1 part in plain text in the first UPDATE packet to the correspondent host. The K2 part is sent in the same packet but encrypted with a session key that the mobile host had established with the correspondent host.

Figure 11.4 illustrates the process of key splitting. After receiving the UPDATE packet, the middlebox closest to the mobile host stores the K1 key, zeros the K1 field in the packet, and forwards it further on. The correspondent host decrypts the K2 key and sends it back as plain text in its reply UPDATE packet. The middlebox obtains the second part of the split key K2 and verifies that the whole key obtained by an XOR operation between K1 and K2 keys matches to its HMAC from the UPDATE packet. If the keys mismatch, the UPDATE packet is dropped but otherwise the middlebox forwards the UPDATE packet after setting the K2 field to zero. Therefore, it remains the only middlebox that knows the split key of the mobile host.

To attack the presented security model, an adversary can position itself between the mobile host and the adjacent middlebox. To prevent this attack, the adjacent middlebox can check if UPDATE packets from the mobile host contain a zero K1 key. In this case there is an attacker and the adjacent middlebox zeros the K2 key when forwarding the UPDATE packets to the mobile host. The attacker would not be able to establish a secret key with the mobile host and the mobility update would fall back to macromobility involving the correspondent host. A more complicated attack scenario would involve an attacker intercepting packets on both sides of the adjacent middlebox and modifying hash values.

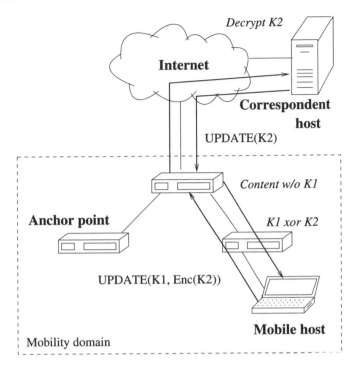

Figure 11.4 Secure attachment to a mobility domain.

11.2.3 Micromobility handover

Figure 11.5 presents a local handover that is handled with a micromobility mechanism. The handover does not involve the correspondent host and is only possible after a first macromobility handover, shown in Figure 11.4, that established the state with the middleboxes. Both middleboxes on the path learn the hash chain and the middlebox adjacent to the mobile host also shares a secret key with it. After the mobile host moves to a new anchor point in the middlebox hierarchy, the host performs a mobility update. The mobile host generates a new split key and sends the new K1 to its correspondent host. The adjacent middlebox stores K1 and forwards the UPDATE packet upward to the next middlebox, which can verify the continuity of the hash value in the packet. The root middlebox forwards the packet further to the previous anchor point of the mobile host and replaces the source IP address with its own. The second and third UPDATE packets also travel through the root middlebox, which verifies their hash values.

The old adjacent middlebox can decrypt the new K2 with a secret key it has previously established with the mobile host. K2 is sent in plain text back to the mobile host via the new adjacent middlebox, which obtains the new secret key by performing an XOR operation on K1 and K2. After completing the update processing, the old adjacent middlebox deletes the state associated with the mobile host. The new adjacent middlebox becomes the only entity sharing a new secret key with the mobile host.

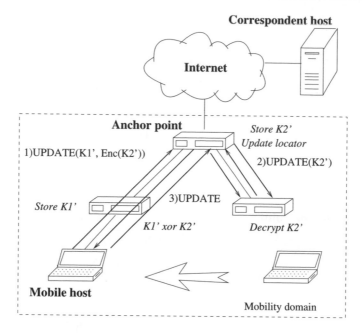

Figure 11.5 Secure intra-domain handover.

The prototype of the presented micromobility solution was implemented with FreeBSD HIP. Its performance was evaluated by simulating network latencies to distant web servers. In an experiment with a distance equivalent to a server located in the USA, the micromobility update is 2.6 times shorter than the macromobility handover. The improvement is directly determined by the latency to the correspondent server.

11.3   Network mobility

In this section, we discuss extensions of HIP for network mobility. A typical example of a mobile network is a WLAN located within a moving train. While the users are attached to the same router within the train, the router can change its attachment point to the Internet as the train moves between stations.

The section starts by introducing the general idea of signaling delegation, which enables the hosts to move the task of mobility updates to the close-by router. Then we present the design and implementation of a HIP mobile router capable of simultaneous multiaccess over several links. Finally, we describe the hybrid Mobile IPv6 and HIP architecture for network mobility.

11.3.1   Delegation of signaling

When hosts are identified using their public key, they can delegate certain signaling rights, such as mobility updates to other hosts. Such delegation can reduce the cost of signaling over an expensive last-hop link without weakening security. This section describes the use of delegation in various scenarios, including host mobility and multihoming, and network mobility (Nikander and Arkko 2002). For example, to authorize a mobile router to send updates for the mobile host within a certain network area, the mobile host creates a certificate $\{K_{MN}^+, K_{MR}^+, region = 10.0.0.0/20\}_{K_{MN}^-}$. The certificate includes public keys of the mobile host and mobile router, and a network prefix that limits the mobility area of the mobile host. The certificate is signed with the private key of the mobile host.

When a mobile host is multihomed, the certificate can include a specific link where the updates can be sent. This prevents a mobile router from claiming that the mobile host is completely unreachable when in fact it has connectivity outside of the mobile router. Then, the mobile router is only able to send updates related to connectivity changes on the direct link to the mobile host and not other links.

A specific application of signaling delegation was proposed to enhance security in network mobility (Ylitalo 2005). There, a mobile host creates a certificate of the form <Issuer, Subject, Delegation, Tag, Lifetime>, which authorizes the mobile router to send mobility updates on its behalf. The certificate typically allows further delegation of rights from the mobile router to the next router in the hierarchy or even to a signaling proxy located somewhere in the Internet. After a handover, the mobile router can send a single message to a correspondent host to update all security associations that mobile hosts established with it. This optimization can reduce the amount of signaling as multiple mobile hosts can be connected to the same correspondent host.

11.3.2 Mobile router

An Ericsson implementation of the HIP mobile router was demonstrated in several events since 2006. The proposed architecture supports nested mobile networks, where a separate subnetwork (e.g., a personal area network of a passenger) can change the location within a parent mobile network (e.g., by walking inside a moving train). The mobile router implementation is integrated with service discovery and simultaneous multiaccess extensions for HIP. To resolve possible SPI collisions during mobility, the mobile router implements SPINAT extensions.

Mobile hosts are located in a network with a private address space that is connected to the Internet via the mobile router. Mobile hosts delegate their signaling rights to the mobile router by signing a certificate containing own HIT and the HIT of the mobile router. The mobile router sends location updates on behalf of mobile nodes when the mobile router re-attaches to a different location in the Internet. The mobile nodes move together with the mobile router and use the same private address space independently of their location in relation to the Internet. Therefore, the network mobility process is transparent to mobile hosts.

Figure 11.6 illustrates a scenario where a new mobile node attaches to the mobile router. The mobile host obtains a private IP address from the DHCP server that can be combined with the mobile router. The mobile node simultaneously performs the registration procedure and updates its correspondent hosts of the new location. The I2 packet during the registration to the mobile router includes a certificate to authorize the mobile router to update the mapping between a HIT of the mobile host and its IP address.

The mobile host can activate a base exchange with a correspondent host via the mobile router. If the mobile hosts has established HIP association with correspondent hosts in the

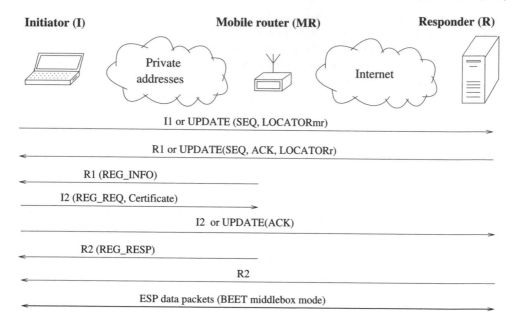

Figure 11.6 Base exchange and registration with HIP mobile router.

Internet, it sends each correspondent host an UPDATE message containing the public IP address of the mobile router. The mobile router performs address translation and forwards packets between the mobile and correspondent hosts. After completing the base exchange or mobility update, the mobile host establishes an ESP Bound End-to-End Tunnel to the correspondent host. The standard BEET ESP mode is modified for convenience of the mobile router and referred to as the BEET middlebox mode.

The mobile router performs a mobility update on behalf of mobile hosts as shown in Figure 11.7. After the mobile router attaches to a new place in the Internet, it sends UPDATE messages to all correspondent hosts that have active security associations with mobile hosts registered to the mobile router. To prove its authorization to perform a mobility update, the mobile router provides a certificate signed by the mobile host. After completing the update, correspondent hosts send all data packets destined to mobile hosts to the new address of the mobile router. Private IP addresses of mobile hosts do not change.

The actual demonstration set up by Ericsson NomadicLab is shown in Figure 11.8. Two mobile nodes communicate with a correspondent host located in the public Internet. The mobile router connects to the Internet via a cellular 3G link (IPv4) or WLAN link (IPv6). Two links can be active one at a time or simultaneously. Mobile host 1 is attached only to the mobile router. Mobile host 2, in addition to the mobile router, has a wireline Ethernet connection to the Internet (IPv6). The mobile and correspondent hosts support simultaneous multiaccess extensions that enable parallel data transmission over two or more available links.

During the demonstration, a video clip is streamed from the correspondent host to mobile hosts. When the 3G or WLAN link is disconnected from the mobile router, the data traffic is automatically redirected to use the other available link. The links are used in parallel if both

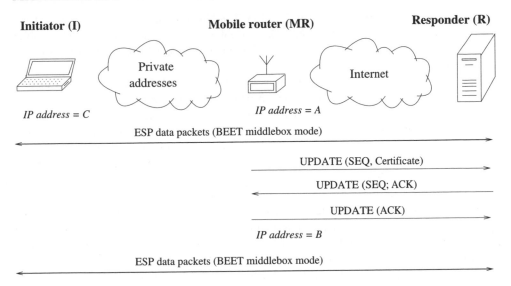

Figure 11.7 Mobile router performs an update on behalf of a HIP host.

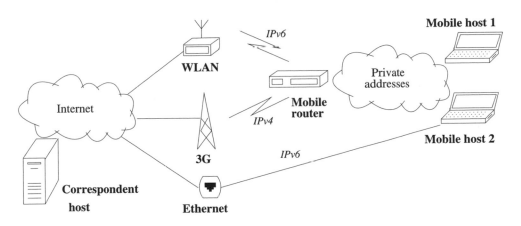

Figure 11.8 A demonstration of the mobile router and SIMA.

are available. Mobile host 2 is able to move between the wireline IPv6 access to the Internet and the link provided by the mobile router.

11.3.3 HarMoNy

An alternative proposal for HIP network mobility is HarMoNy (HIP and NEtwork MObility) (Herborn *et al.* 2006). HarMoNy focuses on a hybrid mobility solution combining Mobile IPv6 routers implementing Network Mobility (NEMO) extensions and HIP mobility. The mobile host implements context-aware handover between a NEMO router and public

Internet connectivity available at a new location. The NEMO router provides a private IPv6 address while the public connectivity provides a topologically correct IP address. The mobile host can use some clues, such as the speed of moving, to select between NEMO or direct Internet connectivity.

The advantage of higher-level mobility solutions such as HIP is the ability to function without special infrastructure, integration with IPsec, and multihoming support. Lower-level mobility with NEMO requires infrastructure components such as a home agent and support from the local router, but can be more efficient and require less involvement from end hosts. HarMoNy does not introduce new infrastructure requirements but enables simultaneous use of several mobility solutions. Such an approach enables co-existence of several mobility mechanisms deployed in various network spots.

The hybrid design of HarMoNy is suitable for a moving vehicular scenario. When the vehicle is moving fast and the rate of handovers is high, a mobile host allows the mobile router located in the same vehicle to handle handovers on its behalf using NEMO with MIPv6 (Figure 11.9). When the vehicle is slowing down, in addition to mobile router connectivity the mobile host also obtains a care-of address from the nearby access router. When the vehicle is stationary, the mobile host establishes a HIP association directly to the access router (Figure 11.10). When making these decisions, the number of mobile hosts in the vehicle should also be taken into account. It is likely that the mobile router can handle update signaling more efficiently than each mobile host individually.

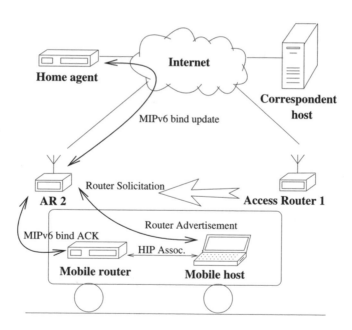

Figure 11.9 A HarMoNy mobile network when moving.

A prototype implementation of HarMoNy combines an OptiNets mobile router with HIP on Linux (HIPL). Mobile and access routers announce their presence using modified

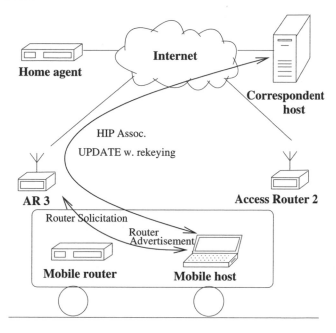

Figure 11.10 A HarMoNy mobile network when stationary.

router advertisement messages. The mobile host configures its IPv6 addresses using stateless address configuration. Selection of the address to use depends on the emulated speed of the vehicle. For NEMO, a MIPv6 rebind packet must reach the home agent and be acknowledged. The handover delay is the RTT between the mobile router and home agent or between the mobile router and correspondent host if the route optimization is used. The HIP mobility update can require up to two RTTs between the mobile host and the correspondent host to complete. Therefore, it could be more efficient to use NEMO handovers when the latency toward the correspondent host is high.

In experiments, the handover latency of HIP was compared with NEMO MIPv6. It was found that without rekeying, HIP latency is half that of NEMO, but with rekeying HIP latency is about double. The use of rekeying is a local policy issue and is typical when the mobile host moves to another network, i.e., its IP address prefix is changing. The experiments also reveal packet losses of 0–10% during handovers. The study concludes that it is preferable to use NEMO in scenarios when rekeying is required to reduce the handover latency. However, the security properties of NEMO are weaker than those provided by HIP with rekeying.

12

Communication privacy

In this chapter, we cover identity and location privacy aspects of HIP architecture. Section 12.1 describes the functions of SPINAT that can translate IP addresses between public and private domains based on SPI values. At the same time, SPINAT hides the exact user location by replacing the user's host IP address with own public IP address. In Section 12.2, we present a modification of HIP base exchange based on the idea of blinding host identities. This mechanism hides the identity of communicating hosts from the third party. Section 12.3 considers the range of identifiers on different protocol layers and describes a privacy mechanism for the simultaneous change of identifiers on multiple layers.

12.1 SPINAT

IPsec packets cannot directly traverse legacy NAT devices as the Authentication Header or ESP prevents modification of the packet or encrypts its content. With ESP, there is no transport port information visible to the NAT, only the IP addresses and the SPI values. To traverse legacy NATs, IPsec packets are typically encapsulated into UDP datagrams. A device specially designed to support IPsec address translation is called Security Parameter Index multiplexed NAT (SPINAT) (Ylitalo et al. 2005).

As the HIP data packets carry no endpoint identifier, the SPI acts as an index to the host identities in middleboxes. A SPINAT translates packets between address domains (private and public addresses, or IPv4 and IPv6 addresses) based on the destination IP address and the SPI value from the IPsec header. The middlebox establishes the mapping state by listening to the HIP base exchange and storing SPI values in both directions. If a SPINAT is located off the base exchange path of HIP hosts, a separate signaling protocol can be used to establish the mapping state. As an example, a trigger insertion can establish the state on i3 servers acting as a SPINAT.

A mobile host can change its network attachment point so that a different SPINAT device appears on the path. In this scenario, the HIP UPDATE exchange through the new SPINAT establishes the mapping state. Afterward the data IPsec packets can be forwarded by the SPINAT. The delay in packet forwarding can cause a performance penalty through lower throughput and packet losses; it prevents the use of Credit-Based Authorization (CBA), which otherwise enables limited packet transmission before the UPDATE exchange completes. The SPINAT authenticates IP address and SPI updates to prevent DoS and redirection attacks.

When a large number of hosts are located behind a single SPINAT, collisions of SPI values are quite likely. The SPI values are chosen randomly by the host and two or more hosts can accidentally choose the same value. In that case SPINAT cannot demultiplex IPsec packets arriving to its public IP address with the same SPI. The SPINAT can either drop an incoming packet that has an SPI already in use, or translate the SPI value on the fly. SPINAT requires that the SPI value in IPsec data and control packets is not encrypted. This requirement is met with the HIP base exchange packets, but not with IKE and IKEv2 protocols. Additionally, if SPI collisions are resolved through translation, the SPI value must not be included into a signature in control nor data packets to enable its modification by the SPINAT.

If SPI collisions are solved through dropping packets with an SPI value already in use, the source host needs to generate a new random SPI value and retransmit the packet. In fact, several retransmissions may be required in a busy network before a value without collisions is found, especially if several SPINAT devices are present on the path. A SPINAT can inform the source host of an SPI collision and a consequent packet drop by sending an ICMP packet. However, the ICMP packet is unsigned and can also be sent by a malicious host. Therefore, the source host should not directly trust the ICMP packet but treat it only as a hint and wait for a retransmission timer to expire before re-sending the packet.

When SPINAT implements on-the-fly SPI translation, it replaces the colliding SPI value in HIP control packets and in IPsec data packets. This is similar to NAPT (Network Address Port Translation) replacing IP addresses and TCP or UDP ports in packets, except that SPI translation is less intrusive. A packet with a spoofed port number through a NAT can reach the host application but packets with spoofed SPI would be dropped by the IPsec layer. Essentially, the security does not suffer if SPI is excluded from the signature.

To enable on-the-fly translation, a modification of the Bound End-to-End Tunnel (BEET) model for IPsec is needed that permits changing the SPI value. The Stripped End-to-End Tunnel (SEET) mode is the same as BEET, except the SPI value is not included into ESP header integrity protection. The SEET mode only changes the details of IPsec processing in hosts, not the packet headers. A more significant change would extend the IPsec header to increase the SPI size to 128 or more bits. Then SPI collisions would become unlikely, removing the need for SPI translation, and use IP addresses for SPI disambiguation.

12.2 BLIND

This section presents BLIND, a framework for identity protection of hosts that are identified with public keys (Ylitalo and Nikander 2004a). The framework extends the HIP base exchange to hide identity by not revealing the real public keys of the host. The protection is robust against passive and active Man-In-The-Middle attacks. In addition, location privacy is achieved with forwarding agents.

12.2.1 Location and identity privacy

Using public keys for identifying hosts creates a privacy problem as third parties can determine the source host even if attached to a different location in the network. Various transactions of the host could be linked together if the host uses the same public key. Furthermore, using a static IP address also allows linking of transactions of the host. Multiplexing multiple hosts behind a single NAT or using short address leases from DHCP can reduce the problem of user tracking. However, IPv6 addresses could eliminate NAT translation and cause additional security issues related to the use of MAC addresses in IPv6 address autoconfiguration.

DNS records can provide information combining host identity and location information, the host public key, and host IP address. Therefore, identity and location privacy are related and should be treated in an integrated approach. The goal of the BLIND is to provide a framework for identity and location privacy. The identity protection is achieved by hiding the actual public keys from third parties so that only the trusted correspondent hosts can recognize the keys. Location privacy is achieved by integrating traffic forwarding with NAT translation and decoupling host identities from locators. The use of random IP and MAC addresses also reduces the issue of location privacy shifting the focus to protecting host identifiers from third parties.

All two-round-trip variations of the Diffie–Hellman key exchange using public keys for authentication are vulnerable to identity theft. The Responder must not generate the shared session key before receiving two messages from the Initiator to avoid DoS attacks. If the Responder sends its public key in the first reply message to the Initiator, the Responder's identity will be revealed to third parties. Therefore, the public key is sent encrypted in its second message of the base exchange. The Initiator cannot determine the identity of the Responder after receiving the last message of the key exchange. As the result, an active attacker can find out the public key and identity of the Initiator by pretending to be a trusted correspondent host. The Initiator's public key is sent encrypted in the third message of the Diffie–Hellman key exchange and can be decrypted by an attacker based on the established session key.

To prevent revealing the identity, the host public key and its hash (HIT) can be encrypted with a secret key known beforehand to both Initiator and Responder. However, this a requirement that cannot be easily implemented in practice. The BLIND framework provides protection from active and passive attackers using a modified two-round-trip Diffie–Hellman key exchange protocol. If the host avoids storing their public keys in the reverse DNS or DHT repository, the framework achieves full location and identity privacy.

12.2.2   Protecting host identity

BLIND requires that the Responder's HIT is known to the Initiator beforehand. The Initiator can typically find out the HIT of the Responder from a directory service or store it from a previous communication session. The identity protection is achieved by scrambling HITs of Initiator and Responder during the Diffie–Hellman key exchange in a way that each scrambled version is used only once. The blinded version of HITs is obtained by applying an SHA hash to the actual HIT concatenated with a nonce. A newly generated nonce is used for each key exchange.

$$HIT_{blind}(i) = SHA1(nonce(i)||HIT_{plain}(i))$$

$$HIT_{blind}(r) = SHA1(nonce(i)||HIT_{plain}(r))$$

The identity protection is based on the idea that if the host knows a list of potential HITs that its peers use, it can test these HITs one-by-one together with a known nonce to check if they match the blinded HIT. However, if only the blinded HIT and the nonce are known, it is infeasible to compute the plain HIT.

Figure 12.1 illustrates the HIP base exchange using BLIND. The I1 packet includes blinded HITs of the Initiator and Responder, a plain-text nonce, and an optional hint disclosing k bits of the actual Responder's HIT. In most cases, the Responder has only one or a few public keys and does not need the hint. However, a busy server with many identities can drop I1 messages without a sufficiently long k to protect itself against DoS attacks.

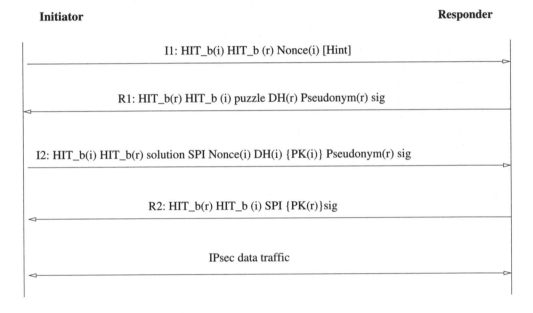

Figure 12.1 Modified base exchange with BLIND.

The Responder generates a random number and includes it into the R1 as a pseudonym for its public key. The public key itself is not sent in R1, but is bound to the pseudonym by a signature with the private key. The pseudonym only helps the Responder to locate its own public key when the I2 packet arrives and does not reveal the Responder identity. The R1 also includes the puzzle and other HIP fields not shown in Figure 12.1.

After receiving the R1 packet, the Initiator computes the keying material using the actual HIT of the Responder and own blinded HIT:

$$KEY_1 = SHA1(KEY_{DH}||HIT_b(i)||HIT(r)||1),$$

$$KEY_n = SHA1(KEY_{DH}||KEYn - 1||n),$$

$$KEY_{MATERIAL} = KEY_1|| \ldots ||KEYn.$$

The Initiator cannot verify the identity of the Responder since its public key is not present in the R1 packet and only the Responder's HIT is known to the Initiator. However, if after receiving the R2 packet, the Initiator finds out that the signature in R1 was invalid, the Initiator can terminate the base exchange without revealing its own identity. The third party does not know the actual Responder's HIT and therefore cannot compute the secret session key.

After receiving I2, the Responder computes the keying material the same way as the Initiator. I2 includes the same nonce from I1 and pseudonym from R1. Using the computed secret key, the Responder can decrypt the Initiator's public key and verify that the blinded HIT is correct. The base exchange completes successfully, after the Initiator receives R2, and decrypts and verifies the Responder's public key.

12.2.3 Protecting location privacy

A forwarding agent is similar in function to a SPINAT, except that it does not have to reside on the path between the Initiator and Responder. While SPINAT is located on the border between the public Internet and the private address space, the forwarding agent can be located anywhere in the Internet. A HIP host can lease a virtual interface from the forwarding agent to be virtually present in that Internet location. The forwarding agent replaces the IP address in packets from the Initiator host to the Responder, and the destination address from the Responder to the Initiator.

The forwarding agent assigns IP addresses from a pool common to all its clients. For IPv4, the forwarding agent is likely to overload the IP addresses as typical NATs do. For IPv6, the address mappings are likely to be one-to-one. The lease contains blinded HITs of Initiator and Responder, their addresses, and lifetime of the lease:

$$HIT^b_{dst}, IP_{dst}, HIT^b_{src}, IP_{src}, lifetime$$

Both the Initiator and Responder can use forwarding agents to secure their location privacy. Then, two forwarding agents, one on the Initiator side, and one on the Responder side, can participate in delivering data packets between the Initiator and the Responder. The forwarding agents can be trustworthy and untrustworthy. The trustworthy forwarding agents are likely to authenticate a host before leasing the interface, while untrustworthy agents allow anonymous leases.

A HIP host relies on a trustworthy forwarding agent not to reveal its identity while the agent trusts the host not to establish too many leases through it. A trustworthy forwarding agent can know the real identity of the host and therefore could reveal both identity and location of the host to an attacker. The untrustworthy agent can reveal only the location of the host with a blinded identifier.

When both Initiator and Responder use BLIND with forwarding agents, a complete identity and location privacy is achieved given the following assumptions. First, the hosts should not keep public mappings from their IP addresses to public keys, for example using reverse DNS. Although BLIND does not require the hosts to know each other's public keys, the second requirement is that the Initiator knows the HIT of the Responder beforehand.

An alternative proposal to BLIND was made in the IETF (Matos *et al.* 2006). Its architecture is similar and based on the use of rendezvous agents (RVA), which are an extended version of rendezvous servers. An RVA acts as a NAT translating addresses between own private address space and public Internet. Hosts outside of the RVA address space would not be able to learn the exact location of a HIP host within the RVA area, although they can determine that the host is behind a certain RVA. Overall, the privacy level is the same as that provided by the micromobility solutions described in Section 11.

12.3   Anonymous identifiers

The privacy of the user is achieved if its location and identity cannot be tracked as the user moves around the network. Unfortunately, security mechanisms, such as the use of HIP, can introduce new security risks with the use of persistent host identifiers. While several telecommunications networks, such as WLAN and GSM, involve the use of pseudonym identifiers on the link layer, it is not sufficient for preserving privacy. The user can be tracked by identities present on several protocol layers, starting from the link layer up to application data (Arkko *et al.* 2005).

The following properties are desirable for privacy-preserving networking. Anonymity refers to the conditions where neither the communicating peer nor any third party is able to find the real identity of the user. Pseudonymity is a weaker form of anonymity where the user has a persistent identifier that is not connected to the real identity, but individual actions by the user can be linked together. Unlinkability is the property of lack of possible connections between separate user actions. Unobservability is a stronger form of unlinkability where separate user actions cannot even be detected or separated from the background actions of other users.

12.3.1   Identifiers on protocol layers

Figure 12.2 shows identifiers on different protocol layers. The MAC address is assigned to the Network Interface Card at the factory and is guaranteed to be unique as each manufacturer uses a unique prefix combined with an incremental sequence number. An adversary that is located in the same local network as the user can clearly identify the user by its MAC address. Fortunately, most modern NICs allow overriding the default MAC with an arbitrary software-generated address.

Even temporal identifiers, such as the IPsec SPI value, can be used to track the user. After moving and changing the IP address, the SPI values remain the same and visible to an adversary outside of the local area network of the user. With 32-bit SPI values, observing the same SPI value as before leaves only a 1 to 4 billion chance that the user is a different one. The problem is aggravated by deficiencies of the IKE protocol for negotiating new SPI values. The protocol transmits old and new SPI values together thus enabling the attacker to correlate them.

The use of public keys on the HIP layer strongly identifies the user. The attacker can track the user by examining the HIT, public keys, and signatures from HIP control packets. The packets can be shown to belong to the user by verifying packet signatures with its public key. Even the Lightweight HIP without the use of public keys is susceptible to privacy-targeted attacks as hash chains are easily linkable by following the chain in the forward direction.

Even protocol sequence numbers at IPsec and transport layers can be used to track the user, although the number is different in each packet. Indeed, if the attacker notices that the sequence number in the packet is just a little over the sequence number seen in packets at

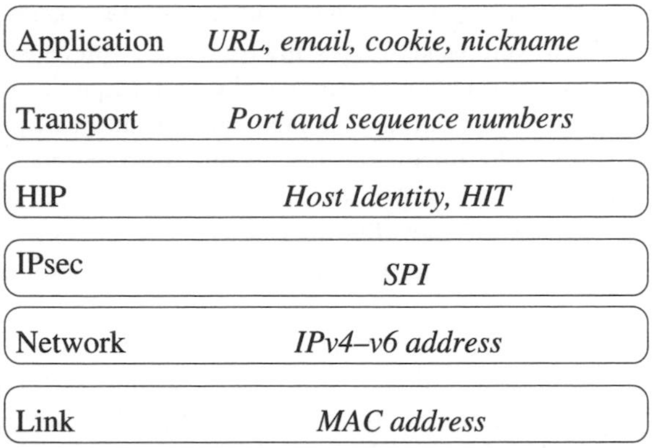

Application	URL, email, cookie, nickname
Transport	Port and sequence numbers
HIP	Host Identity, HIT
IPsec	SPI
Network	IPv4–v6 address
Link	MAC address

Figure 12.2 Privacy-sensitive identifiers on different protocol layers.

another location, it is quite probable that the packets belong to the same user. The probability depends on the transmission rate of the links, and 32-bit sequence numbers offer a sufficient correlation possibility in many scenarios.

At the transport layer, port numbers can reveal the user. Servers use well-known ports for communication, such as port TCP port 80 for HTTP traffic and ephemeral source ports are used sequentially. The attacker can track the user by noticing packets with the same or close port number. At the application layer, various identifiers can be sent in the payload data, such as email addresses or HTTP cookies. Fortunately, IPsec encrypts all protocol headers and data from the network layer upwards. Therefore, for the HIP architecture the most serious privacy threats are limited to MAC and IP addresses, SPI and HIT values.

12.3.2 Changing identifiers

A natural approach to reducing privacy threats of persistent identifiers is to replace them with short-lived identifiers that are changed regularly to prevent user tracking. Furthermore, identifiers must be changed simultaneously at all protocol layers, otherwise an adversary could still link the new identifier through looking at the identifier at another protocol layer that remained the same after the change. The HIP privacy architecture that simultaneously changes identifiers on MAC, IP, and HIP/IPsec layers was developed in TKK (Takkinen 2006). The default frequency of changes is every 6–10 minutes. Unfortunately, each change causes a delay of 3 seconds, and possibly loss of data packets, which might be unacceptable for real-time applications.

For changing identifiers without explicit signaling, two hosts can rely on shared pseudo-random sequences (Arkko *et al.* 2005). When initiating the association, hosts agree on sequences of pseudo-random values to be used for identifiers. When the host moves or a periodic timer expires, the host just starts to use the next value in the agreed sequence of pseudo-random values. To derive pseudo-random values, a hash function such as SHA-256 can be applied, the i-th identifier of a given type would be $SHA(k|i|ID_{type})$ where k is the shared secret key between hosts. To ensure unlinkability, identifiers at all layers need to be changed simultaneously.

The peer host notices the change in identifiers and verifies that the new identifier belongs to the agreed sequence. However, packet losses and reordering can cause gaps in the values, especially if the identifier is itself changing, such as the TCP sequence numbers. Therefore, the host must be prepared to accept not just the next value from the sequence, but any value from a window. A possibility of collisions, when a value from one identifier type falls into the range of acceptable values for another identifier type, does exist, but is relatively low and collision resolution is not difficult. Fortunately, using HIP with IPsec ESP encrypts all transport and application-layer identifiers. The only non-constant identifier that requires changing is ESP sequence numbers.

Frequent change of public keys to avoid tracking of a user can be computationally expensive as generating new keys can take up to a few seconds, depending on hardware. While it is possible to pre-generate keys in advance, a preferable solution is a hybrid with symmetric cryptography. Public keys can be used only to agree on a secret key that is later used to authenticate Diffie–Hellman key exchange instead of public key signatures. Since other parameters in each Diffie–Hellman key exchange are different, there is no information available to the attacker to track the user.

Hash chains that replace public keys in Lightweight HIP can be also used to track the user. Plain hash chains are easily verifiable by a third party in the forward direction. Therefore, an attacker observing a hash value can attempt to track the user by repeatedly applying the known hash function to a previously observed hash value and checking if the result matches the currently observed value. A keyed version of a hash chain $h_i = H(h(i-1)||k)$ is verifiable only by parties sharing the secret k.

Periodic change of identifiers on different protocol layers can affect middleboxes trying to follow the data flow. For instance, changing transport protocol port numbers or IP addresses is likely to cause packet drops in NATs and firewalls that do not know how to link new identifiers to existing state for old identifiers. The change of IPsec SPI would require informing SPI-NAT of the mapping between old and new values. Finally, the bridge spanning tree protocols can get confused by the frequent change of MAC addresses.

Part IV

Applications

13

Possible HIP applications

In this chapter, we describe several applications whose use has the potential to bring about widespread adoption of HIP. Section 13.1 outlines the use of HIP as a VPN solution that enables traveling users to access their home or enterprise networks remotely. In Section 13.2, we describe the peer-to-peer architecture for sharing Internet access with HIP. Section 13.3 presents HIP as a solution to interoperate IPv4 and IPv6 applications. Section 13.4 presents the Secure Mobile Architecture, which is an integrated network solution including HIP among other components. Its prototype is deployed in a testbed at a Boeing airplane factory. Section 13.5 describes migration of virtual machines, preserving ongoing data transfers with HIP. In Section 13.6, we summarize the main issues that a telecommunication network operator can face when deploying HIP in its network.

13.1 Virtual Private Networking

The Internet has become a part of everyday life for many people who prefer to stay online when traveling. When on the move, a user needs a local Internet connectivity and a way to access data and services in the home network or company's intranet. Users who attempt to work on the move are sometimes referred to as Road Warriors (Nikander 2004). Traditional Virtual Private Network (VPN) solutions use various forms of tunneling, such as Layer 2 Tunneling Protocol (L2TP), that add significant amounts of header overhead and are difficult for middleboxes to process. The user is required to enter a VPN password for authentication each time when establishing the tunnel. Most VPN solutions do not support user mobility or multihoming, thus the user is required to re-establish the tunnel and re-authenticate when changing the IP address. Often even small disruptions in, e.g., WLAN connectivity terminate the VPN connection, which annoys the user.

An ideal VPN solution would enable a Road Warrior to instantly connect to their intranet and use all the same services as they would within the intranet. The VPN should not require manual authentication but provide secure and fault-tolerant connectivity to the intranet. The user should be able to access email, the local file server, and other intranet applications such as a personnel database and payroll information. Fortunately, most modern intranet

Host Identity Protocol (HIP): Towards the Secure Mobile Internet   Andrei Gurtov
© 2008 John Wiley & Sons, Ltd

applications are web-based, operating over HTTP. The user's Internet connection can be a slow cellular link, therefore VPN should rely on locally cached data whenever possible and only transfer data that has changed recently.

Road Warriors face various obstacles when trying to access their intranets with VPN: widespread presence of NATs, misconfigured firewalls, disruptive authentication methods for network access, and legacy intranet applications. Many existing VPN products do not work when the user is located behind middleboxes, or are forced to use UDP encapsulation to traverse the NAT and local firewalls. To gain the Internet connectivity the user is often redirected to a captive web page requiring registration and entering user name and password information. Such pages unnecessarily consume the user's time and often create problems for network protocols and applications. Finally, upgrading the corporate application servers to support new VPN software can take years.

Using HIP as a VPN solution is an attractive option for Road Warriors. HIP provides IPsec encryption, support for mobility and multihoming, and enables automatic authentication both to a visiting network and to an intranet firewall. The use of HIP enables Single Sign-On (SSO) functionality to visiting networks; the operator is only required to obtain a list of HITs authorized to use the network and later verify the HIT of a visiting user during the HIP base exchange or a mobility update. For small companies, the administrator can manually configure the list of authorized HITs into the corporal firewall that monitors traffic from external HIP hosts to the intranet. For larger companies, an automated solution involving PKI integrated with a firewall Access Control List (ACL) is likely.

A HIP VPN solution is likely to integrate the HIP aware firewall described in Section 9.4, UDP encapsulation for legacy NAT traversal (Section 9.2), and a HIP proxy located in the intranet (Section 15.6). As shown in Figure 13.1, the proxy serves as an endpoint of the HIP UDP tunnel that decapsulates payload from IPsec packets and forwards them to the corporate servers providing email, file access, and web services. The proxy is likely to be necessary because of the legacy nature of corporate servers that cannot be updated to support HIP directly. The user accessing an intranet with HIP gets other associated benefits, including mobility and multihoming support that other VPN solutions lack.

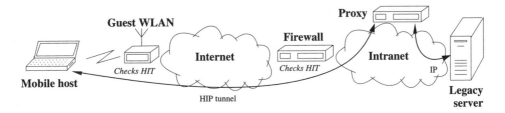

Figure 13.1 Virtual Private Networking with HIP.

Next we discuss how to configure a simple Access Control List for accessing a home Linux server over HIP. Today's ACLs are based on IP addresses and port numbers that are stored in /etc/hosts.allow and /etc/hosts.deny. By adding IP addresses to these files, the server administrator can control access to such services as SSH, HTTP, or FTP. A keyword "ALL" can be used to give the default handling for allowing or denying connections. Access control with IP addresses is inconvenient as users frequently move, e.g., between home and work and their laptop's IP address changes. Using the HITs instead of IP addresses alleviates this problem. The server administrator can add HITs to hosts.allow and hosts.deny files to control the access. Since HITs have the format of IPv6 addresses, such use is backward-compatible. If the HIP base exchange successfully completes between a host and the server, and the host's HIT matches one in hosts.allow file, the host can use the services.

13.2   P2P Internet Sharing Architecture

Proliferation of flat-rate Internet connectivity in homes and the low prices of WLAN equipment have created a good opportunities for getting Internet access when traveling. Many users just leave their WLAN networks completely open, either intentionally because they want to enable other users to benefit Internet connectivity, or unintentionally because of low technical competence, just leaving the default factory settings in WLAN base stations. However, depending on the country, using such WLAN access points can have serious legal consequences for both parties. For instance, just connecting to an open access point without obtaining a prior permission from the owner can be a criminal offense. Likewise, the owner of the access point can be charged with illegal activities executed using his IP address.

Users of open wireless networks also often risk their traffic being intercepted and explored for passwords or other sensitive information. Some access points can even modify web pages by inserting advertisements or viruses, infecting the user's equipment. In summary, the users of WLAN need to be authenticated to protect the owner of the access point from legal prosecution and the user traffic needs to be encrypted to avoid eavesdropping and being tampered with.

Commercial companies such as FON and Wippies in Europe promote the use of WLAN access by giving free access points in exchange to offering other users network connectivity. Their authentication model is quite weak and disruptive, requiring the user to enter the password on a web page. The user's traffic is not encrypted nor could it be separated from the traffic of other users thus creating ambiguity in legal disputes. A malicious user could eavesdrop MAC and IP addresses of an authenticated user and use them to transfer data through the access point. Despite these issues, the service is very popular, involving hundreds of thousands of installed access points worldwide.

The P2P Wi-Fi Internet Sharing Architecture (PISA) (Heer *et al.* 2007) solves the problem of traffic security and access control by applying HIP. In particular, the WLAN access points only allow authenticating HIP traffic to the home router of the user that has been certified by the community. This shifts the legal responsibility from the owner of the access point as all user Internet traffic appears to come from the home router of that user. Furthermore, the user is protected from possible attacks on the traffic since everything sent over WLAN is encrypted. Figure 13.2 illustrates the PISA architecture.

To authenticate itself, the user initiates a base exchange from his host to the home router. The WLAN access points know the list of authorized home routers and only forward packets that are HIP control packets to IP addresses from the list. The access point acts as a middlebox verifying signatures in HIP base exchange packets and thus authenticates the hosts. To prevent a malicious host from simply replaying previously recorded base exchange, the access point inserts a nonce into HIP control packets that both hosts are required to sign.

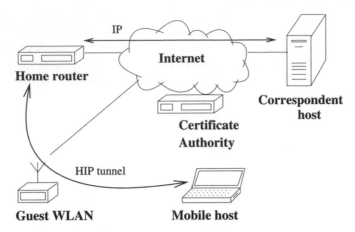

Figure 13.2 P2P Wi-Fi Internet Sharing Architecture.

To verify that the user belongs to the Wi-Fi sharing community, PISA requires the user's host to send a certificate in a CERT parameter during the HIP base exchange. The certificate includes the host identity of the home router that is signed by a Certificate Authority (CA) determining membership in the community. The access point can verify certificate validity by checking for expired and revoked certificates from the CA.

After authentication succeeds, the user's host has an IPsec tunnel to the home router. All data packets are sent inside the tunnel up to the home router, which extracts payload from HIP packets and forwards them to the actual destination. Likewise, the home router receives response packets from the correspondent host and sends them to the user host inside the IPsec tunnel. Therefore, the home router acts as a HIP proxy.

In addition to basic authentication and encryption, PISA supports data compression to compensate for asymmetric speeds of many home ADSL lines. In particular, the direction towards the home router can be the bottleneck for user traffic. The PISA architecture also supports seamless user mobility between WLAN access points. The new access point can authenticate the user from the exchange of UPDATE messages between the mobile host and the home router, without the need to repeat the base exchange. The drawback of PISA is potential routing inefficiency especially if the correspondent host is located close to the user while the home router is far away.

The PISA prototype and demo included a modified version of WLAN access points running OpenWrt Linux distribution. The HIP on Linux implementation was extended with middlebox authentication functionality. During the demo, the user was able to browse the web by transparently authenticating through a local access point, while his traffic was tunneled to a remote home router. The attacker is only able to observe HIP control and encrypted IPsec packets by snooping on the local WLAN.

13.3 Interoperating IPv4 and IPv6

The shortage of IPv4 addresses gradually advances the deployment of IPv6 networks in the Internet. In particular, China has converted its university backbone network CERNET to route only IPv6. However, the transition process is likely to take years and some networks can remain IPv4. Both IP versions will co-exist for a long time raising the issue of interoperation between hosts and applications. A host can have only IPv4 or IPv6 connectivity, or both. Likewise, an application running on the host can be IPv4- or IPv6-only or understand both protocols. Most modern applications support both protocols, although there is a significant number of legacy IPv4-only applications.

In the classic TCP/IP architecture, the transport and networking layers are not independent but connected through checksum calculations and address information. Interdependency between the layers prevents their independent evolution. As an example, implementations of transport protocol such as TCP and UDP need modifications as the result of introducing IPv6.

The use of HIP decouples networking and transport layers making the Internet stack closer to the Open Systems Interconnection (OSI) protocol model. The multihoming extensions enable a HIP host to use IPv4 and IPv6 addresses simultaneously bound to a single or several network interfaces. The mobility extensions enable cross-family handovers between IPv4 and IPv6 addresses. The handover process is similar whether the protocol family changes or not. Cross-family handovers were demonstrated by Ericsson with the FreeBSD HIP implementation in 2003 (Jokela *et al.* 2003).

In the rest of this section, we discuss interoperation between IPv4 and IPv6 applications. To enable cross-family communication, the transport layer checksums need to be computed identically at both communicating hosts. In a classic TCP/IP stack, TCP and UDP pseudo-header checksums include IP addresses from the network layer. With HIP, the checksums are computed over the 128-bit source and destination HITs and IPv6-format pseudoheader is used both for IPv4 and IPv6. This approach enables IPv4 and IPv6 applications to communicate as shown in Figure 13.3. In the Ericsson implementation of HIP, data packets always pass through the IPv6 version of TCP and UDP, even for IPv4 sockets (Ylitalo and Nikander 2004b). The HITs are converted to LSI on the system call level just before passing data to an IPv4 application.

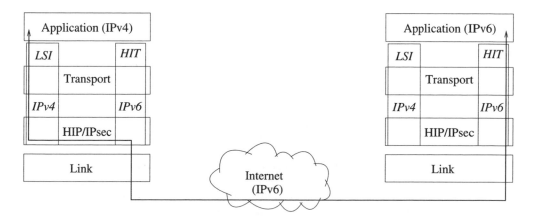

Figure 13.3 IPv4-only application talking to IPv6 application via HIP.

The applications that explicitly use IP addresses or pass them in messages to remote applications can experience problems with cross-family communication using HIP. Applications explicitly using the IP addresses (such as ping) are used mostly for network diagnostics. The user of such applications typically knows what the application is trying to accomplish. Typically such applications operate on the IP layer without involving HIP.

Applications that pass IP addresses in their messages operate normally if the IP address family is the same at both communication endpoints. For IPv4 applications, the LSI is sent if the application passes an IP address to its peer. For IPv6 applications, HIT is passed to the peer application. If an IPv4 application passes LSI to the IPv6 application, it would likely succeed since most IPv6 applications understand IPv4-format addresses. However, if an IPv6 application passes HIT in place of an IPv6 address to an IPv4 application, the communication can fail as legacy IPv4 applications do not understand IPv6 address format.

Figure 13.4 illustrates the scenario where two IPv6 applications communicate over an IPv4-only network over HIP. Likewise, a scenario where legacy IPv4 applications communicate over an IPv6 network is possible.

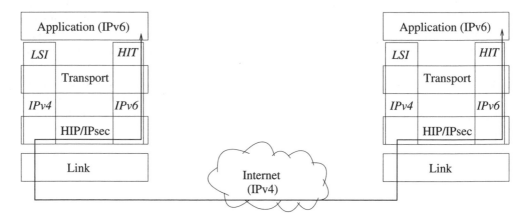

Figure 13.4 Two IPv6 applications talking over IPv4 network via HIP.

13.4   Secure Mobile Architecture

Secure Mobile Architecture (SMA) is an integrated network architecture standardized by the OpenGroup (The Open Group 2004). The OpenGroup is a vendor- and technology-neutral organization that aims at providing integrated solutions for information flow within and between organizations. Its platinum members are Capgemini, EDS, HP, IBM, and NEC. In addition, many companies, government bodies, and academia are part of the consortium. The OpenGroup is composed of ten forums focusing on various aspects of IT systems.

13.4.1 Components of SMA

OpenGroup published the Secure Mobile Architecture (SMA) Vision and Architecture in 2004 by joint work of Security and Mobile Management Forums. SMA combines Public Key Infrastructure (PKI), HIP, Network Directory Services (NDS), and Location-Enabled Network Services (LENS). Boeing was the primary contributor to including HIP in the architecture.

The SMA name has the following meaning. The "Secure" part refers to having user and host authentication through enterprise directory as well cryptographic identifiers in each data packet. "Mobile" refers to the ability to change network location while being backwards compatible with legacy applications. "Architecture" implies an integrated approach to augment the enterprise network with end-to-end encryption, delegation-based authentication, and enabling secure access of hosts outside of the security perimeter.

SMA addresses the business needs of having secure network access using various available networks (LAN, WLAN, cellular) with the capability to switch between them, multi-level security that works on top of any MAC layer, backward compatibility, and automatic configuration.

The PKI part of SMA consists of company PKI servers, an employee's badge card with built-in smart chip, and Registration Authority (RA) connected by SSL tunnels to the card reader. An employee requests a temporal certificate valid for 8 hours by supplying the badge certificate stored on the card chip when inserting a card into the card reader.

Network Directory Service provides such information as mapping between host identity and address for internal users. The client communicates with NDS over Secure Lightweight Directory Access Protocol (SLDAP). NDS is integrated with the DNS system that enables updates using the Dynamic DNS protocol. The location server that maintains geographical coordinates of a user and a policy decision daemon, as well as various middleboxes, are also part of NDS. To authenticate, new clients need to pass a two-stage provisioning process. First, the client performs regular authentication with the DHT and AAA server (RADIUS) to obtain an IP address. Second, the client contacts the Registration Authority (RA) to obtain a temporal certificate over TLS. Then, the client can update its identity to IP address mapping in the directory using the Secure Lightweight Directory Access Protocol.

In the first prototype of directory service at the SMA testbed, clients were receiving status updates directly from the LDAP server. The server queried the decision daemon whether the client was authorized based on the location information and existing policies. The LDAP server also updated the client's current IP address to the DNS server. Further extension of the directory deployment is based on the publish–subscribe paradigm. The Message Broker Infrastructure handles message delivery in the new architecture. Based on incoming data from a location server, sensors and RFID tags, the decision daemon, LDAP server, and HIP daemons are notified according to their subscriptions.

Location-Enabled Network Service enables tracking of workers and equipment within the factory. The system is a combination of passive tag gates that detect an object passing through and a WLAN positioning system that determines user location based on signal strength and time-of-flight triangulation from several access points. AeroScout active radio-frequency identification tags (RFID) possess a unique 802.11 MAC address and can be carried in a pocket or attached to mobile equipment. The LENS system is connected to the Boeing intranet and forms an integral part of the overall SMA testbed.

13.4.2   SMA testbed at Boeing

According to Richard Paine of Boeing (Co-chair of Secure Mobile Architecture in Open-Group, Chair of IEEE 802.11k Task Group), the US Federal Aviation Administration requires that every step in assembling an airplane can be identified with the person and equipment used in the process. In the traditional approach, a worker uses his personal stamp to mark the parts attached to an airplane. In the case of an accident, the cause could be tracked to the part and the person installing it at the factory.

Making the identification of airplane construction and parts automatic and secure is the driving force behind the SMA in Boeing. The airplane should be gradually moved through the assembly process at the factory to keep up with the schedule. With the current practice, mechanics working on the airplane go to the stationary PC located nearby to them to log their operations. The use of WLAN with portable devices has not been successful because going out of network coverage results in loss of VPN connectivity. Workers get annoyed by such disconnections and prefer to use a PC connected to a LAN instead. With the deployment of SMA, no manual authentication nor reconnection is necessary thus improving the manufacturing process considerably. Other crucial properties that SMA provides are end-to-end packet authentication, prevention of spoofing, and seamless WLAN routing.

An SMA testbed is currently deployed at the Everett manufacturing site assembling Boeing 777 airplanes with expansion plans to two other sites. Figure 13.5 illustrates the location of the testbed within the factory. Robots known as Crawlers are used on the factory to move components of the Boeing 777 aircraft during the assembly process using the Supervisory Control And Data Acquisition (SCADA) system. A device called SMA/HIP Endbox is installed on Crawlers and connected to FactoryNet network. Figure 13.6 presents a photograph of the Crawler with an Endbox installed. The Endbox contains a Subscriber Identity Model (SIM) card to provide strong authentication over the WLAN network to the Registration Authority and Directory. The Endbox securely communicates to the Robot Controller over HIP Security Associations. In addition to the Endbox, Crawlers include a HIP Bridge, a device that enables legacy Ethernet equipment to connect to SMA components in the factory.

The secure handover between cellular and WLAN over HIP was demonstrated on an SMA testbed in 2005. While WLAN access points were connected to the Boeing intranet, traffic to the cellular network traveled through the public Internet before entering the intranet through a Netscreen firewall. The presence of HIP encryption and authentication enabled moving the hosts outside of the security perimeter, making the home, office, and public hotspot connectivity equally secure.

The initial testbed version was deployed in 2004. It had evolved gradually by including more advanced components in following years. Figure 13.7 shows internal components of the SMA testbed at Boeing as of 2007. It is organized into two subnetworks, one at Bellevue and one at Everett. In addition it includes a company-wide AAA server (RADIUS) and a PKI server. Each subnetwork includes a separate Message Broker, Directory server, DNS server (namespace mobile.tl.boeing.com), Location information server, Wireless access switch, a Temporal certificate Registration Authority, and SMA access point. The Bellevue subnet includes a WiMAX wireless network where towers are connected to a switch in the subnet. The Everett site includes a Wi-Fi switch connected to several WLAN access points. Nokia N770 Internet Tablets are used as user terminals that can connect to the intranet over the WLAN. Factory Robots (Endbox) and Controllers also connect over the WLAN.

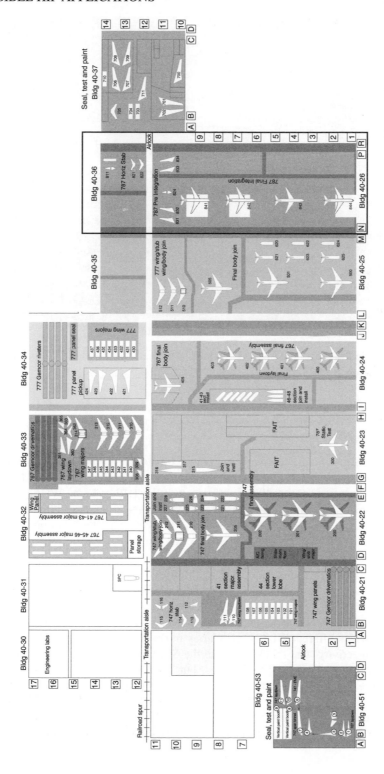

Figure 13.5 Location of the SMA testbed in the Everett airplane factory. (Reproduced by permission © Boeing.)

Figure 13.6 HIP Endbox installed on the Crawler platform under the plane. (Reproduced by permission © Boeing.)

In the testbed, SMA mobile terminals can connect over HIP to the SMA server in each intranet subnetwork and to each other. They can also use public Internet access to pass to the intranet. The Robot and Controller communicate using HIP over the WLAN network. The latest development includes Voice over WLAN (VoWLAN) using Nokia Internet Tablets (N770), which is secured using HIP.

In summary, the deployment of SMA within Boeing brings several advantages that improve secure mobility and at the same time reduce complexity and management costs. The presence of secure identifiers in every packet allows hosts outside of the security perimeter to access the intranet services. SMA is backwards compatible and does not require changes to routing infrastructure. Legacy hosts that do not support HIP can be still allowed in the deployment, depending on how strict the security policy is. The user of HIP allows intranet hosts to move between different WLAN boundaries and even use the public cellular network as a fall-back connectivity. The middleboxes in the SMA network are able to authenticate the traffic not based on changing IP addresses, but with a permanent identity associated with hosts and users. The SMA provides seamless integration of IPv4 and IPv6 through the use of DNS for connections and HIP to support cross-family handovers. The presence of PKI secures host and user identities from spoofing. The delegation mechanism enables certain operations, such as setting a new host, without evolving into a complete bureaucratic process.

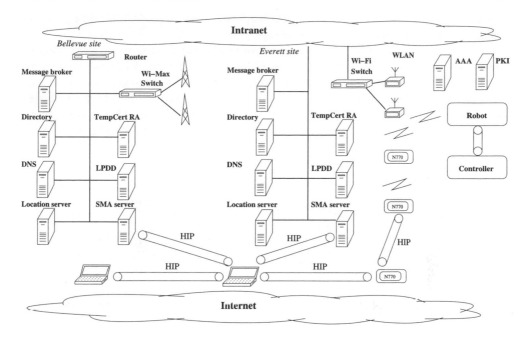

Figure 13.7   Components of the SMA testbed at Boeing. (Reproduced by permission © Boeing.)

13.5   Live application migration

A *virtual machine* (VM) is an instance of an operating system such as Linux or Windows running on virtual hardware that is mapped to the real hardware of the host machine. Several virtual machines even with different operating systems can run on the same host machine. Virtual machines are widely used to reduce costs of hosting WWW servers or for the convenience of running Linux in a separate window after booting from Windows.

Recently, several proposals address the *migration* of virtual machines from one host machine to another. The migration is useful, for example, when the host machine needs to be shut down for maintenance. The VM can move to another host machine and continue operation without a noticeable service break to the users. VM can also move for load balancing if the host machine becomes overloaded. The migration process can introduce a *residual dependency* if the VM on the new host machine still needs the old host to operate. As an example, the network traffic to the VM can still arrive to the old host machine, which then needs to forward it to the new host machine. Therefore, the old host machine cannot be brought down until all active sessions have terminated.

The *Virtual Machine Monitor* manages execution of virtual machines on the host machine. In a full virtualization approach, the VM sees a virtual hardware replica and can run unmodified operating system code. Such an approach is convenient but rather inefficient in performance. Instead, the para-virtualization approach requires small modifications to the operating system but provides better performance. The Xen VMM is suggesting the use of a hypervisor that serves as an interface for VM hardware (Barham *et al.* 2003). The guest OS is modified to use hypercalls to access hardware.

Xen has demonstrated migration from one host machine to another within a few hundred milliseconds within a single LAN. To avoid residual dependencies, VM migration moves the MAC address from the old to the new host, which limits the possible migration area to a LAN. This section describes the extension of migration of Xen VM to wide-area networks (Koponen *et al.* 2005). HIP is used to provide mobility of VM data connections from the old to the new host machine. The migration process is illustrated in Figure 13.8.

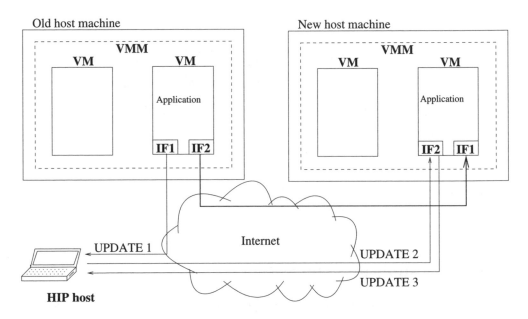

Figure 13.8 Wide-area migration of Xen VM.

A simple approach to wide-area VM migration would be break-before-make when a VM would redirect data packets destined to the old host machine only after moving to the new host machine and acquiring a new IP address. Such an approach would introduce a long break in network connectivity of a VM causing a delay and packet loss to applications. Therefore, the make-before-break approach is used when a VM obtains an IP address at the new host machine before moving. A VM has two virtual Ethernet interfaces, one connected to the old and one to the new host machine. Therefore, VM is multihomed during the migration with link-layer tunnels connecting it to the network of the second host machine.

The timing of the migration of network connections of a VM can impact the outage experienced by applications. If the connections are migrated before VM migration, the link layer tunnel would be loaded with traffic redirected from the new to the old host machine. The tunnel capacity would be reduced thus slowing down the VM migration itself. Migrating connections after VM migration would increase the latency and packet loss for applications. Therefore, data connections and VM migration should occur synchronously.

During migration, the peer hosts perform the return routability test that is required by HIP mobility protocol to verify the ownership of the new IP address. By delaying the last reply message, the migrating VM can control when the peer would redirect data traffic to the new address.

Migration of a Xen VM involves the following steps.

1. *Preparation.* VM running on VMM host A opens a tunnel to VMM on host B.

2. *Address configuration.* VM obtains an IP address in network B.

3. *Migration starts.* VMM A starts copying memory pages to VMM B.

4. *Binding update.* The HIP daemon executes a mobility update so that data traffic will go to VMM B.

5. *Suspending.* VM on host A is stopped and remaining dirty memory pages are copied to VMM B.

6. *Activation.* VM is activated on host B.

7. *Traffic switch.* VM starts receiving new traffic on host B.

8. *Closing.* Link-layer tunnels close between VMM A and B.

When a VM moves, its Host Identity associated with public and private keys has to be relocated to the new host machine. Sometimes, the keys are located on a separate security device or external flash memory and thus are not easily movable. Migrating a VM in this case would introduce residual dependency as the VM on a new host machine would be dependent on the old host machine for signing its packets. An alternative approach is based on the use of delegation certificates, where the VM uses pseudonym public/private keys to identify itself. The machine storing the permanent keys can issue a certificate that binds the pseudonym to the permanent identity.

Migrating a VM in this case would introduce residual dependency as the VM on a new host machine would be dependent on the old host machine for signing its packets. An alternative approach is based on the use of delegation certificates, where the VM uses pseudonym public/private keys to identify itself. The machine storing the permanent keys can issue a certificate that binds the pseudonym to the permanent identity.

The migration process starts with pre-copying when the VM memory image is transferred to the new host machine. This is a bandwidth-hungry operation that can take tens of seconds to execute. Memory pages that have recently been modified are called dirty and are also transferred. The pre-copying stage of VM migration ends when Xen determines that the dirtying rate of memory pages exceeds the transfer speed of memory pages to the new host machine. Afterward, the VM is suspended and remaining dirty pages are transferred to the new location. The HIP daemon also receives a signal to begin the UPDATE procedure and change its current IP address in DHT to redirect connecting clients to the new host machine. The Xen VM communicates with the HIP daemon using a TCP and a raw IP socket.

The prototype implementation of VM wide-area migration is based on OpenHIP and uses OpenDHT for storing mappings of host identity to IP address. Two servers connected with Gigabit Ethernet were used to test the migration process. The wide-area network was emulatcd by the Linux kernel. A set of clients running on PlanetLab was generating workload connections to the migrating VM. The clients connected to the VM over HIP. Two types of traffic, of TCP and UDP, were tested. The TCP window had to be increased to several megabytes to sustain the high transmission rate necessary for a memory image transfer.

In TCP experiments, the VM was prepared for migration in 9 seconds, while the suspend interval was 170 milliseconds. However, since TCP acknowledgments were lost during the suspended period, the TCP sender on VM experienced a retransmission timeout that increased the overall outage in the connection to 370 milliseconds. UDP traffic is not sensitive to losses and the outage is less than 200 milliseconds.

13.6   Network operator viewpoint on HIP

The organized use of HIP in a telecommunication network, such as 3G UMTS, can cause a number of issues for the network operator. This section summarizes an IETF draft by Vodafone and NEC describing the aspects of traffic accounting and charging, lawful interception, integration with SIP, and network interface selection (Dietz *et al.* 2005). These issues mostly arise when the operator likes to provide a centralized and supported deployment of HIP for its own customers. The users are, of course, able to use HIP themselves without touching those issues.

Network operators traditionally make their profits from providing data connectivity to users. *Flat rate* is the simplest charging scheme where users pay a fixed price per month for unlimited use of the network. Flat rate is easy to implement, popular with users, and does not incur any issues with the use of HIP. Many users are accustomed to flat rate charging from broadband Internet connections to homes, provided by ADSL and cable modems. Unfortunately, most cellular and WLAN hotspot networks still do not use flat rate charging due to the high cost of their installation and operation.

The time-based model charges the user based on the time spent online. It is popular with WLAN hotspots and functions well with HIP. The data volume model charges the user based on the amount of data transmitted and received. It is often applied in cellular networks, such as 3G, and works in the presence of HIP.

The most difficult charging model from the HIP point of view is content-based charging, when the operator likes to price data differently depending on its type. The reason for such an approach is that various data have vastly different per-byte values to the customer. Users might not be willing to pay the same rate for downloading large video files as they do for short messages. Some operators are also worried about customers using VoIP software such as Skype instead of paying per-time in a conventional phone call. When HIP is applied end-to-end, from user equipment to the service provider, all data traffic including transport protocol and port information is encrypted with IPsec ESP. The operator cannot use the content-based charging model and therefore can be less willing to promote the use of HIP for its customers.

Another issue related to end-to-end encryption is lawful interception of user traffic. In most countries, network operators are required to provide a capability to eavesdrop user communication if requested by police. If the operator provides HIP support in a way that the private keys of the user are not revealed to the operator, providing lawful interception is not possible. *Key escrow* is an arrangement whereby the user provides its private keys to the operator to be used for traffic decryption. Otherwise, the operator can drop any encrypted packets. Another alternative can use a different approach and terminate the HIP association in the proxy located in the operator network. The proxy can perform lawful interception, collect charging information, and establish HIP association to the traffic destination.

Many modern telecommunication networks, such as the IP Multimedia Subsystem (IMS) in 3G, rely on SIP protocol for locating users and placing calls. Therefore, from the operator viewpoint, useful integration of HIP and SIP is important. Section 15.2 discusses how HIP can be useful for SIP and its considerations apply here as well. However, IMS SIP has differences from IETF SIP that can impact its use with HIP.

In a managed environment, the operator might need to centrally create and maintain Host Identity for a user. In 3G telecommunication systems, the user identifier is stored in a Subscriber Identity Module (SIM) card protected by a Personal Identification Code (PIN). The user identity is bound to the SIM card, not the phone. The user can be still reached by others after moving the SIM card to another phone. Operators are interested in centrally creating user identities and storing private keys securely on a SIM card because most users are not sufficiently educated in nor willing to involve themselves with details of HIP software. This is something of a shift in the original concept of having Host Identifiers bound to the host, not the user.

Many modern smart phones have both 3G cellular and WLAN interfaces. HIP naturally helps there with switching between interfaces and even using them simultaneously. When both networks are managed by the same operator, user authentication is easily done. However, mobility and multihoming in a multi-operator environment is still a challenging area because of involved accounting and agreement issues. Interface selection could either be done by the user phone or controlled from the operator network elements. Factors affecting the interface selection are connection cost, coverage, speed, delay and jitter, loss rate, and type of connectivity (public IPv4, NAT, or IPv6).

14

Application interface

Two types of applications can use HIP. Section 14.1 describes *legacy* applications unaware of HIP. Section 14.2 describes *native* applications written to use HIP using an extended socket API. The advantage of the first approach is that existing applications do not need modifications to benefit from HIP. The disadvantage is that a legacy application cannot convince the user of a secure service.

14.1   Using legacy applications with HIP

The HIP interface for legacy applications is discussed in an IETF document (Henderson *et al*. 2007). It describes three possible methods to use HIP with applications that employ the standard Berkeley socket API using IPv4 or IPv6 addresses. The first method continues to use IP addresses in applications while mapping between the host address and identity with the HIP layer. The second method deploys DNS resolution to provide an address-compatible representation of host identity to the application. In particular, HIT is returned to IPv6 applications and LSI to IPv4 applications. The third method uses a HIT or LSI in the application directly without prior DNS resolution.

The advantage of using IP addresses is better support for *referrals*. A referral is an attempt to pass a locator to an application on another host. As an example, the FTP protocol has messages passing a host IP address to its peer. The problem of referrals is well-known for NAT users as many applications pass their internal address to the peer application. Such addresses are not routable outside of the NAT and disrupt the application connectivity.

The advantage of using the HIT versus an IP address in the application is the concept of *channel binding*. The calling application is bound to the cryptographic host name and the ESP tunnel created by HIP. Therefore, either an application connect() call would connect to a host owning the private key corresponding to the HIT, or the call fails.

Host Identity Protocol (HIP): Towards the Secure Mobile Internet   Andrei Gurtov
© 2008 John Wiley & Sons, Ltd

14.1.1   Using IP addresses

In this scenario, an application is using a connect call with a destination IP address to establish a connection. Since the standard socket API does not allow to indicate whether the use of HIP is desired, an external mechanism should be used to configure HIP associations. For ESP HIP encapsulation, an entry to the security policy database is added to invoke HIP for a given pair of IP addresses. The mapping between the host identity and IP address can be implemented by wrapping the socket API or by patching the operating system.

The HIP association establishment can be triggered in several ways after an application calls connect() with a destination IP address:

- *Static configuration.* Policy rules are pre-configured in the host. The security level corresponds to the current use of IPsec policies.

- *Reverse DNS lookup.* The HIP layer can perform the reverse DNS mapping from an IP address to the FQDN for the destination host. If the destination host has the HIT stored in its forward DNS record, it can be retrieved and put to the I1 packet. Unless secured by DNSSEC, such queries can be the subject of spoofing attacks. The reverse and forward DNS queries introduce additional delays to the application. If DNSSEC is used and the zone signatures are trustworthy, this method provides a high security level.

- *DHT lookup.* If a DHT stores HITs indexed by the host IP address, the HIP layer can lookup the destination HIT after the connect call but before sending the I1 message. The procedure has similar security and delay concerns as the DNS lookup.

- *Opportunistic mode.* The HIP layer tries to send an I1 packet to all IP addresses that an application connects. The security level is similar to plain IP – vulnerable to Man-In-The-Middle attacks, except that HIP associations are robust to hijacking attempts.

The use of IP addresses in applications does not provide complete mobility support. If an application stores the destination IP address and uses it when calling the sendto() function, connectivity will break when the destination host changes the IP address. On the other hand, if a connect() call is used for a socket, the change of an IP address can be transparently implemented as socket re-addressing. Further drawbacks of using IP addresses in the application interface include weaker security binding between the host identity and host locator. The binding in this case is implemented internally at the HIP layer and is not visible to the application or the user.

The primary benefit of IP addresses for applications is better support for referrals, when an application, such as FTP, passes its IP addresses to a peer application.

14.1.2   Using DNS resolution

Typical Internet applications rely on DNS resolution for obtaining arguments for socket calls from a FQDN supplied by the user. This section discusses an approach to support HIP for legacy applications using the DNS resolution. The resolver library can be modified to support HIP in the following ways:

- *Use of a suffix*. This method uses a suffix such as ".hip" for an application to request HIT or LSI. Thus, a request for "www.example.com.hip" would return only HIT or LSI, depending on the request family, but no IP addresses. A request to the "www.example.com" would instead return IP addresses as usual. The suffix method provides finer control of application needs in a user-friendly way to trigger the use of HIP. However, it requires the user to be aware of HIP suffixes and remember to add them to DNS names.

- *Returning HIT and LSI instead of IP addresses*. This method transparently removes IP addresses from a DNS reply and returns only HIT or LSI to the application. It does not require the user to know anything about HIP. If neither HIT nor LSI is found for a particular DNS name, IP addresses can be returned for backward compatibility with non-HIP hosts.

The main drawback of the DNS-based approach is lack of support of referrals. Additionally, not all legacy applications use DNS so cannot necessarily be supported this way. The strong side of this approach is closer binding of the peer host identity in the application.

Security properties of the DNS-based approach depend on the use of DNSSEC. If DNSSEC is properly used, the security level is fairly high and corresponds to the current use of TLS with certificates. Without DNSSEC the level of authentication is similar to plain IP, except that an established HIP association is robust to hijacking attempts.

14.1.3   Directly using HIT

The third possible approach for a legacy application is to use HIT directly in a connect() call. This approach provides best channel binding properties and the highest security level. However, the random-looking HIT format makes it cumbersome for human users to input the HIT in the application. The HIP layer has to distinguish between IPv6 addresses and HITs using an ORCHID prefix. For HITs, the HIP layer has to lookup the IP address using a pre-configured local mapping (e.g., in the /etc/hip/hosts file) or using a DHT.

14.2   API for native HIP applications

This section describes an extension of the standard Berkeley socket Application Programming Interface (API) for HIP-aware applications, also known as the Native HIP API. Its specification enables the use of any shim-layer protocols, although the main focus is on HIP (Komu 2007).

14.2.1   Overview of the design

The standard Berkeley API can be used with legacy applications with LSI, HITs, or IP addresses as described in 14.1. However, applications that are written with HIP support can take full advantage of HIP functions, such as specifying a custom host identity (Komu 2004). Furthermore, using HITs in the interface may not be always possible in the future for the following reasons:

- When HIP is used in the opportunistic mode, the peer HIT is unknown for the application. Therefore, some local identifier is needed for the application.

- The current HIT length of 128 bits selected for compatibility with IPv6 addresses could be insufficient in the future if more attacks on hash functions are found.

- Multiple HITs can be used for a single HIP association in the future, when HIP multicast extensions are specified.

- For implementing session mobility or process migration, the binding between the identifier in an application and the host identifier requires changes.

To overcome these limitations, the Native API uses Endpoint Descriptors (ED) in the application interface instead of HITs. EDs resemble file handles similar to the UNIX file system interface and have only local significance. In that aspect, EDs are similar to LSI. However, LSIs have an IPv4-compatible format with a prefix to separate them from routable IPv4 addresses and can be used only with the AF_INET address family, while an ED is a small integer number. The number serves as an index to the host identity database at the host and cannot be passed to other hosts as a referral.

The Native API hides the transport port numbers from the application as the numbers are often changed by NATs in the network. The EDs are passed to the application using a new address family PF_SHIM that complements existing AF_INET and AF_INET6 families. From an architecture viewpoint, the API is placed between the application and transport layers in the Internet protocol stack model. The shim layer is located between the transport and network layers.

With Native API, applications continue to use human-friendly FQDNs as identifiers, although communication using HITs or IP addresses is still possible. Within the application, different connections are identified by source ED, destination ED, source port, destination port, and the transport protocol number. At the transport layer, the protocol number is already fixed and the EDs are replaced by Host Identities. At the HIP shim layer, associations are identified by a pair of source and destination Host Identities. The network layer still uses the IP addresses for routing packets.

Each PF_SHIM socket is bound to one source and destination ED, one source and destination port, and the transport protocol. However, a single HI can be associated with multiple EDs above it and multiple network interfaces beneath it. The binding between the ED and HI can be established after the first connect call by the application in the opportunistic HIP mode. In advanced scenarios, such as multicast, a single ED can associate with several HIs. The socket listening on an ED can accept connections from any local HI.

The application resolves FQDN to EDs as illustrated in Figure 14.1. The application calls the resolver library with FQDN as the argument. The local resolver stub sends a query to a DNS server, which returns a set of Host Identifiers and locators available for the given FQDN. The resolver passes the DNS response to the HIP layer for processing. The HIP layer stores the mapping between Host Identities and their locators to the local host identity database and sends the Endpoint Descriptor back to the resolver. The resolver returns the ED to the application.

14.2.2 Interface specification

The endpoint descriptor is passed to the application in a structure sockaddr_ed listed below. The ed_family field is set to PF_SHIM (AF_SHIM is an alias to PF_SHIM and

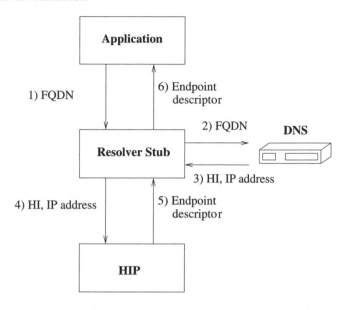

Figure 14.1 Native HIP API resolves a FQDN to an Endpoint Identifier.

can also be used) in host byte order. The port number ed_port is two bytes. The ed_val field contains the 4-byte value of ED in network byte order. The value is opaque for applications and should only be handled, e.g. compared with other EDs, through special interface functions. In addition, three macros are provided for server applications to set ed_val field to any, public or anonymous ED: SHIM_ED_ANY, SHIM_ED_ANY_PUB, and SHIM_ED_ANY_ANON. Those macros are similar to INADDR_ANY and IN6ADDR_ANY_INIT, and can be used only by a single process per port to prevent ambiguity.

```
struct sockaddr_ed {
        unsigned short int ed_family;
        in_port_t ed_port;
        sa_ed_t ed_val;
}
```

The endpoint structures below store application-specific endpoint identifiers containing private keys in a uniform format. The structures endpoint and endpoint_hip represent a generic endpoint and a HIP endpoint. For a HIP endpoint, the family field is set to EF_HI. The flags field can have the SHIM_ENDPOINT_FLAG_ANON flag set to indicate that the endpoint is anonymous. If a flag SHIM_ENDPOINT_FLAG_PRIVATE is set, the host_id field in the union contains a private key of the endpoint in the format given in the HIP base specification. Alternatively, the host_id can store a HIT if a flag SHIM_ENDPOINT_FLAG_HIT is set.

```
struct endpoint {
   se_length_t    length;
   se_family_t    family;
```

```
};
struct endpoint_hip {
  se_length_t length;
  se_family_t family; /* EF_HI in the case of HIP */
  se_hip_flags_t flags;
  union {
    struct hip_host_id host_id;
    hit_t hit;
  } id;
};
```

The HIP interface resolution function reuses the same data structure as the existing getaddrinfo() resolver with ai_family set to PF_SHIM.

```
struct addrinfo {
  int     ai_flags;          /* e.g. AI_ED */
  int     ai_family;         /* e.g. PF_SHIM */
  int     ai_socktype;       /* e.g. SOCK_STREAM */
  int     ai_protocol;       /* usually just zero */
  size_t  ai_addrlen;        /* length of the endpoint */
  struct sockaddr *ai_addr;  /* endpoint socket address */
  char    *ai_canonname;     /* canon. name of the host */
  struct addrinfo *ai_next;  /* next endpoint */
};
```

The resolver requires that the AI_ED flag is set to prevent compatibility problems with legacy applications. The resolver returns found endpoint addresses in a linked list. If found, sockaddr_ed structures are listed first followed by other address types. However, HITs in the sockaddr_in6 format are returned with sockaddr_ed structures and not separately on the linked list. If AI_ED_RVS flag is set in the argument to the resolver, only the rendezvous service addresses are resolved for an endpoint identifier.

Applications can set AI_ED_NOLOCATORS to prevent the resolver from looking up the locators, if the application is using some external mechanism for resolving locators. The flag AI_ED_ANON reverses the default resolution order of first looking up public and then anonymous endpoint identifiers.

The flags AI_ED_ANY, AI_ED_ANY_PUB, and AI_ED_ANY_ANON make the resolver return a single socket address. For local addresses, the flags allow late binding to a group of endpoint identifiers by a server application. When used for peer endpoints, the flags enable the HIP association establishment in an opportunistic mode.

The Native API uses the sockaddr_ed structure in place of sockaddr_in or sockaddr_in6 structures in the standard socket API. Otherwise the interface of standard API functions bind, connect, send, sendto, sendmsg, recv, recvmsg, and recvfrom is unchanged.

The sockaddr_ed structure is produced by the getaddrinfo() function listed below. The function accepts the node name or the service name as arguments. One of the arguments, but not all, can be NULL. The nodename argument contains the host name for resolution and the NULL value refers to the local host. The servname argument contains the port number that is set in the resulting socket structures. The hint argument limits the type of socket addresses

returned by the resolver. As an example, setting the SHIM_ENDPOINT_FLAG_ANON flag
in hints would make the resolver return only anonymous socket structures.

```
int getaddrinfo(const char *nodename,
                const char *servname,
const struct addrinfo *hints,
struct addrinfo **res)
void free_addrinfo(struct addrinfo *res)
```

The function getaddrinfo() returns zero on success and places the socket addresses to the
pointer passed as the res argument. A non-zero return value denotes an error. The memory
for the linked list of socket addresses is dynamically allocated by the resolver and should be
freed by the application afterward using the free_addrinfo() function.

The functions shim_endpoint_pem_load() and shim_endpoint_pem_load_str() load the
endpoint identity from a file or from a string. The function loads a public or a private key from
a source encoded with the Privacy Enhanced Mail (PEM) standard. The result is returned
through the endpoint pointer using dynamically allocated memory. The application should
use the free() call to release memory later. Functions return zero on success and non-zero
value on failure.

```
int shim_endpoint_pem_load(const char *filename,
                           struct endpoint **endpoint)
int shim_endpoint_pem_load_str(const char *pem_str,
                           struct endpoint **endpoint);
```

Most Native API functions operate on endpoint descriptors and not the endpoint identities. The function getlocaled() creates the endpoint descriptor for a local and getpeered()
for an external endpoint identity. The endpoint descriptor structure is internally allocated in
the functions and should be freed later by the calling application. The endpoint argument
specifies the endpoint identity and the NULL value makes the function create a random
identity itself. The servname argument defines the port number that is copied to the
sockaddr_ed structure. The addrs argument is a pointer to the current IP addresses of the
peer or local host. The NULL value for addrs argument is possible but not recommended.

```
struct sockaddr_ed *getlocaled(const struct endpoint *endpoint,
                               const char *servname,
                               const struct addrinfo *addrs,
                               const struct if_nameindex *ifaces,
                               int flags)
struct sockaddr_ed *getpeered(const struct endpoint *endpoint,
                               const char *servname,
                               const struct addrinfo *addrs,
                               int flags)
```

The getlocaled() function accepts a list of interfaces as an argument *ifaces*. If the
argument is NULL, the interface selection is unlimited. Otherwise, the ED is bound to the
specific interfaces from the argument. The function if_nameindex() can be called to obtain a
list of interfaces in the system. The getlocaled() function accepts SHIM_ED_REUSE_UID,
SHIM_ED_REUSE_GID, and SHIM_ED_REUSE_ANY flags. These flags allow the use of

endpoint identity for other processes with the same User ID (UID), group ID (GID) or any processes. Flags SHIM_ED_IPV4 and SHIM_ED_IPV6 limit the use of local addresses to IPv4 or IPv6 protocol families.

Two functions, getlocaledinfo() and getpeeredinfo(), provide a reverse mapping from EDs to Host Identities. Their interface is given below. The first argument my_ed or peer_ed contains the ED that functions look up. The result is returned in endpoint and addrs pointers using dynamically allocated memory. The getlocaledinfo() also returns the list of interfaces associated with the ED.

```
int getlocaledinfo(const struct sockaddr_ed *my_ed,
                    struct endpoint **endpoint,
                    struct addrinfo **addrs,
                    struct if_nameindex **ifaces)

int getpeeredinfo(const struct sockaddr_ed *peer_ed,
                    struct endpoint **endpoint,
                    struct addrinfo **addrs)
```

The application operating on multiple EDs can use these functions to check the properties of each ED and select the most suitable one for a given task.

14.2.3   Socket attributes

Several HIP policy attributes can be set for a Native API socket using getsockopt() and setsockopt() calls. The IPSEC_ESP_TRANSFORM attribute sets the preferred ESP transform algorithm. IPSEC_SA_LIFETIME defines the lifetime of IPsec security association in seconds. SHIM_PROTOCOL is set to PF_HIP although different protocols could be supported in future. SHIM_CHALLENGE_SIZE defines the challenge size of the puzzle. SHIM_SHIM_TRANSFORM is the preferred SHIM transform algorithm and SHIM_DH_GROUP_IDS is the Diffie–Hellman key group. SHIM_AF_FAMILY contains the preferred locator family, by default AF_ANY. SHIM_FAST_FALLBACK is set to enable extensions for opportunistic HIP and SHIM_FAST_HANDSHAKE for TCP piggybacking.

In addition to socket attributes mentioned above, there are more generic attributes specified for SHIM protocols, such as HIP and SHIM6 (Komu *et al.* 2007a). These attributes enable the application to handle the following tasks:

- *Enabling the SHIM layer.* The application should be able to query if the SHIM layer is used by the given socket and be able to enable or disable it.

- *Management of locators.* The application may need to query the list of all local and peer locators, and the pair of preferred locators. The actual pair of source and destination locators for each packet is selected by the SHIM layer based on path availability, application preferences, and path characteristics.

- *Event triggers.* The transport protocol such as TCP or the SHIM layer may need to inform the application of path reachability problems based on timeouts or arriving ICMP messages. The application can also get positive notifications that the path works well.

- *Configuring timeouts.* The application should be able to set the timeout values for detecting path reachability and whether keepalive packets are sent by the SHIM layer.

- *Hot-standby configuration.* An application requiring low recovery time in the case of a path failure can request the SHIM layer to keep a pair of hot-standby locators. These locators use an alternative tested path that can be immediately utilized if the current locator pair becomes unreachable.

- *Path exploration.* The application can set the aggressiveness of new path exploration in the case of a path failure. As an example, the SHIM layer could send several discovery packets concurrently over multiple paths.

- *Delaying context establishment.* The application should be able to control if the context establishment, such as the base exchange, is started immediately or later when the data is sent.

- *Locator queries.* For each incoming or outgoing packet, the application should be able to check its locator pair. As an example, after giving its preferred locators, the application can check what were the actual locators set in the outgoing packet.

Below we list the attributes currently defined for the SHIM layer. For more complex attributes we give their data structure and an example of their use with getsockopt() or setsockopt() calls.

- The SHIM_ASSOCIATED attribute can be used only with getsockopt() to check if the socket is using the SHIM layer. An example below shows how the attribute can be used. The result is returned via a reference to the optval integer, with non-zero indicating an associated SHIM context and zero otherwise.

```
int optval;
int optlen = sizeof(optval);
getsockopt(fd,SOL_SHIM,SHIM_ASSOCIATED, &optval, &optlen);
```

- The SHIM_DONTSHIM attribute defines if the SHIM layer provides multihoming for a socket. It can be queried and set by the application with the getsockopt() and setsockopt() functions.

- The SHIM_HOT_STANDBY attribute can be queried and set by the application to enable a secondary location pair at the SHIM layer that can be immediately used in the case of primary path failure.

- The SHIM_LOC_LOCAL_PREF attribute queries or sets the local preferred locator for the socket. The structure shim_locator that stores the locator is listed below. The lc_family contains the address family of the locator, AF_INET or AF_INET6. The lc_ifidx is the read-only index of the network interface that has the locator. The lc_flags can mark special locators, such as hash-based or cryptographically generated addresses. The lc_preference is a preference level of the locator compared with other locators. Finally, lc_addr contains the locator itself.

```
struct shim_locator {
  uint8_t    lc_family;      /* address family */
  uint8_t    lc_ifidx;       /* interface index */
  uint8_t    lc_flags;       /* flags */
  uint8_t    lc_preference;  /* preference value */
  uint8_t    lc_addr[16];    /* locator */
}
```

An example below performs the SHIM_LOC_LOCAL_PREF set operation.

```
struct shim_locator lc;
struct in6_addr ip6;
// ip6 = preferred IPv6 address
bzero(&lc, sizeof(shim_locator));
lc.lc_family = AF_INET6;   /* IPv6 */
lc.lc_ifidx = 0;
lc.lc_flags = 0;
lc.lc_preference = 255;
memcpy(lc.lc_addr, &ip6, sizeof(in6_addr));
setsockopt(fd, SOL_SHIM, SHIM_LOC_LOCAL_PREF, &lc,
           sizeof(optval));
```

- The SHIM_LOC_PEER_PREF attribute works similarly to SHIM_LOC_LOCAL _PREF but queries or sets the preferred locator for the peer host.

- The SHIM_LOC_LOCAL_RECV attribute determines if the SHIM layer stores the destination locator of arriving packets as ancillary data that can be obtained with a recvmsg() call. The attribute value is 0 or 1 and can be queried by getsockopt() and set by setsockopt() calls.

The format of a POSIX message header data structure from socket.h file is listed below. The structure serves as an argument to sendmsg() and recvmsg() calls. The msg_control field contains a pointer to ancillary data that can contain SHIM-specific parameters. The msg_controllen is the length of the ancillary data.

```
struct msghdr {
  caddr_t msg_name;        /* optional address */
  u_int   msg_namelen;     /* size of address */
  struct  iovec *msg_iov;  /* scatter/gather array */
  u_int   msg_iovlen;      /* # elements in msg_iov */
  caddr_t msg_control;     /* ancillary data, see below */
  u_int   msg_controllen;  /* ancillary data buffer len */
  int     msg_flags;       /* flags on received message */
};
```

The ancillary data is an array of cmsghdr structures. Their cmsg_level is set to SOL_SHIM; the cmsg_data field contains a single locator in a sockaddr_in or sock-addr_in6 structure. The cmsg_type is one of the following. For recvmsg(), it is SHIM_

LOC_LOCAL_RECV or SHIM_LOC_PEER_RECV. For sendmsg(), it is SHIM_
LOC_LOCAL_SEND, SHIM_LOC_PEER_SEND, SHIM_FEEDBACK_POSITIVE, or
SHIM_FEEDBACK_NEGATIVE. The types described are only applicable to UDP or
raw sockets, as there is no direct mapping between send or receive calls and actual
TCP segments.

- The SHIM_LOC_PEER_RECV attribute determines if the SHIM layer stores the source
  locator of arriving packets as ancillary data that can be obtained with a recvmsg() call.
  The use of this attribute is similar to SHIM_LOC_LOCAL_RECV.

- The SHIM_LOCLIST_LOCAL attribute controls the list of local locators associated
  with Endpoint ID of the socket. The attribute can be queried and set by getsockopt()
  and setsockopt() calls with a buffer for the locator list as an argument. The buffer size
  should be sufficient to store all locators.

- The SHIM_LOCLIST_PEER attribute is similar to the SHIM_LOCLIST_LOCAL
  attribute, except it controls the list of remote locators associated with a socket.

- The SHIM_APP_TIMEOUT attribute determines the application timeout to detect a
  path failure. The timeout can be queried and set with getsockopt() and setsockopt()
  calls with an integer argument containing the timeout value in seconds. The Reachability Protocol sends the keepalive messages according to the configured timeout.

- The SHIM_PATHEXPLORE attribute defines the interval and the maximum number
  of path exploration probes when a primary path fails. By default, the interval is 0.5
  seconds and the maximum retry number is 4. The probe message is sent per locator
  pair. The attribute can be queried and set by the application.

  The shim_pathexplore structure stores the parameters for path exploration.

```
struct shim_pathexplore {
  uint8_t    pe_probenum;      /* # of initial probe */
  uint8_t    pe_keepaliveto;   /* Keepalive Timeout */
  uint16_t   pe_initprobeto;   /* Initial Probe Timeout */
  uint32_t   pe_reserved;      /* reserved */
};
```

  Below is an example of setting the SHIM_PATHEXPLORE attribute.

```
struct shim6_pathexplore pe;
pe.pe_probenum = 4;        /* times */
pe.pe_keepaliveto = 10;    /* seconds */
pe.pe_initprobeto = 500;   /* milliseconds */
pe.pe_reserved = 0;
setsockopt(fd, SOL_SHIM, SHIM_PATHEXPLORE, &pe,
           sizeof(pe));
```

- The SHIM_CONTEXT_DEFERRED_SETUP attribute is an integer that can be queried
  and set by getsockopt() and setsockopt() calls. The attribute determines if the SHIM
  context is setup before or after initial packet exchange between the peer hosts.

Out of the listed attributes, SHIM_LOC_LOCAL_PREF and SHIM_LOC_ PEER_PREF are similar to the IPV6_PKTINFO attribute in the standard socket API. Furthermore, SHIM_LOC_LOCAL_RECV and SHIM_LOC_PEER_RECV are similar to IP_RECVDSTADDR and IPV6_PKTINFO. For legacy applications using the standard API, the SHIM layer returns Endpoint IDs rather than locators in the socket calls.

In the standard socket API, getsockname() and getpeername() calls return the local and remote IP address and port data. In the SHIM API, these calls return Endpoint ID instead of the address and port.

15

Integrating HIP with other protocols

In this chapter, we describe possible synergies that can be created by combining HIP with other protocols, such as SIP, SRTP, and Mobile IP. The chapter starts with Section 15.1, which considers how the HIP architecture could be made more general so that it is suitable for a wider range of applications. In Section 15.2, running SIP over HIP is discussed in detail. Section 15.3 considers the use of SRTP as an alternative encapsulation format instead of ESP. Section 15.4 discusses the experimental proposal of using IKEv2 as HIP base exchange instead of I1-R2 packets. Section 15.5 compares properties of Mobile IP versus HIP. The chapter concludes with Section 15.6, which presents the design of a proxy for interfacing HIP-aware and legacy hosts.

15.1   Generalized HIP

The HIP architecture implements the identifier–locator split in the Internet by introducing public keys as host identifiers. Not all network engineers agree that a new name space is needed in the Internet. However, the community appears to have a consensus that some level of indirection is needed in the Internet.

Several other proposals implement indirection in the Internet at a more general level. Mobile IP, multi6, i3, FARA, TRIAD, IPNL, and MOBIKE are examples of such proposals. Mobile IP is the only standard-track solution at the time of writing. Mobile IP uses IP addresses as host identifiers instead of a new name space.

If the HIP architecture were made more general, it could be used by other indirection proposals to facilitate experimentation (Henderson 2005). Although these HIP revisions have not yet been implemented, it is interesting to see what would be needed for other protocols to benefit from HIP. Three main changes to HIP were proposed. First, a requirement that specific HIP messages must carry certain parameters is relaxed. Second, a definition of profiles for the HIP base exchange is proposed. Third, type-length vector encoding is made extensible.

Host Identity Protocol (HIP): Towards the Secure Mobile Internet   Andrei Gurtov
© 2008 John Wiley & Sons, Ltd

This section describes and classifies network indirection proposals providing some degree of location independence at the networking (IP) layer, or between the networking and transport layers. It assumes that some identifier is provided on the networking layer, rather than in other layers above the protocol stack. Implementing identifiers at the network layer has advantages of security, generic support for upper protocols, traversal of middleboxes, multihoming, and mobility.

15.1.1 Classification of proposals

- *Upper-layer identifier* is an identifier used at the application, session, or transport layer. For compatibility with existing socket API and legacy applications, upper-layer identifiers should have an IPv4 or IPv6 length of 32 or 128 bits, respectively. Upper-layer identifiers can be routable at the existing IP layer (e.g., specially formed IP addresses having a topologically correct prefix) or not-routable (e.g., HIT as a hash of a public key) at the IP layer but routable at the overlay layer on top of IP. For applications supporting enhanced socket API, identifiers of a format incompatible with IP addresses, such as public keys, are possible.

- *Resolving the identifier.* If the upper-layer identifier is not the same as the destination address of a host, the identifier requires resolution or binding to the proper address. The binding can occur "early", i.e. in the sending host, or "late" in an overlay node in the network.

- *Establishing context.* The context establishment protocol sets up the mapping between upper-layer identifiers and locators in communicating hosts. The context can define common cryptographic keys and demultiplexing rules for data traffic.

- *Managing locators.* While the identity of a host typically remains stable, its active set of locators can change dynamically. Some proposals support only the change of one active locator (mobility), others require pre-defined set of locators (multihoming), and some solutions can add and remove locators combining mobility and multihoming. Some mechanisms choose the locator of the destination host (preferred address) and the sending host (default source address). To prevent locator hijacking a secure binding mechanism between identifier and locator is required.

- *Tagging packets.* When a packet arrives to the receiver, some tag is needed to locate a correct context for processing the packet. In HIP with ESP encapsulation, the SPI serves as the tag, while other proposals may use IP extension headers.

Table 15.1 compares HIP, Mobile IP with Routing Optimization, i3, and shim6 proposals against described characteristics.

Several proposals have been presented to combine the individual proposals for greater benefits. A combination of HIP and Internet Indirection Infrastructure (Hi3) achieves efficient end-to-end transport of HIP with overlay routing capabilities of i3. The HIP rendezvous server can be seen as a combination of STUN server and Mobile IP home agent. There has been a proposal to use HIP to secure route optimization of Mobile IPv6. The use of IKEv2 as a control plane for HIP has been considered. In the multihoming for IPv6 (multi6) IETF working group, a proposal has been made to use a lighter version of HIP to avoid heavy public-key cryptography.

Table 15.1 Comparison of proposals for decoupling identity and location.

Characteristic	HIP	MIP with RO	i3	shim6
Upper-layer identifiers	128-bit hash of public key used instead of IP addresses by transport protocols.	IP address of the home agent.	256-bit trigger ID.	A specially-formed IP address (e.g., a local network prefix and cryptographically generated host suffix).
Resolving the identifier	Can be performed using DNS, Distributed Hash Tables, or opportunistically by sending a packet to a known locator (early binding).	Done by the Care-of Address at endpoints (early binding).	Done by the i3 servers and is decoupled from the act of sending a packet (late binding).	At the end host (early binding).
Establishing context	Base exchange of four packets performing a Diffie–Hellman key exchange. With ESP encapsulation, also sets up Security Associations.	Binding update sets up the tunneling context.	Context for packet forwarding is created by receivers by inserting a trigger in one of the i3 servers. No end-to-end packet context exists, except when shortcuts are deployed.	
Managing locators	Mobility and multihoming HIP extensions can add and remove locators to active Security Associations with authenticating updates.	The return routability test generates a key to authenticate Binding Updates that change the locator binding at corresponding node. Multihoming is not supported.	Receivers can update their locator set by changing the trigger in the i3 infrastructure. Mobility and multihoming, as well as anycast and multicasts are naturally supported.	May require the set of locators to be known before context establishment. Alternatively, some end-to-end update protocol can be used.
Tagging packets	With ESP encapsulation, 32-bit SPI in each packet represents a compressed version of 128-bit HIT.	Each packet contains the home address in a Destination or Routing Header.	Each packet contains a destination trigger ID in UDP or TCP payload.	No explicit tagging may be needed when locators are cryptographically bound to identifiers. Alternatively, a new extension header or a flow label can be used.

15.1.2 HIP implications

It has been suggested that HIP can be used as a general experimentation framework for multiple proposals towards separation of identifiers and locators (Henderson 2005). Practical deployment of such proposals in the Internet is still an open issue and it may be beneficial to combine the efforts on initial experiments. To support the approach of generalizing HIP, the IETF working group has removed the ESP encapsulation from the base specifications, and is publishing it as one possible HIP encapsulation. Other possible data packet encapsulation methods may include SRTP (Tschofenig *et al.* 2006).

The HIP header can be made more generic at the expense of additional overhead in new fields. Figure 15.1 shows a more general HIP header using Type-Length encoding for representing host identifiers. It allows HIP to use HITs of a different size than 128 bits, which may become necessary, e.g., if new exploits against SHA hash function are found in future. Additionally, the use of other identifiers instead of HITs becomes possible.

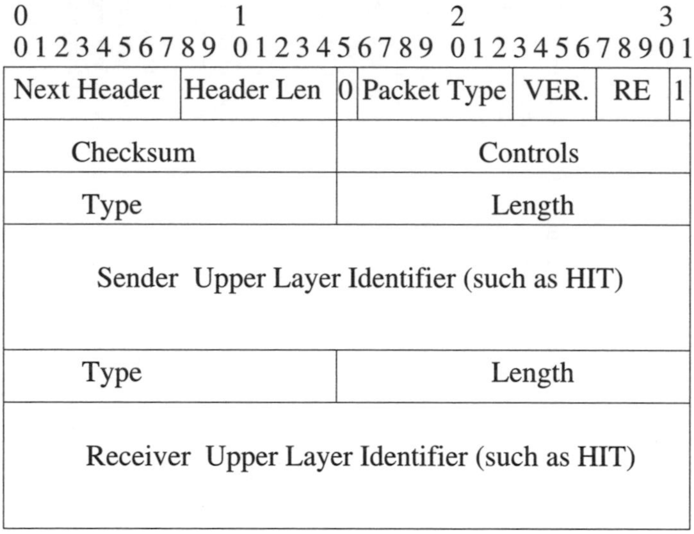

```
0                   1                   2                   3
0 1 2 3 4 5 6 7 8 9 0 1 2 3 4 5 6 7 8 9 0 1 2 3 4 5 6 7 8 9 0 1
```

Next Header	Header Len	0	Packet Type	VER.	RE	1
Checksum			Controls			
Type			Length			
Sender Upper Layer Identifier (such as HIT)						
Type			Length			
Receiver Upper Layer Identifier (such as HIT)						

Figure 15.1 Generalized HIP packet header.

Existing HIP specifications could be mapped to the generalized version using an implementation profile. The HIP SIGMA-compliant Diffie–Hellman key exchange would remain the core of the HIP profile. However, other options such as delaying the context establishment handshake until later during a connection (shim6) are useful. During the handshake, the R2 packet could return the list of supported options using a TLV encoding or by setting control bits in the HIP header. One possible option is Lightweight HIP where public-key operations are replaced with hash chains.

The base header of generalized HIP should have few compulsory fields. It appears that source and destination identities encoded in TLV format are the only compulsory fields. Each separate profile, such as the base HIP, can include additional compulsory parameters, such as a puzzle in R2. However, the puzzle may be optional or unsupported in other profiles.

Practical use cases for extended HIP profiles include NAT traversal and micromobility. The current NAT traversal extensions for HIP use UDP encapsulation to pass NATs. Instead, a common NAT-friendly profile could be defined for generalized HIP that can be used by other solutions in addition to base HIP. Furthermore, the base HIP mobility model only supports end-to-end mobility updates or a single rendezvous server. Efficient micromobility can require a hierarchical set of mobility anchor points. Generalized HIP can include a micromobility profile common for Mobile IP and base HIP.

Another possible application of generalized HIP could be securing the Mobile IPv6 binding updates. The present MIP specifications suggest creating a binding management key through a return routability test from the home network to the mobile node. The key is necessary to create security context with arbitrary correspondent nodes. The key becomes invalid if the mobile node change its IP address. With generalized HIP, a mobile node could initiate base exchange to the correspondent node using the home address as the source identity and correspondent node address as the destination identity in I1 message. Such a "Mobile IP" profile would include Diffie–Hellman parameters as mandatory components, while the puzzle, nonce, and signatures would be optional.

The Site multiHoming by IPv6 interMediation (shim6) IETF working group produces a multihoming solution using routable specially-formed IP addresses as upper-layer identifiers. Shim6 defines policies for selecting source and destination addresses that could be used by HIP. One requirement that shim6 plans to meet is the context establishment protocol that could be executed in parallel with data packets even in a late phase of a data transfer. The generalized HIP profile permitting using routable addresses as upper-layer identifiers in combination with hash-chain security context (Lightweight HIP) could become a context-establishment protocol for shim6.

15.2 The use of Session Initiation Protocol

This section briefly describes the Session Initiation Protocol (SIP), the use of SIP infrastructure proxies as a rendezvous mechanism for HIP hosts, the use of HIP to secure SIP traffic, and the extensions to Session Description Protocol to exchange HITs between HIP hosts (Henderson 2004).

At the time of writing, an active area of development in IETF is the use of HIP for the design of a peer-to-peer version of SIP (P2PSIP). Various proposals suggest either to use HIP directly (Hautakorpi et al. 2007) or just borrow ideas from HIP (Cooper et al. 2007). Since this area is still under active development, this section focuses only on the traditional SIP with proxies.

15.2.1 SIP as a rendezvous service

The Session Initiation Protocol is an application-layer protocol for setting up sessions between users identified by Uniform Resource Locators (URI) similar to email addresses (user@example.com). SIP is often used for voice and video telephony, chat messaging, and other interactive applications. The SIP protocol enables User Agents (UA) to locate the party and negotiate the session parameters. The actual data is delivered using a different protocol, such as RTP.

While HIP is primarily used for user-to-machine communication (e.g., opening a telnet connection to a server), SIP is mostly used for user-to-user communication (Camarillo 2004). SIP provides a rendezvous service, when a given user with a permanent identifier can be located at different places in the Internet. The UA registers the user identifier (e.g., user@example.com) at the current location (e.g., user@voipphone.example.com) to the SIP registrar server.

The SIP proxy server forwards the session initiation request to the UA at the registered location (Address of Record, AoR). The SIP redirect server, on the other hand, returns the current user location to the Initiator, which can directly contact the UA. HIP hosts can use the SIP rendezvous mechanism to locate each other. To accomplish this, a HIP host generates a SIP URI and publishes it in a directory or sends it beforehand to the peer hosts. Before establishing a HIP association, two HIP hosts can set up a SIP session exchanging their HITs and current IP addresses.

The SIP session setup consists of request-replay messages negotiating the session parameters. The initiating UA sends an INVITE message containing the session parameters such as the IP address, UDP ports, and HITs. The parameters are presented with the Session Description Protocol (SDP) extensions, which are explained in detail in Section 15.2.4. The peer replies with a "200 OK" message to confirm the INVITE and send its own HIT.

The use of HIP can be negotiated during the SIP session establishment. When HIP hosts are using SIP URI to locate peers with SIP servers, some contacted hosts may not support HIP. Furthermore, even if both hosts support HIP, they may prefer not to use HIP for certain types of session, such as RTP voice traffic. When RTP is used over slow data links, the header compression is necessary to reduce the header overhead. A typical RTP packet can carry only 30 bytes of useful payload data together with over 50 bytes of IP, UDP, and RTP headers.

If HIP is enabled for RTP sessions, whole data packets, including most of the headers, are encrypted by ESP IPsec. The compressor and de-compressor located at the ends of a slow data link would not be able to process the headers. Therefore, the use of SRTP protocol instead of ESP IPsec for encapsulating data packets is preferable in this scenario. SRTP only encrypts the data payload and not the headers, which enables the compression. An experimental use of SRTP as the HIP data encapsulation protocol is described in Section 15.3.

In SIP, the end-to-end security is provided with S/MIME similar to email. The user messages and some header fields are encrypted and integrity-protected and can be read only by the receiver and not the SIP proxies on the path. SIP uses self-signed certificates to implement opportunistic session security. Both SIP and HIP have the practice of establishing a first association using the leap-of-faith when the destination identity is assumed to be correct. After association establishment, the identity is stored locally and can be verified on subsequent connections.

When SIP and HIP are used together, the leap-of-faith needs to be executed only at the SIP layer. The SIP layer is preferable for opportunistic security over HIP because once the user identity is verified it can be applied to several device identities that the user has. After the SIP UAs have exchanged their self-signed session descriptions, each host has another HIT and can exchange host certificates using the HIP base exchange. The host certificates are signed by the user's private key instead of using self-signed certificates.

Naturally, if a HIP or SIP identity can be reliably confirmed using a trusted source such as DNSSEC, Public Key Infrastructure or a web of trust it is preferable to use that identity as a trust anchor instead of executing the opportunistic exchange of identities.

15.2.2   Complementary mobility

SIP and HIP provide mobility, but on different layers. Thus, the joint use of protocols is not conflicting but is complementary. The SIP provides mobility to users and sessions, while HIP provides host mobility. When a user moves from one device to another, the new host will have a different host identifier and the HIP association cannot be moved from the old host. Although there are methods to overcome these limitations (e.g., placing the host public/private keys on a portable storage device such as a USB flash memory stick or using delegation certificates) user mobility is better handled by SIP.

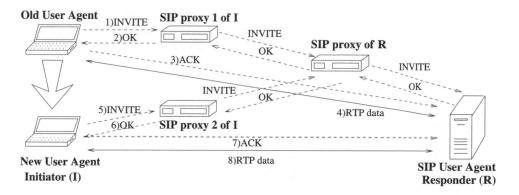

Figure 15.2 Session mobility with the SIP Replaces mechanism.

SIP also supports session mobility when a user can change UA without loosing their sessions. The change is implemented with the basic SIP mechanisms (if the proxy is not changing) or with a special mechanism called Replaces. After moving to a new device (and UA), the user can trigger Replaces to restore its existing sessions from the old device. For this, the new UA sends an INVITE message with a Replaces header that identifies sessions established with the old UA. Figure 15.2 illustrates how a user can switch devices with the Replaces mechanism. In this scenario, the new UA needs to re-establish a HIP association with the Responder UA and SIP proxies. Naturally, it would be a new association using a different Initiator host identity than from the old UA.

Many of the SIP services use UDP where a datagram can be sent from any IP address. However, the use of SIP mobility mechanisms does not enable maintaining TCP sessions after the change of the IP address in a UA. The use of HIP would enable the SIP UA to keep all data connections during network mobility. Furthermore, it would enable the use of multihomed UAs for reliability and performance, something that the basic SIP does not provide.

In Figure 15.2, the old UA informs the peer UA that it is now reachable at the new UA using a referral. A referral represents an identifier that has meaning outside of local content and can be routed to the correct destination when given to a third party. SIP implements a REFER message that contains a Globally Routable UA URI that can be always routed to a correct UA. After receiving a REFER message, the new UA sends INVITE and ACK messages to the peer UA and confirms the referral with a 202 Accepted message.

Recall that HIP has a problem with referrals when a legacy application passes a HIT that it believes to be an IPv6 address to another application (see Section 14.1). As SIP already has a mechanism to handle the referrals, it can be used by HIP hosts instead of directly passing HITs to each other. After a SIP referral, the new UA can obtain peer HIT using the DNS resolution of the URI and establish a HIP association with the peer UA.

HIP can be useful to prevent exploiting the SIP protocol to generate Denial-of-Service attacks as follows. The request-reply session establishment mechanism by SIP can be misused to send unsolicited traffic to a third party. The INVITE message includes an IP address that the Responder sends a reply to. A malicious host could put the address of a victim instead of its own in the INVITE message and cause the Responder to send, e.g., RTP over UDP packets to the victim. The ICMP "Non Reachable" messages may not stop the attack if dropped by firewalls or not understood by the Responder implementation. The use of HIP would prevent this attack as the Responder can verify the IP address of the Initiator.

15.2.3   Securing SIP control traffic

In this section, we discuss the use of HIP with SIP control messages between UAs and proxies. The use of HIP for the data traffic after establishment of a SIP session is orthogonal to the use of HIP for control traffic. End-to-end data security can be always provided by HIP as described in the previous section, subject to performance issues such as the header compression. This section focuses on hop-by-hop security in the SIP infrastructure and support of host mobility.

Figure 15.3 shows separate data and control planes of a SIP session. While the control signaling is performed through a sequence of SIP proxies, the data flows directly between communicating hosts. Hop-by-hop communication with SIP proxies is secured by Transport Layer Security (TLS). Authentication between proxies is typically performed using certificates during a TLS exchange. However, as clients typically do not have the certificates, they use a different authentication method known as Digest.

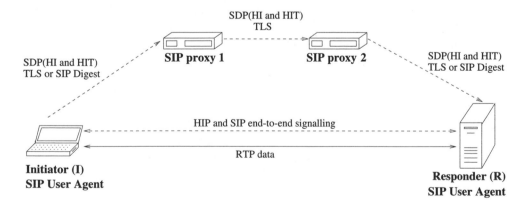

Figure 15.3 Combining HIP and SIP protocols.

The use of IPsec for security between proxies was a possible design alternative to TLS. However, with IPsec the application cannot be certain that a given message was indeed sent over a protected connection. With TLS, a UA sending a message to a SIP URI is sure that the message is delivered encrypted and integrity protected to another proxy.

Using HIP makes the IPsec option more attractive for SIP. HIP allows the applications to be certain that a message is delivered protected if sent to a HIT. The use of a HIT as a message destination instead of an IP address implies that the message is secured by HIP. SIP proxies would locate the HIT and IP address of a next hop proxy using the DNS. Since HIP also supports certificate authentication, it can be used for agent–proxy and proxy–proxy hops. A HIP host can send a CER packet containing the certificate after the R1 packet.

Another aspect where HIP can be useful for the SIP control plane is UA mobility. SIP provides UA mobility by informing all associated UAs about the new location. This is done with a hop-by-hop approach where an UPDATE message from the moving UA is relayed by several proxies until it reaches the destination UA. When using HIP, a mobile UA can just inform its own proxy about its new location using a HIP UPDATE message. This way, only the last proxy that really needs to know the location of the UA is informed of its current address. Such an approach can reduce the load on other SIP proxies and the network, which would not forward unnecessary updates.

A modeling study of a hybrid mobility solution using HIP and SIP was made (So *et al.* 2005). The handover delay as well as the signaling overhead of plain SIP, HIP/SIP, plain Mobile IP, and Mobile IP/SIP were compared. The handover duration is dominated by a large DHCP address assignment delay. Overall, the plain SIP delay was the highest and solutions based on HIP or Mobile IP produced less delay. The signaling overhead of the HIP/SIP mobility solution was found to be half that of plain SIP and a third that of Mobile IP/SIP.

HIP can help SIP UA to resolve the problem of simultaneous mobility. When both UAs move at the same time, their SIP UPDATE messages would not be useful because they are sent to invalid destinations. If HIP were used in this scenario, each UA would update its own proxy with the current location information and the problem would not arise. Although using the SIP Replaces mechanism with Globally Routable UA URI (GRUU) can also resolve the problem of simultaneous mobility, such a solution is more complex and less efficient than the use of HIP.

The other area where HIP could be useful for SIP is the instance identifier for UAs. While Address of Record identifies a particular user, the instance identifier gives the location of a particular UA. A forking proxy uses instant identifiers to probe several UAs sequentially or in parallel. A UA registers to the Registrar by providing a unique instance identifier, the Registrar returns the GRUU that routes to that UA. If the same UA re-registers from a different location with the same instance identifier, the Registrar will supply the same GRUU.

The use of HITs as instance identifiers is suitable for certain UAs. It prevents registration hijacking because the Registrar can verify that the registration comes from the UA that owns the instance identifier, i.e., HIT. Additionally, the last-hop proxy can authenticate the UA just before relaying a session initiation message to it. Currently, authentication is implemented by maintaining a TLS connection between UA and the proxy for the whole registration period. Maintaining the idle TLS connection causes overhead to the UA and the proxy.

Unfortunately, HITs cannot be used by software UAs that change their location device or run simultaneously on the same device. Indeed, a HIT identifies a host and if the UA moves to another host its instance identifier would change. Furthermore, several UAs running on the same device would not get unique instance identifiers from the same HIT. Fortunately, several types of instance identifier can be used by SIP. Hardware UAs bound to the same host can use HITs as instance identifiers, while mobile software UAs can use the standard SIP identifiers.

There are possible synergies between HIP and SIP protocols for middlebox traversal. Both protocols adopt Interactive Connectivity Establishment (ICE) for traversing legacy NATs. SIP traffic would benefit from using HIP to traverse HIP-aware NATs that provide additional security by verifying the Host Identity of communicating hosts.

15.2.4 Session Description Protocol extensions

Apart from the HIP opportunistic mode, establishment of a HIP association requires prior secure exchange of Host Identities. This section describes the extensions to the Session Description Protocol (SDP) to be used during the SIP initial exchange to carry Host Identities as shown in Figure 15.3 (Tschofenig *et al.* 2007a).

During the initial SIP exchange, TLS security is used to protect the exchange of Host Identities without the need for a global Public Key Infrastructure (PKI). Furthermore, additional security benefits come from the fact that control and data packets follow different network paths. Therefore, an attacker must be able to intercept both the signaling and data traffic, which is more difficult than intercepting one path. Although, the successful exchange of Host Identities requires only the integrity and not the confidentiality of SDP messages, TLS encryption provides both mechanisms.

The Host Identities can be carried in SDP parameters "k" or "a". The "k" parameter has the structure:

```
k=<method>:<encryption key>.
```

Two new fields are defined for the "k" parameter to carry the Host Identity or a HIT:

```
k=host-identity:<Host Identity>
```

```
k=host-identity-tag:<Host Identity Tag>
```

The Host Identity includes a public key and cryptographic parameters as defined in the HIP base specification. It is encoded with the base64 algorithm.

The "a" parameter can be used as an alternative to "k" and has the following structure:

```
a=<attribute>:<value>.
```

Two new fields are defined for the "a" parameter to carry the Host Identity or a HIT:

```
a=key-mgmt:host-identity <Host Identity>
```

```
a=key-mgmt:host-identity-tag <Host Identity Tag>
```

The SDP only deals with media streams and not with users or UAs. Several media streams can be involved with a single SIP session when, e.g., one device is used for receiving video and another for receiving an audio stream. Then, several Host Identities, one per device, need to be exchanged. As an alternative to using SDP for exchanging host identities, the SIP signaling header could carry HI or HIT in the "Contact:" field protected by S/MIME encryption.

The SIP integrity mechanism protects the delivery of Host Identity from the Initiator UA to the Responder UA. However, the Response Identity Mechanism is still under development by the SIP community in the IETF. When finalized, this mechanism should be used to secure delivery of Host Identity from the Responder to the Initiator. Currently, several existing mechanisms can be used to assert the identity of the Responder. After the Responder replies with the SDP message containing its Host Identity, it can immediately send an UPDATE message using the SIP identity mechanism to confirm the information sent in SDP. This approach has the drawback of requiring one additional round-trip time but works even if SIP proxies are not trusted.

An alternative approach is the use of the SIP mechanism based on hop-by-hop TLS security. If all SIP proxies on the signaling path are trusted, it provides the authenticity of the Responder Host Identity. However, a malicious proxy could mount a Man-In-The-Middle attack by replacing the Responder identity when relaying an SDP message. The use of S/MIME could provide end-to-end integrity of a message from the Responder to the Initiator, but is not yet deployed for SIP.

15.3   Encapsulating HIP data using SRTP

The HIP base specification (Moskowitz *et al.* 2008) describes the HIP base exchange that creates a HIP association between two hosts. The default mechanism for protecting HIP data packets is the use of IPsec ESP (Jokela *et al.* 2008). This section describes an alternative experimental encapsulation for HIP data packets based on Secure Real-Time Protocol (SRTP) (Tschofenig *et al.* 2006). The security properties of SRTP encapsulation are similar to IPsec ESP, but enable header compression of protocol headers. Header compression is important for real-time applications (such as VoIP) transmitting small packets over a slow (e.g., wireless) link.

SRTP is an application protocol operating over UDP. SRTP is often used together with SIP as a signaling mechanism. SRTP is an extension profile for Real-Time Protocol (RTP). SRTP provides packet integrity, encryption, and protection against replay attacks specifically for real-time data. However, SRTP does not include a mechanism to exchange keys and protocol parameters necessary to establish an SRTP security context. This section discusses how the HIP base exchange can be extended to carry SRTP-specific parameters. In particular, the R1 packet needs to contain a list of available transforms at the Responder. The Initiator selects one of the transforms and informs the Responder using the I2 packet.

Figure 15.4 illustrates the setup of an SRTP association between two hosts. Compared with the standard base exchange, R1 and I2 packets are extended with several SRTP-specific parameters. The SRTP_T parameter contains a timestamp to prevent replay attacks. SRTP_PARAM defines the mapping between Cryptographic Sessions and Synchronization sources (SSRC). KEY_PARAM contains SRTP_RAND (a random string for key generation), SRTP_SP (list of security algorithms and transforms), and SRTP_MKI (master key). In R1, SRTP_KEY includes the list of all transforms supported by the Responder. In I2, SRTP_KEY includes a single transform selected by the Initiator. The Responder generates the session keys from the master key and salt, and the selected security policy for each SRTP stream.

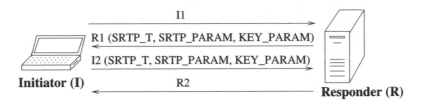

Figure 15.4 HIP base exchange with SRTP.

The rekeying process using the UPDATE messages is shown in Figure 15.5. Rekeying can be started either by the Initiator or Responder by sending new SRTP parameters and a DIFFIE_HELLMAN parameter for generating a new key. The host starting the rekeying can also continue using the current master key and only generate a new session key. In that case, the DIFFIE_HELLMAN parameter is not sent in the UPDATE packet.

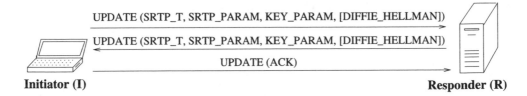

Figure 15.5 HIP mobility update with SRTP.

The whole rekeying process takes three UPDATE messages protected by the retransmission mechanism. The second UPDATE performs a return routability check to the source of the first UPDATE packet. The third UPDATE packet is an acknowledgment of successful rekeying that triggers the switch from the old to new key material.

Next we describe the format of SRTP-specific parameters, SRTP_PARAM, SRTP_T, SRTP_RAND, SRTP_SP, and SRTP_MKI. Out of five parameters, three are grouped together as a complex parameter KEY_PARAM. Figure 15.6 shows the format of SRTP_PARAM, which maps Crypto Sessions (CS) to SRTP associations. The proposed type for SRTP_PARAM is 40000 and the length depends on the number of Crypto Sessions included.

The Crypto Session Bundle (CSB) field contains an identifier that is randomly generated by the Responder. The identifier need not be globally unique but must be unique between the Initiator and Responder. The next field in SRTP_PARAM gives the number of Crypto Sessions combined in the Crypto Session Bundle. The maximum number of sessions is 255 and the zero value can be used in base exchange to denote no existing sessions. Security Associations are only created for Crypto Sessions selected in the CS map info field.

0	1	2	3
0 1 2 3 4 5 6 7 8 9	0 1 2 3 4 5 6 7 8 9	0 1 2 3 4 5 6 7 8 9 0 1	

Type		Length	
Crypto Session Bundle ID			
CryptoSession #	CryptoSession map info		RESERV.
Policy–1	SSRC–1		
SSRC–1 (cont)	ROC–1		
ROC–1 (cont)	Policy–2	SSRC–2	
SSRC–2 (cont)	ROC–2		
ROC–2 (cont)	Padding		

Figure 15.6 The HIP mapping parameter for SRTP.

For each Crypto Session in a Bundle, three fields are present in SRTP_PARAM: a policy number, Synchronization Source (SSRC), and the Rollover Counter (ROC). The policy number specifies a security policy in use for the following SSRC. The rollover counter is used by SRTP cyphers for protection against replay attacks.

The SRTP_T parameter contains a timestamp, as shown in Figure 15.7. It has a fixed length of 4 bytes and has the type of 40001. The timestamp is needed to prevent replay attacks.

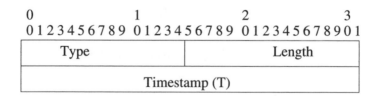

Figure 15.7 The HIP timestamp parameter for SRTP.

Figure 15.8 shows the format of SRTP_RAND parameter containing a master salt for generating a key. The length of the pseudo-random bit string is 112 bits with the remaining part to a 32-bit boundary being in reserve. The type of this parameter is 40002 and the length is 14 bytes.

The SRTP_SP parameter is composed of several security policies, as in Figure 15.9. Each policy defines cryptographic transforms and cypher algorithms supported by a HIP host. A particular policy is selected using the mapping parameter (SRTP_PARAM). The type of SRTP_SP is 40003 and its length depends on the number of included security policies. The policies are concatenated using the Type-Length-Value (TLV) encoding. The policy

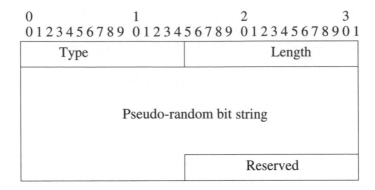

Figure 15.8 The HIP pseudo-random bit string parameter for SRTP.

parameters are preceded by a header containing the type and length of a parameter as well as its policy number. The maximum number of policies is 255 although the actual number of policies in use is limited by the R1/I2 packet size. The existing security policy parameters are summarized in Table 15.2.

0	1	2	3
0 1 2 3 4 5 6 7 8 9	0 1 2 3 4 5 6 7 8 9	0 1 2 3 4 5 6 7 8 9 0 1	

Type		Length	
Type	Length	Policy #	Value
Value			
Type	Length	Policy #	Value

Figure 15.9 The HIP policy parameter for SRTP.

The Master Key Identifier (SRTP_MKI) is an optional parameter that determines the master key and its salt of the session key that encrypts and authenticates the given packet. The SRTP_MKI format is shown in Figure 15.10, with the type value of 40004 and variable length.

The HIP NOTIFY packet is extended to include two error types specific to SRTP. The error NO_SRTP_PROPOSAL_CHOSEN is used when none of the suggested SRTP transforms are suitable for a HIP peer. The error INVALID_SRTP_TRANSFORM_CHOSEN is set when the chosen transform is not one offered.

The processing of incoming and outgoing SRTP-encapsulated packets is described in RFC 3711 (Baugher *et al*. 2004). The packets are encrypted on transmission and decrypted by the receiver. The new HIP parameters SRTP_PARAM, SRTP_T, SRTP_RAND, SRTP_SP, and SRTP_MKI are protected with HMAC and a signature as described in the HIP base specifications (Moskowitz *et al*. 2008).

Table 15.2   Parameters for SRTP security policies.

Type	Length	Name	Value
1	1	Encryption transforms	NULL, AES-CM, AES-F8
2	2	The length of encryption session key	128
3	1	Authentication transforms	NULL, HMAC-SHA-1
4	2	The length of authentication session key	160
5	2	The length of a tag	80
6	4	The SRTP prefix length	Variable (default 0)
7	1	Pseudo Random Function for key generation	NULL, AES-CM
8	8	The rate of key generation	Variable (default 0)
9	8	Maximum packet lifetime for SRTP	Variable
10	8	Maximum packet lifetime for SRTCP	Variable
11	1	Forward Error Correction	2 bits

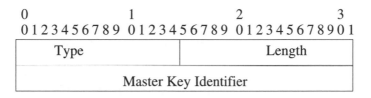

Figure 15.10 The HIP master key identifier parameter for SRTP.

The HIP base exchange establishes the SRTP association between two hosts. However, for processing a data stream, the interface between the HIP daemon and implementation is necessary. The SRTP processing code for data packets is typically integrated to a user application and is not a part of the operating system kernel as IPsec. Therefore, the application needs to receive the session key from the HIP daemon to process data packets. An API is defined for the HIP daemon that enables the application to query the key when receiving a data packet. The details of the API are specific to each particular implementation.

15.4   Replacing HIP base exchange with IKEv2

This section describes an experimental proposal to use an extended Internet Key Exchange protocol version 2 (IKEv2) (Kaufman 2005) to set up ESP Security Associations for HIP (Jian *et al.* 2005). This proposal, named IKE-H, replaces the HIP base exchange consisting of I1, R1, I2, and R2 packets. The use of IKEv2 may provide better security properties than the HIP base exchange and may benefit from a larger implementation and deployment of IKEv2.

The IKE-H extension performs the exchange of keying material and mutual authentication between two HIP hosts. ESP Security Associations between Initiator and Responder HITs are established using the keys from the IKE-H exchange. When a HIP host changes its IP address, a readdressing exchange is performed to inform the peers of the new address.

Figure 15.11 shows the IKE-H exchange consisting of four messages. The first two messages form the IKE_SA_INIT exchange and the last two messages the IKE_AUTH exchange. All messages include a standard IKEv2 header with an H-flag set up to indicate the HIP extension. During the IKE_SA_INIT phase, the hosts exchange their HITs, nonces, negotiate cryptographic transforms, and perform a Diffie–Hellman key exchange.

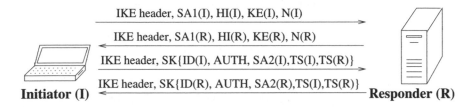

Figure 15.11 The HIP base exchange with IKEv2.

When the IKE_SA_INIT phase completes, the hosts have established a secure tunnel. During the IKE_AUTH phase, HIP hosts identify each other and establish a child SA bound to HITs (not to IP addresses). The payload of IKE_AUTH packets is encrypted using the session key generated during the IKE_SA_INIT phase as shown in curly brackets in Figure 15.11. The ID(I) and ID(R) fields are identifiers of the Initiator and Responder. The AUTH parameter contains data for host authentication and the SA parameter data for establishment of a child SA. TS(I) and TS(R) are the traffic selectors.

The rekeying process for IKE-H is illustrated in Figure 15.12. There, a child SA is created by exchanging new nonces and SA parameters. A host identity can be changed during the process, as well.

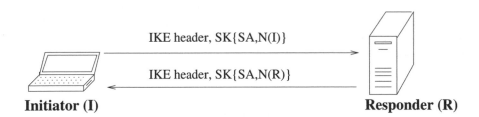

Figure 15.12 The HIP rekeying with IKEv2.

The IKEv2 header is shown in Figure 15.13. To indicate the support of IKE-H, a host must set the H flag, which is the second bit in the flags field. The rest of the header is identical to the standard IKEv2 header, and we describe the meaning of remaining fields for completeness. The Initiator SPI contains the Security Parameter Index (SPI) chosen at the Initiator. The Responder SPI value is set to zero in the first message during IKE exchange, and afterward is set to the value sent by Responder. The next payload field gives the time of payload (such as the Host Identity) that follows the basic header.

MjVer and MnVer are major and minor versions of IKE implementation. The exchange type denotes the kind of IKE exchange that the packet belongs to (e.g., IKE_SA_INIT, IKE_AUTH, CREATE_CHILD_SA). The flags field contains various bit flags to indicate protocol parameters. The length is the total length of the message in bytes including the header.

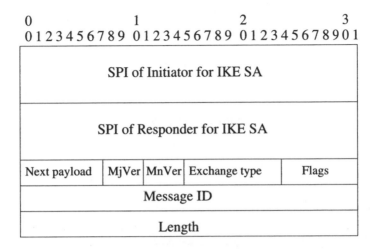

Figure 15.13 The IKEv2 header with HIP extensions.

Figure 15.14 shows the format of the Host Identity payload for IKE-H. The parameter is sent by the Initiator and the Responder in the IKE_SA_INIT phase of IKE-H exchange and during the rekeying procedure. The exchanged HITs serve as identifiers to Security Associations for IKE-H. The parameter contains both the source and destination Host Identities. When the destination identity is unknown (for example, in the first message during the exchange), it is set to zero.

```
 0                   1                   2                   3
 0 1 2 3 4 5 6 7 8 9 0 1 2 3 4 5 6 7 8 9 0 1 2 3 4 5 6 7 8 9 0 1
+---------------+--+-----------+------------------------------+
| Next payload  | C| Reserved  |        Payload length        |
+---------------+--+-----------+------------------------------+
|                                                             |
|                         Source HI                           |
|                                                             |
+-------------------------------------------------------------+
|                                                             |
|                       Destination HI                        |
|                                                             |
+-------------------------------------------------------------+
```

Figure 15.14 The IKEv2 parameter for carrying a Host Identity.

During the IKE_AUTH phase, HIP hosts exchange Identification Payloads for authentication. Figure 15.15 shows the format of HIP-specific payload (type ID_HIP). The reserved fields are set to zero on transmission and ignored on reception. The identification data contains a HIT.

```
0                   1                   2                   3
0 1 2 3 4 5 6 7 8 9 0 1 2 3 4 5 6 7 8 9 0 1 2 3 4 5 6 7 8 9 0 1
```

Next payload	C	Reserved	Payload length
ID type		Reserved	
Identification data			

Figure 15.15 The IKEv2 parameter for carrying identification data.

With IKE-H, the application always uses the IP addresses (not HITs) when sending a packet. The IPsec security policies determine which packets from an application are sent using IPsec. If an outgoing packet requires IPsec processing, IKE-H daemon looks up a HIT corresponding to the destination IP address. If not found, a new IKE-H exchange is initiated to the destination host. If a HIT is found, the packet is encrypted and integrity protected according to negotiated IPsec transforms. The peer host can deliver the incoming packet to the right application by using the SPI value to look up a proper HIT to IP address mapping. The HITs are replaced with IP addresses before passing the packet to the application.

15.5   Mobile IP and HIP

In this section, we compare the pros and cons of using HIP, i3, Mobile IPv6, and Hi3. We use the following evaluation criteria for compatibility with the previous studies (Henderson 2003).

- Mobility

    - simultaneous mobility and multihoming support,
    - fault-tolerance,
    - inverse mapping support.

- Security

    - Denial-of-Service (DoS) resistance,
    - end-to-end security and privacy,
    - accountability,
    - trust model.

- Efficiency

  - routing efficiency,
  - infrastructure cost.

The basic HIP provides efficient and secure end-to-end connectivity. If the HITs and IP addresses of end points are known, it can work without additional infrastructure, thus having no issues with infrastructure cost, accountability, trust, or fault-tolerance. Basic HIP provides limited DoS protection by enabling the Responder to make the Initiator solve a computationally substantial puzzle before creating state in the Responder. Mobility of one end point at a time is supported, but there is no way to perform the reverse mapping support[1]. HIP with a rendezvous server enables mobility of both end points, while preserving accountability and the trust model, since the rendezvous server is chosen by the Responder.

The advantages of i3 include better DoS protection, support for simultaneous mobility, and higher fault-tolerance when using a DHT with data replication. Disadvantages of i3 include reliance on an extensive infrastructure, server scalability, use of UDP, and lack of traffic encryption. Since i3 is an overlay network on top of the Chord DHT, it makes the infrastructure fairly complex. There is limited experience with widespread i3 deployment, thus it is difficult to assess how scalable the servers are. The latency of relayed control traffic will mostly be affected by forwarding and network delays. However, relaying all control and data traffic through i3 infrastructure would likely prove burdensome. By mutual agreement, the client and the server could use i3 only for initial contact and afterward exchange the data directly using shortcuts.

Although the i3 implementation could run on both UDP and TCP, currently only UDP is supported because maintaining many TCP connections between servers is challenging. As a consequence, some i3 features may be disabled in the wide area, because UDP packets often do not traverse firewalls and Network Address Translation devices[2].

The basic i3 system does not provide data encryption, although it could be implemented as an add-on feature. i3 also lacks encryption and privacy for control packets. When a public infrastructure is used, i3's extensive infrastructure requirements bring other serious security issues including the possibility of malicious or misbehaving i3 nodes that do not forward correctly and a lack of trust of arbitrary i3 servers from end points. Note that Secure-i3 introduced several constraints on the structure of triggers to prevent misuse of triggers by third parties and formation of loops in the topology. Finally, diagnosing problems in a distributed Internet system is always challenging, and the added indirection introduced by i3 further complicates the situation.

A combination approach helps to address some of the separate shortcomings of HIP and i3. The advantages of using i3 as a control plane for HIP in Hi3 include protection from DoS attacks, solving the double-jump problem, and providing an initial rendezvous service. By hiding parties' IP addresses until the HIP handshake partially authenticates them, Hi3 provides additional protection against DoS attacks. Although some DoS protection could be provided by a HIP rendezvous server, the client's IP address is revealed to a server in the first control packet. Simultaneous mobility of both hosts in i3 is supported by sending update

---

[1] A reverse lookup from an IP address to HIT (similar to reverse DNS) provides additional functionality, for example, for security purposes.

[2] To be fair, we note that not all firewalls block UDP traffic and many NAT devices will support reverse flows of UDP traffic. This problem equally affects HIP if IPsec traffic is not forwarded.

control packets via i3 when end-to-end connectivity is lost. Hi3 inherits the challenges of the extensive i3 infrastructure, including trust, accountability, and cost issues.

Comparing Mobile IP with HIP, basic Mobile IP does not provide any DoS protection mechanisms, end-to-end security, nor support for multihoming or co-existence of IPv4 and IPv6. Hi3 inherits the benefits of HIP and provides better DoS protection than HIP does. End-to-end security can be added to Mobile IP with IKE (Devaparalli 2005). In Mobile IP, the mobile node and home agent are assumed to have a business relationship, leading to a fairly clear trust and accountability model. While there are proposals for supporting mobility between IPv4 and IPv6 (Soliman and Tsitsis 2004), we are not aware of any serious work to address end-host multihoming with Mobile IP. From a fault tolerance point-of-view, the Mobile IP home agent forms a single point of failure. From an efficiency point of view, Mobile IP adds extra mobility-related headers to all packets while HIP/Hi3 does not[3]. The capacity requirements for Mobile IP and Hi3 appear to be at the same level; however, the administrative models differ considerably[4].

15.6   HIP proxy for legacy hosts

To be useful, HIP requires that both communicating hosts support the HIP protocol. However, HIP is expected to be incrementally deployed in the Internet and initially most hosts would remain HIP-unaware. A HIP proxy can help to obtain some HIP benefits when communicating with legacy hosts in the Internet. In this section, we describe the use scenarios and design challenges of a HIP proxy (Salmela 2005).

The HIP proxy should handle four possible communication scenarios:

- A legacy mobile host contacting a legacy Internet host
- A legacy mobile host contacting a HIP Internet host
- A HIP mobile host contacting a legacy Internet host
- A HIP mobile host contacting a HIP Internet host.

The first and last scenarios are simple, as the proxy does not interfere with the connection but just forwards the packets. The only challenge for the proxy is to determine whether the host being contacted is HIP-aware. Two more interesting scenarios appear when a plain IP packet is sent to an Internet host detected to be HIP-aware, or when a HIP-aware mobile host sends an I1 packet to a legacy host. Scenarios where a mobile host is contacted by a host from the Internet are not currently considered in the proxy architecture.

15.6.1   Legacy mobile hosts

In the first scenario, illustrated in Figure 15.16, a HIP proxy helps legacy hosts in a local network to gain HIP benefits. An example of such deployment can be an intranet of

---

[3]While HIP currently requires ESP encapsulation, this may change in the future.

[4]In Hi3, the infrastructure is distributed and may be controlled jointly by several organizations. Unless some structure is imposed on the trigger identifiers, the parties cannot easily control where an identifier is stored. In Mobile IP, on the other hand, the home agents are configured to the mobile hosts.

a company or 3G telecommunication networks where hosts are mobile phones or other lightweight devices. By introducing a proxy, the network owning company can instantly start benefiting from HIP without upgrading all local hosts. The transition to end-to-end HIP can happen smoothly by gradually upgrading local hosts to support HIP.

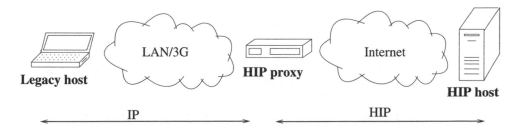

Figure 15.16 A proxy enables HIP benefits for legacy mobile hosts.

The advantage of using a HIP proxy for local hosts is traffic protection in the public Internet. While the local network or operator 3G network is relatively secure, the public Internet harbors a great number of security risks. LANs often use Ethernet switching technology that prevents users from eavesdropping on others' traffic. 3G networks have own security mechanisms that authenticate mobile hosts and encrypt their traffic within the network. Furthermore, 3G networks include own mobility management that can duplicate HIP functionality. 3G terminals can have insufficient computational resources to run HIP.

The HIP proxy generates a public/private key pair for each mobile host that attempts to communicate with a HIP-aware host in the Internet. The HIP proxy notices a DNS query when a mobile host attempts to discover the IP address of the correspondent host. When a reply comes back, the proxy can see if any HITs are present in the DNS records. The AAAA records can contain HITs that are distinguished from the regular IPv6 addresses with an ORCHID prefix. Alternatively, a HIP-specific record can be present containing a host public key and a HIT.

If no HITs are present in the DNS reply nor does the proxy have any internal HIT-IP mapping configuration for the destination host, the proxy concludes that the host is HIP-unaware. The proxy could attempt to establish an opportunistic HIP association with a correspondent host with unknown HIT by transmitting an I1 message to it with NULL destination HIT. However, this approach would increase the communication latency for all legacy correspondent hosts. Thus, the proxy just forwards the packets without further processing if it cannot find the destination HIT from DNS, DHT or local configuration.

When receiving a plain IP packet from a mobile host, the HIP proxy checks if it has a HIT available for the destination IP address of the packet. If not, the packet is forwarded to the destination as is. Otherwise, the proxy looks up existing Security Associations (SA) for the source IP address of the packet. If no SAs are found, the proxy caches the packet internally and initiates a base exchange to the correspondent host. After the SA is established, the proxy encapsulates the IP payload of the cached packet to a newly generated ESP packet and sends it to the destination host in the Internet. Likewise, when receiving an ESP packet from the Internet, the proxy determines the destination mobile host based on internal HIT-IP address mappings, removes ESP encapsulation, and forwards the packet to the mobile host.

A complicated scenario evolves if the DNS record contains an IP address of the rendezvous server together with a HIT of the correspondent host. A legacy mobile host will send packets to the IP address of the rendezvous server. If only a single connection to the RVS server exists, the proxy can successfully forward ESP packets with the HIT of the correspondent hosts. However, if several mobile hosts attempt to contact correspondent hosts using the same rendezvous server, the proxy would have an ambiguous mapping where several destination HITs correspond to a single IP address of the RVS. The proxy can either drop or send the packet to all hosts in the hope that incorrect packets would be rejected by transport or application protocols.

To resolve this conflict, the proxy can modify DNS reply messages containing the HIT and RVS IP address for Internet hosts. The proxy can act as a NAT inserting a private IPv4 address instead of a public IPv4 RVS address, or the destination HIT instead of an IPv6 address. Then, the proxy would know which correspondent host is being contacted by a mobile host by looking at the destination IP address. After locating the real IP address of the correspondent host or its RVS, the proxy would replace the destination IP address of the packet and forward it.

The 3G network includes own access control, encryption, and mobility mechanisms. The mobile hosts in the network are mostly lightweight devices such as smart phones that are running closed-source OS, such as Symbian. Deploying HIP on such hosts is not an easy task given their limited hardware capabilities, difficulty of performing a software update, and lack of HIP implementations for Symbian. On the other hand, when IP packets leave the 3G network they travel unprotected through the Internet. For cellular networks, such as 3G/UMTS, a natural place to implement the HIP proxy is the Gateway GPRS Support Node (GGSN). GGSN acts as a gateway between the 3G network and the Internet.

The proxy prototype for 3G network was implemented in the FreeBSD operating system by Ericsson. The details of the implementation and performance evaluation are described in a Master's thesis (Salmela 2005).

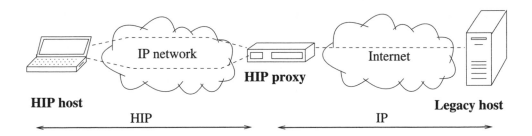

Figure 15.17 A proxy enables HIP benefits for legacy remote hosts.

15.6.2 Legacy correspondent hosts

Figure 15.17 shows the scenario where a mobile HIP host uses a proxy to communicate with legacy servers in the Internet. The advantage of using a proxy in this case is in providing mobility and multihoming for the mobile host and traffic protection on the path between the mobile host and the proxy. Such deployment is particularly useful when the access network does not have built-in mobility management and security mechanisms. As an example, a mobile host using WLAN for Internet access can connect to own HIP proxy in its home network to encrypt all air traffic and be able to move between different WLAN networks. The scenario matches well with the fact that popular servers in the Internet are HIP-unaware. Therefore, to gain some HIP benefits, a mobile user needs a HIP proxy to communicate with the rest of world.

An alternative to developing a HIP-specific proxy is to adapt a generic proxy such as Overlay Convergence Architecture for Legacy Applications (OCALA) (Joseph *et al.* 2006). OCALA proxy runs on multiple platforms including Windows, Mac OS, and Linux, and provides a capturing mechanism for packets that a legacy application sends or receives. OCALA includes a generic overlay-independent sublayer for packet capturing and encapsulation, as well as adaptation modules for several overlay architectures, including Internet Indirection Infrastructure (i3) and HIP. The OCALA implementation is open-source and freely available at http://ocala.cs.berkeley.edu/. The drawback of OCALA is that it must be run both on the HIP host and on the proxy host.

Appendix A

Installing and using HIP

This Appendix presents an overview of current HIP implementations. It also describes the process of setting up and using HIP on Linux (HIPL) implementations.

A.1 Overview of HIP implementations

In this section, we present an overview of existing HIP implementations. First we describe three open-source implementations that are up-to-date with the latest specifications. Interoperability tests between them are run regularly at IETF meetings.

- The *OpenHIP* implementation has been started by the Boeing Phantom Works company in the USA. OpenHIP runs on Linux, Windows, and Mac OS. The implementation is released with a GPL license. Figure A.1 shows the components of the OpenHIP implementation. URL: http://www.openhip.org.

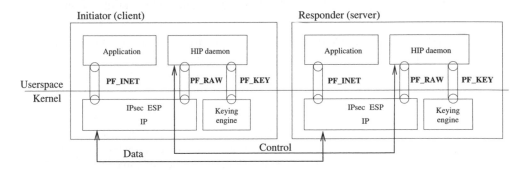

Figure A.1 Internal structure of OpenHIP implementation.

- The *HIP on Linux* (HIPL) implementation is developed by the Helsinki Institute for Information Technology (HIIT) in Finland. HIPL runs on Linux, and there is an ongoing project to port it to Symbian OS. Originally, HIPL was a kernel IPv6-only

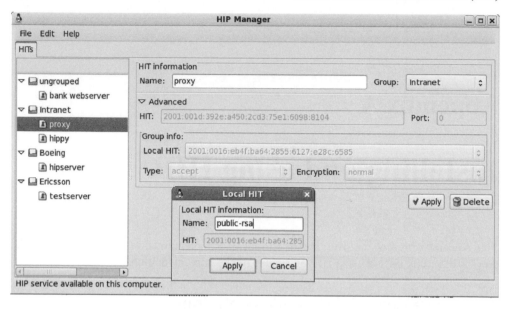

Figure A.2 Graphical User Interface for managing HITs in HIPL.

implementation, but later IPv4 support was added in the user-space. However, because the application interface lacks LSI support, the application must support IPv6. For public use, the implementation is released with the GPL license, but a less restrictive MIT license is available on special request. Figure A.2 shows the interface for classifying and managing HITs in a HIPL implementation. URL: http://infrahip.hiit.fi.

- The *HIP for inter.net* implementation is developed by Ericsson NomadicLab, located in Jorvas, Finland. The primary platform for HIP for inter.net is FreeBSD, although the Linux variant is also available. Both IPv4 and IPv6 protocols are supported. The implementation has a special "Ericsson Finland Public Source" license.

  This implementation consists of a BSD kernel patch and the user-space daemon. The patch implements the IPsec BEET mode and will also appear in the later official kernel releases. The user-space daemon is written in C. The developers have re-written the implementation several times to ensure its briefness and lack of software defects. URL: http://www.hip4inter.net/

In addition to these three implementations, there is an open-source but outdated HIP implementation in Python language by Andrew McGregor. Its source code is available from the book web page.

Several network equipment vendors and operators are known to have internal implementations. However, because of their close-source nature, little is known of their implementation approach or current status.

A.2   HIPL tutorial

This section describes the installation and use of HIPL implementation on Linux. The up-to-date packages for Debian and Fedora Linux distributions are available from the web site (http://infrahip.hiit.fi).

First we recap on the important components of HIP.

BEET (Bound End-to-End Tunnel) is a new IPsec traversal mode that is more efficient than the regular tunnel mode.

HIPD (HIP Daemon) is a HIP software process running in userspace. HIPD handles IPsec key management and mobility events.

HIPCONF (HIP CONFiguration tool) is a tool to configure the HIP daemon manually. Parameters such as the NAT support can be controlled using the HIPCONF.

To install HIP, you need a BEET patched kernel, HIP daemon, and optionally the hipconf tool. HIPL can be installed in two ways:

- Pre-built rpm or deb packages:

    Download and install the kernel package from http://hipl.hiit.fi/hipl/release/. The kernel packages contain pre-compiled kernel images containing the IPsec BEET mode code.

    Download and install the HIPL userspace software packages from http://hipl.hiit.fi/hipl/release/. The userspace packages contain the HIP daemon, the hipconf tool, and some testing software.

- Alternatively you can compile the kernel and userspace software from the source.

    Update your kernel with BEET patches: Download the BEET patches from http://infrahip.hiit.fi/beet. Patch the kernel with BEET patches (might already be available in kernels after version 2.6.19).

    Compile the kernel according to instructions in HIPL readme and reboot your machine to the BEET patched kernel.

    Download the HIPL userspace software from http://infrahip.hiit.fi/hipl/hipl.tar.gz. Compile the source as per instructions in the README.

Before running HIP, make sure that the end-host firewall does not block HIP and ESP traffic as shown below, otherwise HIP traffic would not get through. As a workaround, you can use hipconf to enable the NAT mode. Make sure that no middlebox blocks HIP traffic. If the host is behind a NAT, you need to enable NAT support using hipconf.

```
Rules for iptables to allow HIP and ESP traffic:

iptables -A INPUT -p 139 -j ACCEPT
iptables -A OUTPUT -p 139 -j ACCEPT
iptables -A INPUT -p 50 -j ACCEPT
iptables -A OUTPUT -p 50 -j ACCEPT
```

Once the installation has been completed, hipd daemon can be started like this:

```
$ sudo hipd
```

When you start the hipd for the first time, it generates HIs and corresponding HITs for the host and stores them in the /etc/hip/ folder. HITs are added as IPv6 addresses on a dummy0 device. To see HITs of the host, run the following:

```
$ ifconfig dummy0
```

or

```
$ ip addr show dev dummy0
```

We test HIP between two machines called "crash" (IPv6 address 3ffe::1) and "oops" (IPv6 address 3ffe::2). It is possible to use IPv4 addresses as well. We assume that the machines are located in the same network. You should run these commands on the testing hosts as root.

Do the following on oops:

```
$ ifconfig eth0 inet6 add 3ffe::2/64
## To add ipv6 address on oops
$ hipd
## To start as a background daemon process, add -b flag
$ hipsetup -r
## hipsetup: An example echo server program
written by HIP developers.
## Can be found in test/ folder of the source code.
```

Crash will act as the connection Initiator, so we need to configure the host files of crash (note: use of IPv4 addresses is also possible). Do the following on crash:

In /etc/hosts file add a line to indicate the IP address of oops
```
3ffe::2 oops
```
In /etc/hip/hosts add a line to indicate HIT of oops
```
HIT_OF_OOPS oops
```

```
$ ifconfig eth0 inet6 add 3ffe::1/64
$ hipd
$ hipsetup -i oops
## hipsetup: An example echo client program which
contacts echo server
## Can be found in test/ folder of the source code.
```

Type some text in crash, press enter and ctrl+d and you should see some text appearing in the output of the hosts. This will set up a HIP-enabled connection between the hosts and transmit a character string back and forth.

HITs-IP addresses mappings are usually set up automatically with the host files. Manual configuration is also possible but not necessary:

```
$ hipconf add map PEER\_HIT PEER\_IP
$ hipconf add map abcd::1234 10.0.0.1
$ ping6 abcd::1234
```

With this command, hipconf tells the HIP daemon that the host whose HIT is abcd::1234 is currently located at IP 10.0.0.1. After this, the HIT can be pinged directly.

If your host is behind a NAT then you can configure your hip daemon to support legacy NAT traversal. Run the hipd first and then execute the following:

```
$ hipconf hip nat on
```

This switches on the NAT support in the hipd.

We now show an example of how to use HIP with a legacy application. We describe how to set up a HIP-enabled FTP session between two hosts. Proftpd is used as an FTP server (http://www.proftpd.org). LFTP (http://lftp.yar.ru/) was used as an FTP client.

Run hipd on both client and server. On the server, run proftpd as

```
$ proftpd
```

If you are using deb/rpm packages, proftpd is usually started automatically.

On the client do the following:

```
$ hipconf hip nat on
```

Do this only if the client is behind a NAT. Otherwise, skip this step

```
$ hipconf add map HIT_OF_SERVER  IP_ADDR_OF_SERVER
```

```
$ lftp HIT_OF_SERVER
```

This sets up a HIP-enabled FTP connection between two hosts.

To use HIP in your own applications, you need to override the legacy getaddrinfo function with HIP enabled getaddrinfo. Override the getaddrinfo function from the command line as described in http://www-106.ibm.com/developerworks/linux/library/l-glibc.html

Assuming you are using the bash shell, add to the file .bash_profile

```
LD_PRELOAD= libinet6.so:libhiptool.so
LD_LIBRARY_PATH=/home/username/hipl--main--2.6/libhiptool/.
libs/:/home/username/hipl--main--2.6/libinet6/.libs/
```

Make sure that libinet6 is compiled with HIP_TRANSPARENT_API on in libinet6/ Makefile.in CFLAGS += -DHIP_TRANSPARENT_API. This method assumes two prerequisites. First, the application must be IPv6-enabled in order to support HITs. Second, the application must really use the getaddrinfo interface. At least the following applications use getaddrinfo on FC4: telnet, Firefox, lynx, ssh.

HIP GUI has a "Run" button which allows you to execute an application so that it sets up the LD_PRELOAD options automatically. For example, it can be used for starting up a web browser that will then use the HIP libraries. For detailed instructions, refer to the HIPL manual (http://infrahip.hiit.fi).

Next we describe how to try out mobility with HIPL. Once you have a HIP association set up between two hosts, either of the hosts can change its network location without breaking existing TCP connections.

Suppose you have a HIP-enabled FTP session running between crash and oops. Now if oops changes its network location, changing its IP address from 3ffe::2/64 to 3ffe::3/64 (can be IPv4 addresses too), then you should do the following at oops:

```
$ ip addr del 3ffe::2/64 dev eth0
```

This deletes the old IP address

```
$ ip addr add 3ffe::3/64 dev eth0
```

Note that you can use any other mechanism to delete and add IP addresses.

Once the new address is added, you can see that the FTP transfer resumes from where it had left off. It will not break down. If you try to change the address without HIP, it will break the FTP transfer.

Bibliography

Adkins D, Lakshminarayanan K, Perrig A and Stoica I 2003 Towards a more functional and secure network infrastructure. Technical Report UCB/CSD-03-1242, University of California at Berkeley.

Ahrenholz J 2007 HIP DHT interface: draft-ahrenholz-hiprg-dht-01. Work in progress.

Arkko J and Nikander P 2002 Weak Authentication: How to Authenticate Unknown Principals without Trusted Parties. *Security Protocols*, 10th International Workshop Cambridge, UK, April 17–19, 2002, pp. 5–19. Revised Papers. *Lecture Notes in Computer Science*, vol. 2845. Bruce Christianson, Bruno Crispo, James A. Malcolm, Michael Roe (Eds.) Springer, 2004.

Arkko J, Nikander P and Naslund M 2005 Enhancing Privacy with Shared Pseudo Random Sequences. *Security Protocols*, 13th International Workshop Cambridge, UK, April 20–22, 2005, pp. 1–11. Revised Selected Papers. *Lecture Notes in Computer Science*, vol. 4631. Bruce Christianson, Bruno Crispo, James A. Malcolm, Michael Roe (Eds.) Springer, 2007.

Balakrishnan H, Lakshminarayanan K, Ratnasamy S, Shenker S, Stoica I and Walfish M 2004 A layered naming architecture for the Internet. *Proc. of ACM SIGCOMM'04*, ACM Press, pp. 343–352.

Barham P, Dragovic B, Fraser K, Hand S, Harris T, Ho A, Neugebauer R, Pratt I and Warfield A 2003 Xen and the art of virtualization. *Proc. of the 19th ACM Symposium on Operating System Principles (SOSP '03)*, pp. 164–177. ACM Press, New York, NY, USA.

Baugher M, McGrew DA, Naslund M, Carrara E and Norrman K 2004 The secure real-time transport protocol (SRTP). RFC 3711, IETF.

Camarillo G 2004 Combining the Session Initiation Protocol (SIP) and the Host Identity Protocol (HIP). *Proc. of Research Seminar on Telecommunications Software*, Helsinki University of Technology.

Candolin C and Nikander P 2001 IPv6 source addresses considered harmful *Proc. of Sixth Nordic Workshop on Secure IT Systems*. Technical Report IMM-TR-2001-14, pp. 54–68. Technical University of Denmark.

Cavallar S *et al.* 2000 Factorization of a 512-Bit RSA Modulus. *Proc. of International Conference on the Theory and Application of Cryptographic Techniques*, Bruges, Belgium, pp. 1–18. Lecture Notes in Computer Science 1807. Springer 2000.

Cooper E, Johnston A and Matthews P 2007 A distributed transport function in P2PSIP using HIP for multi-hop overlay routing: draft-matthews-p2psip-hip-hop-00. Work in progress. Expired in December, 2007.

Damgård I 1990 A design principle for hash functions. *CRYPTO '89: Proc. of the 9th Annual International Cryptology Conference on Advances in Cryptology*, pp. 416–427. Springer-Verlag, London, UK.

Devaparalli V 2005 Mobile IPv6 operation with IKEv2 and the revised IPsec architecture. Work in progress.

Dietz T, Brunner M, Papadoglou N, Raptis V and Kypris K 2005 Issues of HIP in an operators networks: draft-dietz-hip-operator-issues-00. Work in progress.

Diffie W and Hellman ME 1976 New directions in cryptography. *IEEE Transactions on Information Theory* **IT-22**(6), 644–654.

Eastlake 3rd D 2005 Cryptographic Algorithm Implementation Requirements for Encapsulating Security Payload (ESP) and Authentication Header (AH). RFC 4305 (Proposed Standard). Made obsolete by RFC 4835.

Eggert L and Laganier J 2004 HIP resolution and rendezvous problem description: draft-eggert-hiprg-rr-prob-desc-00. Work in progress. Expires in April, 2005.

Ellison CM, Frantz B, Lampson B, Rivest R, Thomas B and Ylönen T 1999 SPKI certificate theory. RFC 2693, IETF.

Ford B, Strauss J, Lesniewski-Laas C, Rhea S, Kaashoek F and Morris R 2006 Persistent personal names for globally connected mobile devices. *Proc. of the 7th USENIX Symposium on Operating Systems Design and Implementation (OSDI'06)*. ACM Press.

Freedman MJ, Laskhminarayanan K and Mazières D 2006 OASIS: Anycast for any service. *Proc. of the 3rd Symposium on Networked Systems Design and Implementation*, San Jose, CA. USENIX Association.

Frier A, Karlton P and Kocher P 1996 The SSL 3.0 Protocol. *Netscape Communications Corp.* **18**, 2780.

Harkins D and Carrel D 1998 RFC 2409: The internet key exchange (IKE).

Hautakorpi J, Camarillo G and Koskela J 2007 Utilizing HIP (Host Identity Protocol) for P2PSIP (Peer-to-peer Session Initiation Protocol): draft-hautakorpi-p2psip-with-hip-01.txt. Work in progress.

Heer T 2007 LHIP lightweight authentication extension for HIP: draft-heer-hip-lhip-00.txt. Work in progress.

Heer T, Li S and Wehrle K 2007 PISA: P2P Wi-Fi internet sharing architecture. *Proc. of the 7th International Conference on Peer-to-Peer Computing*, Galway, Ireland, pp. 251–252. IEEE Computer Society.

Henderson T 2004 Can SIP use HIP. *HIP Workshop, 61st IETF meeting*. Unpublished, posted on the web site http://hiprg.piuha.net/workshop/

Henderson T 2007 End-host mobility and multihoming with the host identity protocol: draft-ietf-hip-mm-05. Work in progress. Expires in September, 2007.

Henderson T, Nikander P and Komu M 2007 Using HIP with legacy applications: draft-henderson-hip-applications-03. Work in progress. Expires May 21, 2008.

Henderson TR 2003 Host mobility for IP networks: A comparison. *IEEE Network* **17**(6), 18–26.

Henderson TR 2005 Generalizing the HIP base protocol: draft-henderson-hip-generalize-00. Work in progress. Expires in August, 2005.

Heer T, Götz S, Weingärtner E and Wehrle K 2008 Secure Wi-Fi Sharing at Global Scales. *Proc. of the 15th International Conference on Telecommunication (ICT'08)*. St. Petersburg, Russia. IEEE Communications Society.

Herborn S, Haslett L, Boreli R and Seneviratne A 2006 HarMoNy – HIP mobile networks. *Proc. of the IEEE 63rd Vehicular Technology Conference (VTC '06)*, pp. 871–875. IEEE.

Jian C, Yan R, Hongke Z and Sidong Z 2005 A proposal to replace HIP base exchange with IKE-H method: draft-yan-hip-ike-h-02. Work in progress. Expires in May, 2006.

Jokela P, Melen J and Ylitalo J 2006 HIP service discovery: draft-jokela-hip-service-discovery-00. Work in progress. Expires in December, 2006.

Jokela P, Moskowitz R and Nikander P 2008 Using the Encapsulating Security Payload (ESP) Transport Format with the Host Identity Protocol (HIP). RFC 5202.

Jokela P, Nikander P, Melen J, Ylitalo J and Wall J 2003 Host Identity Protocol: Achieving IPv4–IPv6 handovers without tunneling in *Proceedings of Evolute Workshop 2003: "Beyond 3G Evolution of Systems and Services"*, University of Surrey, Guildford, UK.

Jokela P, Rinta-Aho T, Jokikyyny T, Wall J, Kuparinen M, Mahkonen H, Melen J, Kauppinen T and Korhonen J 2004 Handover performance with HIP and MIPv6. *Proc. 1st International Symposium on Wireless Communication Systems, ISWCS'04*.

Joseph D, Kannan J, Kubota A, Lakshminarayanan K, Stoica I and Wehrle K 2006 OCALA: an architecture for supporting legacy applications over overlays. *NSDI'06: Proc. of the 3rd Conference on 3rd Symposium on Networked Systems Design & Implementation*, San Jose, CA, pp. 20–22. USENIX Association, Berkeley, CA, USA.

Joux A 2004 Multicollisions in iterated hash functions. Application to cascaded constructions. *Lecture Notes in Computer Science* vol 3152. *Proc. of Advances in Cryptology – CRYPTO*, 2004, pp. 306–316. Springer: Berlin/Heidelberg.

Kaminsky D 2004 MD5 to be considered harmful someday. Cryptology ePrint Archive, Report 2004/357. http://eprint.iacr.org/.

Kaufman C 2005 Internet key exchange (IKEv2) protocol. RFC 4306, IETF.

Kent S 2005a IP Authentication Header. RFC 4302 (Proposed Standard).

Kent S 2005b IP Encapsulating Security Payload (ESP). RFC 4303 (Proposed Standard).

Kent S and Atkinson R 1998a IP encapsulating security payload (ESP). RFC 2406, IETF.

Kent S and Atkinson R 1998b Security architecture for the Internet Protocol. RFC 2401, IETF.

Kent S and Seo K 2005 Security Architecture for the Internet Protocol. RFC 4301 (Proposed Standard).

Kivinen T and Kojo M 2003 More Modular Exponential (MODP) Diffie–Hellman groups for Internet Key Exchange (IKE). RFC 3526.

Komu M 2004 *Application Programming Interfaces for the Host Identity Protocol*. Master's thesis, Helsinki University of Technology, Telecommunications Software and Multimedia Laboratory.

Komu M 2007 *Native Application Programming Interfaces for SHIM APIs: draft-ietf hip-native-api-02* IETF. Work in progress. Expires January, 2008.

Komu M, Bagnulo M, Slavov K and Sugimoto S 2007a Socket application program interface (API) for multihoming shim: draft-ietf-shim6-multihome-shim-api-03. Work in progress. Expires in January, 2008.

Komu M, Schuetz S and Stiemerling M 2007b HIP extensions for the traversal of Network Address Translators: draft-ietf-hip-nat-traversal-02. Work in progress. Expires in January, 2008.

Komu M, Tarkoma S, Kangasharju J and Gurtov A 2005 Applying a Cryptographic Namespace to Applications. *Proc. of the first ACM workshop on Dynamic Interconnection of Networks (DIN 2005)*. ACM Press, Cologne, Germany.

Koponen T, Gurtov A and Nikander P 2005 Application mobility with Host Identity Protocol. *Proc. of NDSS Wireless and Security Workshop*. Internet Society, San Diego, CA, USA.

Korzun D and Gurtov A 2006 On scalability properties of the Hi3 control plane. *Computer Communications* **29**(17), 3591–3601.

Kovacshazi Z and Vida R 2007 Host identity specific multicast. *International Conference on Networking and Services (ICNS '07)* pp. 1–9. IEEE Computer Society.

Krawczyk H 2003 SIGMA: The 'SIGn-and-MAc' approach to authenticated Diffie–Hellman and its use in the IKE protocols. *Proceedings of Advances in Cryptology – CRYPTO*, October 2003, Santa Barbara, CA, USA, pp. 400–425. Springer: Berlin/Heidelberg.

Krawczyk H, Bellare M and Canetti R 1997 HMAC: Keyed-Hashing for Message Authentication. RFC 2104 (Informational).

Laganier J and Eggert L 2008 Host Identity Protocol (HIP) Rendezvous Extension. RFC 5204.

Laganier J, Koponen T and Eggert L 2008 Host Identity Protocol (HIP) registration extension. RFC 5203.

Lakshminarayanan K, Adkins D, Perrig A and Stoica I 2005 On securing forwarding infrastructures: Protecting the data plane from an untrusted control plane. Unpublished manuscript.

Lamport L 1981 Password authentication with insecure communication. *Communications of the ACM* **24**(11), 770–772.

Li J, Stribling J, Gil TM, Morris R and Kaashoek MF 2004 Comparing the performance of distributed hash tables under churn. *Proc. of the 3rd International Workshop on Peer-to-Peer Systems*, pp. 87–99. Springer: Berlin/Heidelberg.

Liben-Nowell D, Balakrishnan H and Karger D 2002 Observations on the Dynamic Evolution of Peer-to-Peer Networks. *1st Workshop on P2P Systems and Technologies*, Cambridge, MA, USA, pp. 22–33. Springer-Verlag: London, UK.

Lindqvist J 2006a Establishing Host Identity Protocol opportunistic mode with TCP option: draft-lindqvist-hip-opportunistic-01. Work in progress. Expires in September, 2006.

Lindqvist J 2006b Piggybacking TCP to Host Identity Protocol: draft-lindqvist-hip-tcp-piggybacking-00. Work in progress. Expires in January, 2007.

Manral V 2007 Cryptographic Algorithm Implementation Requirements for Encapsulating Security Payload (ESP) and Authentication Header (AH). RFC 4835 (Proposed Standard).

Matos A, Santos J, Girao J, Liebsch M and Aguiar R 2006 Host Identity Protocol location privacy extensions: draft-matos-hip-privacy-extensions-01. Work in progress.

Medina A, Allman M and Floyd S 2005 Measuring the evolution of transport protocols in the internet. *ACM Computer Communication Review* **35**(2), 37–52.

Merkle RC 1988 A digital signature based on a conventional encryption function. *CRYPTO '87: A Conference on the Theory and Applications of Cryptographic Techniques on Advances in Cryptology*, pp. 369–378. Springer-Verlag, London, UK.

Merkle RC 1989 One way hash functions and DES. In *Advances in Cryptology – CRYPTO*, G Brassard (ed.), volume 435 of Lecture Notes in Computer Science, pp. 428–446. Springer-Verlag, 1990.

Miller G 1975 Riemann's Hypothesis and tests for primality. *Proc. of 7th Annual ACM Symposium on Theory of Computing*, pp. 234–239.

Mironov I 2005 Hash functions: Theory, attacks, and applications. Technical Report MSR-TR-2005-187, Microsoft Research.

Moskowitz R and Nikander P 2006 Host Identity Protocol architecture. RFC 4423, IETF.

Moskowitz R, Nikander P, Jokela P and Henderson T 2008 Host Identity Protocol. RFC 5201.

National Institute of Standards and Technology 1995 *FIPS PUB 180-1: Secure Hash Standard*. pub-NIST.

Nikander P 2004 HIPpy road warriors jumping hoods over road blocks. *Workshop on HIP and Related Architectures*. Unpublished, posted on a web site http://hiprg.piuha.net/workshop/

Nikander P and Arkko J 2002 Delegation of signalling rights. *Proc. of the 10th International Workshop on Security Protocols*, pp. 203–212. Springer, Cambridge, UK.

Nikander P, Arkko J and Ohlman B 2004 Host identity indirection infrastructure (hi3) in *Proc. of The Second Swedish National Computer Networking Workshop (SNCNW2004)*, Karlstad University, Karlstad, Sweden, November 23–24, 2004.

Nikander P, Henderson T, Vogt C and Arkko J 2008 End-Host Mobility and Multihoming with the Host Identity Protocol. RFC 5206.

Nikander P and Laganier J 2008 Host Identity Protocol (HIP) Domain Name System (DNS) Extension. RFC 5205.

Nikander P, Laganier J and Dupont F 2007b An IPv6 prefix for overlay routable cryptographic hash identifiers (ORCHID). RFC 4843, IETF.

Nikander P and Melen J 2006 A bound end-to-end tunnel (BEET) mode for ESP: draft-nikander-esp-beet-mode-06. Work in progress.

Nikander P, Tschofenig H, Fu X, Henderson T and Laganier J 2006 Preferred alternatives for tunnelling HIP (PATH): draft-nikander-hip-path-01.txt. Work in progress.

Nikander P, Ylitalo J and Wall J 2003 Integrating security, mobility, and multi-homing in a HIP way. *Proc. of Network and Distributed Systems Security Symposium (NDSS'03)*. Internet Society, San Diego, CA, USA.

Novaczki S, Bokor L and Imre S 2006 Micromobility support in HIP: Survey and extension of host identity protocol. *Proc. of the 13th IEEE Mediterranean Electrotechnical Conference – MELECON 2006*, pp. 651–654. IEEE.

Orman H 1998 The OAKLEY key determination protocol. IETF RFC 2412.

Papadoglou N and Zisimopoulos H 2005 Host Identity Tags (HIT) in presence information data format (pidf): draft-papadoglou-hiprg-hit-presence-00. Work in progress. Expires in April, 2006.

Perrig A, Song D, Canetti R, Tygar JD and Briscoe B 2005 Timed Efficient Stream Loss-Tolerant Authentication (TESLA): Multicast Source Authentication Transform Introduction. RFC 4082 (Informational).

Pierrel S, Jokela P and Melen JM 2006 Simultaneous Multi-Access extension to the Host Identity Protocol: draft-pierrel-hip-sima-00. Work in progress. Expires in December, 2006.

Rabin M 1980 Probabilistic algorithm for primality testing. *Journal of Number Theory* **12**, 128–138.

Rhea S, Godfrey B, Karp B, Kubiatowicz J, Ratnasamy S, Shenker S, Stoica I and Yu H 2005 OpenDHT: A public DHT service and its uses. *Proc. of ACM SIGCOMM'05*. ACM Press, Philadelphia, PA, USA.

Rivest R, Shamir A and Adleman L 1978 A method for obtaining digitial signatures and public-key cryptosytems. *Communications of the ACM* **21**, 2, ACM, pp. 120–126.

Rivest RL 1992 The MD5 message digest algorithm. RFC 1321.

Rivest RL and Shamir A 1996 Payword and micromint: Two simple micropayment schemes in *Proceedings of 1996 International Workshop on Security Protocols*,pp. 69–87 Mark Lomas (ed.) Lecture Notes in Computer Science, 1189, Springer, 1997.

Rosenberg J, Weinberger J, Huitema C and Mahy R 2003 STUN: Simple traversal of user datagram protocol (UDP) through network address translators (NATs). RFC 3489, IETF.

Salmela P 2005 *Host Identity Protocol Proxy in a 3G System*. Master's thesis, Helsinki University of Technology, Department of Electrical and Communications Engineering.

Saltzer JH 1993 On the naming and binding of network destinations in local computer networks. RFC 1498, IETF.

Sander T and Ta-Shma A 1998 Auditable, anonymous electronic cash. *Advances in Cryptology– CRYPTO* **99**, 555–572.

Schmitt V, Pathak A, Komu M, Eggert L and Stiemerling M 2006 HIP extensions for the traversal of network address translators: draft-schmitt-hip-nat-traversal-01. Work in progress.

Schneier B 1996 *Applied Cryptography*. Wiley, New York, NY, USA.

Shannon C 1949 *Communication Theory and Secrecy Systems*. Bell Telephone Laboratories.

So JYH, Wang J and Jones D 2005 SHIP – mobility management hybrid SIP-HIP scheme. *Proc. of The 6th International Conference on Software Engineering, Artificial Intelligence, Networking and Parallel/Distributed Computing and First ACIS International Workshop on Self-Assembling Wireless Networks (SNPD/SAWN'05)*, pp. 226–230. IEEE Computer Society.

Soliman H and Tsitsis G 2004 Dual stack Mobile IPv6v4: draft-soliman-v4v6-mipv4-01.txt. Internet draft, IETF. Work in progress. Expires in April, 2005.

Standard A 2001 Federal Information Processing Standard Publications (FIPS PUBS) 197.

Stiemerling M, Quittek J and Eggert L 2008 NAT and Firewall Traversal Issues of Host Identity Protocol (HIP) Communication. RFC 5207.

Stoica I, Adkins D, Zhuang S, Shenker S and Surana S 2002 Internet indirection infrastructure. *Proc. of ACM SIGCOMM'02*, pp. 73–88. Pittsburgh, PA, USA. ACM.

Stoica I, Morris R, Karger D, Kaashoek MF and Balakrishnan H 2001 Chord: A scalable peer-to-peer lookup service for Internet applications *Proc. of ACM SIGCOMM'01*. San Diego, CA, USA.

Takkinen L 2006 *Host Identity Protocol Privacy Management*. Master's thesis, Helsinki University of Technology, Telecommunications Software and Multimedia Laboratory.

The Open Group 2004 Secure mobile architecture (SMA) vision and architecture. Technical Report E041, The Open Group.

Torvinen V and Ylitalo J 2004 Weak context establishment procedure for mobility management and multi-homing. *Proc. of the 8th IFIP TC-6 TC-11 Conference on Communications and Multimedia Security (CMS'04)* September 15–18, 2004, Windermere, UK, pp. 111–123. Springer.

Tschofenig H and Shanmugam M 2006 Traversing HIP-aware NATs and Firewalls: Problem Statement and Requirements: draft-tschofenig-hiprg-hip-natfw-traversal-04.

Tschofenig H, Ott J, Schulzrinne H, Henderson TR and Camarillo G 2007a Interaction between SIP and HIP:draft-tschofenig-hiprg-host-identities-05. Work in progress. Expires in December, 2007.

Tschofenig H, Shanmugam M and Muenz F 2006 Using SRTP transport format with HIP: draft-tschofenig-hiprg-hip-srtp-02. Work in progress. Expires in September, 2006.

Tschofenig H, Shanmugam M and Stiemerling M 2007b Traversing HIP-aware NATs and firewalls: Problem statement and requirements: draft-tschofenig-hiprg-hip-natfw-traversal-06. Work in progress. Expires in January, 2008.

Vehmersalo E 2005 *Host Identity Protocol Enabled Firewall: A Prototype Implementation and Analysis*. Master's thesis, Helsinki University of Technology, Department of Computer Science and Engineering.

Walfish M, Balakrishnan H and Shenker S 2004 Untangling the web from DNS. *Proc. of the 1st Symposium on Networked Systems Design and Implementation (NSDI '04)*, San Francisco, CA. pp. 17–17. USENIX Association.

Wang X and Yu H 2005 How to break MD5 and other hash functions. *Proc. of Advances in Cryptology – EUROCRYPT*, 2005, pp. 19–35. Springer: Berlin/Heidelberg.

Ylitalo J 2005 Re-thinking security in network mobility. *Proc. of NDSS'05 Wireless and Mobile Security Workshop*, San Diego, CA, USA. The Internet Society.

Ylitalo J and Nikander P 2004a BLIND: A complete identity protection framework for end-points. *Proc. of the 12th International Workshop on Security Protocols*. Cambridge, UK, April 26–28, 2004. Revised Selected Papers. *Lecture Notes in Computer Science*, vol. 3957. Bruce Christianson, Bruno Crispo, James A. Malcolm, Michael Roe (Eds.) Springer, 2006.

Ylitalo J and Nikander P 2004b A new name space for end-points: Implementing secure mobility and multi-homing across the two versions of IP. *Proc. of the 5th European Wireless Conference, Mobile and Wireless Systems beyond 3G (EW2004)*, pp. 435–441. VDE Verlag.

Ylitalo J, Jokikyyny T, Kauppinen T, Tuominen AJ and Laine J 2003 Dynamic network interface selection in multihomed mobile hosts. *Proc. of the 36th Annual Hawaii International Conference on System Sciences (HICSS'03)*. IEEE Computer Society.

Ylitalo J, Melen J, Nikander P and Torvinen V 2004 Re-thinking security in IP based micro-mobility. *Proc. of 7th Information Security Conference (ISC04)* LNCS 3225, Springer Verlag, pp. 318–329.

Ylitalo J, Salmela P and Tschofening H 2005 SPINAT: Integrating IPsec into overlay routing. *Proc. of the 1st International Conference on Security and Privacy for Emerging Areas in Communications Networks (SecureComm '05)*, pp. 315–326. IEEE Computer Society.

Zhang K 1998 Efficient protocols for signing routing messages. *Proc. of the Symposium on Network and Distributed Systems Security (NDSS'98)*, Sand Diego, California, USA. The Internet Society.

Index